# ENVIRONMENTAL SCIENCES

# ENVIRONMENTAL SCIENCES

*Edited by Frances S. Sterrett*

*The New York Academy of Sciences*
*New York, New York*
*1987*

*Copyright © 1987 by the New York Academy of Sciences. All rights reserved. Under the provisions of the United States Copyright Act of 1976, individual readers of the* Annals *are permitted to make fair use of the material in them for teaching or research. Permission is granted to quote from the* Annals *provided that the customary acknowledgment is made of the source. Material in the* Annals *may be republished only by permission of the Academy. Address inquiries to the Executive Editor at the New York Academy of Sciences.*

Copying fees: *For each copy of an article made beyond the free copying permitted under Section 107 or 108 of the 1976 Copyright Act, a fee should be paid through the Copyright Clearance Center, 21 Congress Street, Salem, Mass. 01970. For articles of more than 3 pages, the copying fee is $1.75.*

**Library of Congress Cataloging-in-Publication Data**

Environmental sciences.

(Annals of the New York Academy of Sciences, ISSN 0077-8923; v. 502)
"The papers in this volume were presented at meetings of the Environmental Sciences Section of the New York Academy of Sciences, New York, during the years 1984–1985"—P.
Bibliography: p.
Includes index.
1. Air—Pollution—Congresses.  2. Radioactive  pollution—Congresses.  3. Water—Pollution—Congresses.  I. Sterrett, Frances S.  II. New York Academy of Sciences. Section of Environmental Sciences.  III. Series.

Q11. N5 vol. 502                 500 s                        87-15392
[Q11.N5]                         [363.7′3]
ISBN 0-89766-399-3
ISBN 0-89766-400-0 (pbk.)

CCP
*Printed in the United States of America*
**ISBN 0-89766-399-3** (cloth)
**ISBN 0-89766-400-0** (paper)
**ISSN 0077-8923**

*Volume 502*
*July 2, 1987*

## ENVIRONMENTAL SCIENCES[a]

*Editor*
FRANCES S. STERRETT

## CONTENTS



[a] The papers in this volume were presented at meetings of the Environmental Sciences Section of the New York Academy of Sciences, New York, New York, during the years 1984–1985.

The New York Academy of Sciences believes it has a responsibility to provide an open forum for discussion of scientific questions. The positions taken by the participants in the reported conferences are their own and not necessarily those of the Academy. The Academy has no intent to influence legislation by providing such forums.

# Preface

FRANCES S. STERRETT

*Department of Chemistry*
*Hofstra University*
*Hempstead, New York 11550*

This volume, *Environmental Sciences*, contains most of the papers that were presented at the monthly meetings of the Environmental Sciences Section of the Academy during the years 1984 and 1985. At these meetings distinguished speakers were invited to lecture on topics concerned with prospects for the environment and the human predicament that results therefrom. These lectures were presented before an audience of members of the Academy as well as the general public.

This *Annal* encompasses a wide range of subject matter, many of the papers dealing with exposure to radiation and to toxic or hazardous materials and their risks to human health and welfare. I would like to emphasize here that we realize that in most cases more research will be needed for definite answers to emerge. This volume considers a variety of topics that could have been arranged in many different ways. I chose as the first paper a general discussion on environmental problems. Richard J. Benoit, in "What's Wrong With The Environment," raises some questions about the decision making process with reference to a global system rather than a local one. Several environmental issues are discussed, such as acid rain, asbestos and polychlorobiphenyls (PCBs), and the author concludes that we must strive for a harmonious economic, social and ecologic global equilibrium giving benefits to all.

This general view is followed by four papers dealing with radiation: Jack Geiger, in his excellent treatise "The Effects of Nuclear Winter on Human Social Systems," describes the atmospheric consequences of nuclear war, the destruction of physical, biological and social systems, and the possible extinction of all humans. This is a highly debated issue, since it is impossible to carry out tests and computer simulations are only as good as the input. Leonard R. Solon in "Health Aspects of Low-Level Ionizing Radiation," discusses the cancer rate that might be attributed to low-dose radiation and the problems connected with making evaluations. In this case the difficulty in decision making is due to scientific uncertainties, for only very-large-population sampling can be conclusive. This uncertainty causes polarization of different interest groups, and it is suggested that radiologic health may be better achieved by democratic dialogue. The last two articles in this part deal with nonionizing radiation. Two scientists, Nancy Wertheimer and Sol Michaelson, express opposite views on the controversial subject of health effects caused by exposure to magnetic and electric fields. In their paper Nancy Wertheimer and Ed Leeper report that research indicates that long-term exposure to low and high voltage of alternating magnetic fields from such sources as high-current power lines near homes may act as a cancer promoter. Sol M. Michaelson, on the other hand, states that even extremely high-voltage exposure causes no chronic health effects. In view of the fact that more research is needed, I wonder whether it would not be better to avoid magnetic and electric field exposure wherever possible.

The next five papers deal with air pollution: outdoor air is considered in three papers, indoor air in the fourth, and urban air movement in the last.

James P. Lodge offers a humorous and charming treatise in "Fire and Brimstone." This paper discusses air pollution by sulfur-containing compounds and particulate matter. Lodge notes that the fear of fire and volcanos and of their sulfurous vapors and smoke is associated with Hell. He alludes to the religious imagery of fire and fumes from prehistoric times to today. Lodge suggests that the strict standards set currently be the U.S. Environmental Protection Agency are also influenced by these mysterious beliefs.

Stephen E. Schwartz's article "Both Sides Now: The Chemistry of Clouds," is a superb one to which the author appended more than 200 additional references to cover the findings from 1984, when the paper was presented, to the present. The productive new research carried out in cloud chemistry during this time is thus enumerated. Schwartz's paper deals with the process whereby sulfur- and nitrogen oxides are transformed in the atmosphere and deposited to the surface in precipitation, and it elucidates two complementary approaches in atmospheric science to the study of chemistry that occurs in liquid water clouds.

The chemistry and the free-radical chain mechanism by which photochemical smog is formed are described in Julian Heicklen's paper, "The Formation and Inhibition of Photochemical Smog." In this paper Heicklen suggests an effective free-radical scavenger, diethylhydroxylamine (DEHA), as a safe and practical means to prevent photochemical smog. However, I wonder whether it would be wise to introduce a synthetic chemical into the natural atmosphere, no matter how effective it proves to be during testing.

Amos Turk, in his paper "Air-Cleaning Devices for Home and Office," describes different methods for purifying contaminated indoor air. He discusses the effectiveness of various types of self-contained commercially available units for home or office as well as the different pollutants that they aim to combat. Air is also the subject of the paper by Walter Hoydysh *et al.* In "Air Movement and Vehicular Pollution in Urban Canyons" the authors examine some approaches to dispersion of pollution within urban street canyons using simulation based on mathematical modeling of air flow and mass dispersion.

The next two papers deal with water: one with New York City's drinking water and the other with ground water. Gerald R. Iwan, in "Drinking Water Quality Concerns of New York City, Past and Present," gives excellent and detailed information on all phases of New York City's drinking water supplies and distribution system, and he provides an extensive historical background on the quality and quantity of drinking water in the City from its inception to the present. He describes the means of quality control and the future objectives with respect to monitoring, treatment, and protection. Stephen E. Ragone's presentation, "The U.S. Geological Survey's Toxic Waste–Ground-Water Contamination Program," outlines the research concerning toxic and hazardous waste in ground water, the physical, chemical and microbiological processes affecting contaminants, and the transport and appraisal of ground water quality and mobility. He also discusses the flow and fate of hazardous inorganic and organic chemicals in ground water, and presents guidelines for situating hazardous waste facilities.

The last two papers in this volume stand by themselves. One, by C. P. Wolf, "The NIMBY Syndrome: Its Cause and Cure," discusses the sociological phenomenon of public recalcitrance at allowing environmentally question-able facilities to be sited "close to home" (NIMBY is an acronym for "not in my back yard"). This paper examines the NIMBY syndrome and diagnoses its symptoms. It also prescribes a possible cure and concludes with a construc-tive interpretation, namely, that the local public can be a partner of the fed-eral system in fostering a genuinely cooperative intergovernmental relation-ship of community-building.

The last paper, that of E. Drucker *et al.*, "Exposure of Sheet-Metal Workers to Asbestos: A Study of Workers Engaged in the Construction and Renova-tion of Commercial Buildings in New York City," reports on the systematic removal of asbestos from New York City buildings. This paper assesses the mortality and morbidity of sheet-metal workers in whom exposure was re-duced as a result of programs of health education, safe working practices, and evaluation of personal protection equipment. The article investigates legis-lative and regulatory solutions to the problem of asbestos contamination of commercial buildings.

In closing, I hope that all of the papers in this volume will help to add a scientific dimension to that which is often perceived on an emotional basis.

# What's Wrong with the Environment?

RICHARD J. BENOIT

*Department of Botany*
*Connecticut College*
*New London, Connecticut 06320*

The answer to the question posed in the title of this article obviously depends on the perspective of the person answering; and any and all thoughtful answers are appropriate within the intellectual context. Certain categories of answers are of greater interest to persons concerned with the environmental sciences. As scientists, we have a particular perspective; as citizens we have a responsibility to participate in the governance or decision-making process in our democratic society. Our participation calls for us to contribute as professionals and demands that we cultivate qualities of tolerance, patience, and cooperation.

In the following paragraphs I will try to illuminate the decision-making process as it has operated in three environmental case histories in the hope that as scientist-citizens we can learn to avoid the mistakes of the past and work toward an improvement in the allocation of resources, material and human, in order to correct what is most wrong with the environment.

The catalogue of case histories to choose from is vast: industrial/technological civilization produces byproducts that threaten the life-support system. Sometimes, it is claimed, and not unreasonably, that the *product* threatens the life-support system. In a broad way the question of what is wrong with the environment is answered in *Global 2000*, the 1980 report of the Council on Environmental Quality,[1] and in many other authoritative reports that have appeared in recent years.

In quite another way the future prospects for *Homo sapiens* are examined in studies that followed the path laid down by the Club of Rome's examination of the "human dilemma,"[2] and the very sad state of affairs in the global ecosystem has been very widely commented upon by the international scientific community.[3] W. W. Hutchinson, President of the International Union of Geological Sciences (IUGS), in an address to the 27th International Geological Congress remarked: "Surely, one of the best contributions that the international geological community can make to society is to provide adequately for our present needs without robbing future generations." He listed as universal problems: $CO_2$ and climatic change, the growth of the desert areas, water shortage, soil degradation, and radioactive waste disposal. The new International Geosphere Biosphere Program (IGBP) was described by *Episodes* staff writers in a recent issue. They began their news report with these words: "It is now generally recognized that the solar-terrestrial system can be viewed as a single living organism whose parts are linked by chemical, physical and biological processes. . . . Over millions of years ecosystems have evolved to produce food, fibre and shelter necessary to sustain human life, but anthropogenic impacts have increased and may well now equal in magnitude the natural processes that control the nature of the global life support system." The IGBP has targeted for interdisciplinary study acid rain, ozone depletion, $CO_2$ and methane build-up, and perturbations of biogeochemical cycles. Of course $CO_2$ and methane build-up are perturbations of the biogeochemical cycle of carbon.

In the decision-making context, the threat of $CO_2$ build-up (the green house effect) may prove to be the most difficult and most complex environmental issue that has ever confronted men and nations. There seems to be little doubt remaining concerning the gravity of that problem. The recent NRC-NAS report, *Changing Climate*, clearly documents the trend of increasing atmospheric $CO_2$.[6] Apparently only a fortuitous concomitant increase in atmospheric dustiness, which has a cooling effect, has forestalled the first increment of climatic effects of increased atmospheric $CO_2$. It is ironic that the overdue impetus to solar energy development and fuller utilization of the hydroelectric resource may finally come out of the human dilemma of a fully industrialized world society caught in the environmental vise-like grip of dwindling and climate-spoiling fossil fuels and nuclear energy that threatens the biosphere with fugitive radioactive poisons. In the latter context see especially *Too Hot to Handle*,[4] a recent account of our failure to manage nuclear waste disposal prudently.

One other item deserves comment in the context of public decision-making because of timeliness — the Bhopal tragedy. Because of growing concern over the fact that actually or potentially dangerous substances were being developed and manufactured at a rate exceeding the ability of the public sector to assess and comprehend the threat, the Toxic Substances Control Act (TOSCA)[5] was drafted, debated, and passed in 1976. The administration of the act was assigned to the Environmental Protection Agency (EPA). That agency then promptly promulgated the principle of exempting from the provisions of the act, substances however toxic that were only manufactured and used in the chemical (and other) industries in a "closed and controlled" manner, and transportation subject to ordinary ICC and state regulations was designated as a closed and controlled use. Less than a dozen substances have ever been adjudged to be subject to the full provisions of the act.

Even before Bhopal, however, there was a growing public concern that more stringent and more extensive regulation was called for to improve environmental safety. In announcing a conference for chemical industry executives[5] the sponsors began: "Despite the leadership and initiative taken by the chemical industry for self-regulation of environment, health and safety, the regulations tide continues to rise. . . ." The conference was held in mid October of 1984; the Bhopal episode occurred the first week in December.

The credibility gap *vis-à-vis* the adequacy of self-regulation in the chemical industry will scarcely be bridged by media items like the AP release which appeared around December 9th under the byline of Peter S. Hawes, a writer for the Associated Press. That story went to great length to assure us readers that Mr. Warren M. Anderson, chairman of the Union Carbide Corporation, is concerned, humane, open, and candid according to remarks quoted from some of his friends and colleagues. Would it not be better to ask why a substance like methylisocyanate should be manufactured into inventory at all (there were three 45-ton tanks at Bhopal and other amounts are stored in Institute, West Virginia, Woodbine, Georgia, and France)? Is it not technically feasible to manufacture it only at points of use and only at a rate consistent with its scheduled use so that it need not be stored?

Some other case histories deserve special mention in the context of decision-making in our democratic society. First, let us consider chlorofluorocarbons and the upper-atmosphere ozone layer. Chlorofluorocarbons are inert and stable gases or volatile liquids that are used in refrigeration compressors and as propellants in aerosol products. Freon is the DuPont tradename for them; Genetron is Allied's trademark. Those stable substances released into the air persisted for decades, at least, and eventually found their way into the upper atmosphere, where they underwent reactions with ozone and threatened thereby to breach the protective screen against solar ultraviolet radiation that the ozone layer affords. (Some readers may recall the same issue being debated in the context of high-altitude supersonic airliners). In the case of chlorofluorocarbons and the threat to the ozone layer, at the urging of the National Academy of Science, their use as propellants in consumer products was banned, but other uses were not prohibited. The public benefits of chlorofluorocarbon propellants in aerosol packaging are clearly inconsequential; one cannot think of a single essential or important consumer product that is marketed in an aerosol package; in any case, other safe propellants like carbon dioxide can be used.

In an address in 1976, J. Gribben, writing in *Nature*, expressed some regret concerning the banning of chlorofluorocarbons as a spray-can propellant in the United States because five more years of the existing rate of use would have given atmosphere modellers enough hard data on the behavior of the upper layers to perfect the whole-atmosphere model.[a]

Three case histories will be discussed more fully in the context of public policy decision-making. They are (1) acid rain, (2) asbestos, and (3) polychlorobiphenyls (PCBs).


## ACID RAIN


Acid rain is an environmental phenomenon and a public issue of great current interest. It involves damage and the threat of damage to natural ecosystems, but no known threat to human health. The economic impact of acid rain on buildings, materials, and artifacts has been well documented. Those circumstances—the lack of known health impacts—militate against hasty overreaction in public policy. The Clean Air Act of 1972 with its several amendments

---

[a] In 1985 a comprehensive, up-to-date report on our understanding of anthropogenic impacts on the atmosphere's ozone layer appeared, and soon after the DuPont Company, a major producer of chlorofluorocarbons, issued a position paper urging its customers and others to curb the release of those substances into the environment (see Ref. 6a).

does not of itself provide any adequate mechanism for coping with the problem. The Clean Air Act's provisions are primarily justified on grounds of health. Acid rain is a complex phenomenon involving interactions between the atmosphere, lithosphere (pedosphere), biosphere, and part of the hydrosphere on broad geographical scales.

An enormous base of scientific and technical literature on acid rain has grown up. Several comprehensive literature reviews have been published[7] and many interdisciplinary conferences have been held.[8] It is not my intention here to provide an exhaustive bibliography or a comprehensive summary of facts and hypotheses, but rather to highlight some ideas that have been mostly overlooked and which might be especially important in making prudent public decisions on the subject.

1. Almost everywhere in the northern hemisphere land areas, there are significant levels of anthropogenic contamination of the air. These contaminants invariably deposit on the landscape in rain, snow, and aerosol particles. Two very conspicuous contaminants are sulfur- and nitrogen oxides, which are precursors of strong acids — sulfuric and nitric acids. In the land regions where the effects of deposition are significant, some of the sulfur oxides originate from local industrial and electrical utility fossil fuel combustion but at least half, and usually more, originate from outside the affected region. Nitric acid contributes as much as half, and usually less, of the noncarbonic acidity of rain and more than half of it originates from internal combustion engines.

2. In temperate humid regions with noncalcareous bedrock, soil formation is an acidifying process and soils in those regions are quite acid, independent of air contamination or strong acid deposition. The run-off from acid soils can be expected to be acid, but much of the run-off is not in contact with soil long enough to come to chemical equilibrium with it. Run-off, river water, and lake waters are quite different in composition from soil-pore waters for that reason. Soil-pore waters tend toward chemical equilibrium with soil. Carbonic acidity of run-off and seepage flows will tend toward chemical equilibrium with air; such a water will contain carbonic acid only to the extent that its pH is reduced to 5.6. Soil-pore waters can contain as much as 150 times as much carbonic acid as the air-equilibrium level. Sulfuric and nitric acids in run-off and seepage flows do not "blow off" in contact with the air. The levels of sulfuric and nitric acids found in rain and snow-melt run-off in some areas significantly affected by acid rain can reduce the pH to as low as 3.0.

Because of the scale and complexity of acid rain phenomena, direct proofs or short-path inferential proofs of hypotheses are not possible, but some direct experimental demonstrations of effects have been made (see Glass *et al.*[7]). A large number of simple, incomprehensible (at least to this non-soil-scientist) experiments have been done in which water is poured through a soil column and changes in its chemistry have been noted.

There is also an ongoing large-scale program on experimental acidification of lakes in Canada (see Schindler *et al.*[9])

3. The reduction of acid precursor emissions will reduce acid deposition (and will have ancillary air quality benefits). The task of reducing emissions will have to be effected by some regions or nations outside the regions or nations enjoying the benefits of reduced depositions.

4. There is still a lot to learn about acid rain, but there does exist a scientific consensus concerning important aspects (see Hileman[8]). Canada, New York State, and West Germany have new programs for emissions reduction, based on the scientific consensus.[10] The U.S. Federal government has not acted decisively[11]; the administration-sponsored amendments to the Clean Air Act would permit increased $SO_2$ emissions (which emissions have been slightly reduced since 1980; nitrogen oxide emissions have been steadily increasing). The American scientific consensus is embodied in the following quotation from the 1983 report of the President's Office of Science and Technology Policy — Acid Rain Peer Review Panel; "Recommendations based upon imperfect data run the risk of being in error; recommendations for inaction pending collection of all the desirable data entail even greater risk of damage." The U.S.–Canadian Work Group I, whose recommendations the panel was empowered to peer-review, had concluded that a 25 percent reduction in sulfate deposition (down to a target loading of 30 kg $h^{-1}$ $yr^{-1}$) would eliminate damage to all but the most delicate of fresh-water ecosystems. The Panel did not find support for a precise relation between deposition and emissions and could not find a basis for concluding that there exists a loading (deposition) level below which the average sensitive aquatic system will be protected.[12]

Consensus does not mean unanimous agreement on all points of a complex technical issue. Decision-making on environmental matters in a representative democracy is a complex process and the role of technical experts is a limited one.[b]

## ASBESTOS

In contrast to the acid rain issue, which involves an ecological impact only, asbestos in the environment involves only a human health impact. Asbestos is the name for a number of different fibrous siliceous minerals used in the manufacture of heat-resistant materials — fabrics, friction/wear surfacing, and building materials.[13]

As early as 1900, asbestosis, a specific type of pneumoconiosis in asbestos miners and industrial workers, was recognized. Later, as early as 1935, and definitely by 1955, a correlation between a particular type of lung cancer and occupational exposure to asbestos dust was discovered. The detailed facts concerning the etiology and oncology were soon developed along with the epidemiologic data on asbestos exposure.

The first reaction of the asbestos industry was disbelief and shock, per-

---

[b] Most recently the National Academy of Sciences has completed and released its comprehensive review of the acid rain problem (Ref. 12a). Recent news notes in *Science* comment on its contents and on the reaction in various quarters. Of particular importance is a renewed interest in coal cleaning technology. The latest plan for abatement calls for a $400 million Department of Energy program on coal cleaning. The Edison Electric Institute according to its spokesman has already spent $500 million and has plans for spending $580 million over the next three years. I have not found a summary of Bureau of Mines' long-enduring activities in coal cleaning, but in any case, it is important to distinguish between the technology for improving the fuel value from that for reducing sulfur.

fectly understandable reactions. Asbestos is one of the most inert natural substances known; it is also a very cheap and useful material. In spite of our background of knowledge concerning silicosis and pneumoconiosis, it is not reasonable to expect the industry to have anticipated the cancer problem. The industry association promptly developed a methodology for monitoring workspace air for asbestos fibers and cooperated with public health agencies to institute protective measures in the workplace.

Gradually, however, the public sector developed a panicky attitude toward asbestos and the industry collapsed as virtually all uses of asbestos were banned and the court dockets catalogued thousands of damage suits.[13a] The asbestos industry was dominated by a single blue-chip company, the Johns-Manville Company, which has filed for bankruptcy and has closed all operations involving asbestos. At the present time, the insurance industry and most of the companies, except Johns-Manville, who are targets of damage suits are involved in a negotiation with almost all plaintiffs in an attempt to reach an omnibus settlement. The Dean of the Yale Law School is the key arbitrator.

Some curious circumstances (I am tempted to call them facts) surround the asbestos case history.

1. Some of the studies relating cigarette smoking and asbestos exposure to incidence rates for lung cancer have been flawed in particular ways, as discussed in the Stanford Research Institute report.[14] There is a clear synergism between the two, however. In the general population, cigarette smokers have a risk of lung cancer ten to fifteen times that of nonsmokers. Smokers with occupational exposure to asbestos have about an 8-fold increase in lung cancer over smokers in the general population. The combined risks are multiplicative, not additive: a smoker with occupational asbestos exposure has 92 times more likelihood of lung cancer than a nonsmoker with no occupational asbestos exposure, according to a study made in 1968 by Selikoff.[15]

The American Lung Association[15] estimated the annual health cost of smoking for 1979 at $27 billion. According to the Association, smoking is responsible for 80 percent of all cases of lung cancer, 80 percent of all cases of emphysema and other forms of chronic obstructive pulmonary disease, and 25 percent of all cardiovascular disease. In 1979 smoking claimed 310,000 lives prematurely. According to the Lung Association's 1983–84 Annual Report smoking is the single largest preventable cause of death in America. In addition it annually accounts for $25 billion in lost productivity, wages, and absenteeism in the workplace. Smokers have double the absenteeism rate of nonsmokers.

2. Many public buildings such as schools, hospitals, and libraries had asbestos lagging on steam pipes and asbestos insulation on ducts. In most of these buildings the material has been removed at high cost; and some institutions have instituted damage suits against architects, contractors, and manufacturers. At the same time, virtually every home that is still heated by steam has asbestos lagging on the pipes in the cellar. Those homeowners must decide whether to live with the condition, to attempt to wrap and seal the lagging, or to remove it and substitute some other material.

3. The evidence that oral ingestion of asbestos causes any pathologic con-

dition is too weak to take seriously, but nevertheless, the EPA has listed asbestos in the revised national drinking water standards (Federal Register, October 1983) as "under consideration," although no standard has been set. Many communities have replaced, at great cost, asbestos-lined and Transite (asbestos cement) water mains. Some communities used polyvinyl chloride piping and found they had introduced unacceptable levels of halomethanes and other substances for which primary standards have already been set.

4. The problem with asbestos seems to be a problem with *fugitive* asbestos at some reasonably high level, but the link with cancer doubtlessly obviates any feasibility for safe-and-sane risk analysis. There is also no evidence that the Federal government is willing to alleviate the financial disaster of Johns-Manville, as it did so generously with Lockheed and Chrysler.

### *The Dose-Response Relationships*

A number of industrial hygiene practices can be instituted to reduce exposure of workers to asbestos. The extent and scope of the practices that need to be implemented to reduce risk to an acceptable level depends on the dose-response relationship. That relationship is quite straightforward with regard to asbestos-pneumoconiosis, but with regard to cancer, it has been difficult to establish. On theoretical grounds it is not expected to be simple and straightforward for long exposures to low levels of fiber. It is not surprising that the statistics (and other data) provide no evidence of a threshold level of exposure for the cancer response. FIGURE 1 was drawn from the data of Table 5 in the Stanford Research Institute report.[14]

The relationship is quite remarkably linear for the type of data represented. It is especially noteworthy that the line, which is only sketched in, does not

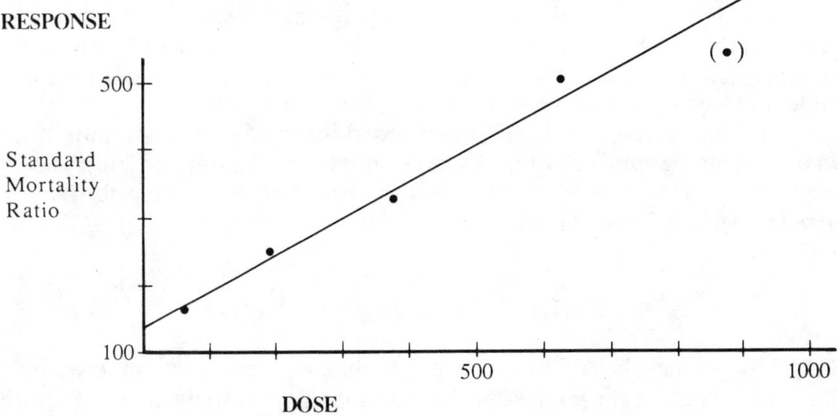



**FIGURE 1.** Dose-response relationship for respiratory cancer (lung, larynx, and pleural mesothelioma) in retired asbestos workers. The dose is exposed-time weighted dustiness of the workspace air; response is the standard mortality ratio. Zero excess deaths over a control cohort is 100 response units. (Data are from Table 5 of the Stanford Research Institute Report.[14])

cross the horizontal axis, and thereby, the data show no evidence of a threshold exposure.

For asbestosic pneumoconiosis, it is commonly accepted that a threshold does exist, judging from the following statement: "Industrial experience indicates that pulmonary fibrosis sufficient to interfere with respiratory and cardiovascular function can be prevented by reducing asbestos dust concentrations to levels that are still far above any likely to be encountered in community air."[16] The National Research Council report[16] finds that the facts reviewed "suggest that there are levels of asbestos exposure that will not be associated with any detectable risk [but that] those levels are . . . not known." The report states further that there is no evidence that persons without occupational, household, or neighborhood exposures have any increased risk of neoplasm, even in the known presence of ferruginous bodies or fibers in their lungs. There was thus, in 1971, a body of informed scientists who were convinced that an exposure threshold existed for the neoplastic response. Even if the threshold does exist, experience with the statistics of dose-response systems signals that it will be embedded in the variance (uncertainty) surrounding the dose-response curve-of-best-fit at the low dose end.

In 1974 the EPA, in reviewing their 1973 standards for allowed levels of asbestos in air, concluded that it was not possible to "identify ambient concentrations of the pollutant which [are] judged to provide an ample margin of safety to protect the public health. . . . For asbestos, it is impossible to prescribe and enforce allowable numerical concentrations or mass emission limitations known to provide an ample margin of safety to protect public health, since no safe level has been identified."

You may have noted that I used the word "allowed" in the first sentence, while the EPA used the word "allowable." The distinction, I submit, is important. The words *allowable* and *permissible* imply a threshold level below which safe exposure prevails, the word *allowed* or *permitted* implies no such thing; rather the implication is that the regulatory agency has set a standard consistent with their authority and responsibility. The whole cause of environmental management would be advanced if we say *permissible* when we mean permissible and say *permitted* when we mean permitted.

The NRC committee[16] in 1971 concluded that "asbestos is too important in our technology and economy for its essential use to be stopped." Is it merely ironic that the subtitle of a 1919 book on asbestos[17] bears the subtitle *The World's Most Wonderful Mineral?*


## POLYCHLOROBIPHENYLS

Polychlorobiphenyls (PCBs) in the environment represent an especially noteworthy case in the context of decision-making, for there are both health and ecological aspects to the problem. The "mussel watch" program[18] has reported high PCB levels in mussels and oysters from all coasts near urban centers.

Polychlorobiphenyls are oily liquids that have ideal properties for certain applications—they are stable, noncorrosive, relatively inexpensive, nonvolatile, and have high dielectric strength. Their principal uses were in oil-filled transformers and other electrical equipment, and in wound foil capacitors. Another important use of PCBs was as a constituent of the inking system in self-inking carbon papers. Other uses of this chemical included its use as a high-temperature heat-exchange fluid. Uses in consumer products derive from PCBs' property as plasticizer and solvent.

In the 1960s PCBs began to appear in the analyses done in environmental surveys for DDT and related chlorinated organic pesticides, but since then it has been thoroughly studied in its own right and it has been shown to be widely dispersed in natural habitats.[18] PCB mixtures invariably contain some chlorinated dibenzofurans, which are extremely toxic. PCBs are themselves reasonably toxic; the dibenzofurans may be degradation products as well as byproducts of manufacture. In any case, most uses of PCBs were forbidden in the 1970s and manufacture ceased. The contamination of mud in the Hudson[19] and Housatonic Rivers is well documented.[20] The sources of contamination in those rivers have been stopped for several years, but fish taken in those rivers still can show dangerous levels of PCBs.

General Electric has joined forces with New York State in a program to clean up the "hot spots" in the upper Hudson River. The most heavily contaminated sediments are to be dredged and disposed of at a specially prepared secure landfill. A similar cooperative program is in planning for the Housatonic River.

In order to minimize the future distribution of PCBs to the environment, a complex and costly regulatory process has been put in place. Waste fuel oils must be tested for PCBs, because under ordinary combustion conditions PCBs are not destroyed and can be spread out over the stack plume area of installation. All PCB-contaminated wastes must be handled and disposed of as hazardous under the Resource Conservation and Recovery Act.

Northeast Utilities Company in Connecticut developed a plan, with the full cooperation of the state's Department of Environmental Protection, to burn PCBs with fuel oil on a controlled scale in the high-stack, oil-fired power plant at Middletown, Connecticut. The plan included a full-scale risk assessment, including trial runs with very thorough monitoring of emissions. The public information aspects of that plan seem to have been poorly handled, and the public reaction was very strongly negative. The risk analysis predicted no more than a one in a million risk of cancer from PCBs, dioxin, or benzofuran emission levels. Compared to ordinary voluntary risks, such a risk level would seem tolerable, but Northeast Utilities and the Connecticut Department of Environmental Protection found themselves in an adversary situation with a polarized press and a very stubborn public sector.[20a] The cost of disposing of the stockpiles of PCB-containing waste is going to be very high indeed.[c]

---

[c] Dr. Elizabeth M. Whelan in her recent book *Toxic Terror*[20b] contends that we have overreacted to the hazards posed by PCBs in the environment, and that "environmentalists" are deluded, hysterical, and pernicious. Dr. Whalen is founder and Executive Director of the American Council

The manufacture of PCBs is prohibited under TOSCA regulations by the EPA. Monsanto voluntarily suspended sales of their products for use in heat-exchangers in food industries and some "open" uses in 1970. Waste oils containing more than a trace of PCBs, other PCB-containing wastes, and even excavated PCB-contaminated soils are subject to RCRA regulations. The costs, mostly borne by the public, of assuring that the public's exposure is minimized, are enormous, but the decisions on PCBs mostly make good sense. Dr. D. Squires, Director of the Connecticut Sea-Grant Program, recently revealed at a seminar there that with the FDA reducing the allowed PCBs residue in fish (and other foods), New York State was going to announce a regulation making it illegal (punishable) to have striped bass in one's possession.

Isn't it fair to ask why Monsanto has not been asked to underwrite some important share of the costs of protection against the residue of PCBs in the environment?

### Safety Engineering, Risk Assessment, and Decision-Making in a Democratic Society

The Middletown, Connecticut PCB disposal facility is a locally conspicuous microcosm of a very broad landscape. The nuclear power debate and, soon to come, the debates over siting of large-scale solid waste management facilities are larger prototypes of the problem of public reaction to environmental risk.

Scientists and engineers have but a poor understanding of that problem. The literature in the social sciences is interesting, useful, and growing. It behooves us natural scientists and engineers who act in the arena of environmental quality management to learn all we can from those studies on public attitudes.

The technological component of environmental decision-making in the public sector has been embodied recently in a process called risk assessment. Risk assessment is a formal methodology derived from and akin to systems analysis and cost-benefit analysis (see Kates, Otway and Pahner, and Rappaport[22]). In the process a quantitative estimate is made of risk in terms of the number of excess fatalities per unit of risk-inducing technological activity.[23]

One reason for the public's slow and still limited acceptance of risk assessment is because the main dimension of risk is the expected number of *deaths*. Because the main dimension of benefit is economic, the dollar value of a human life is sometimes used as a common denominator. One can obviously obtain

---

on Science and Health. In her book she makes a strong case that we have overreacted to all environmental hazards since DDT — and that we dangerously overreacted in that case. It is noteworthy that, on the one hand, she stresses the reality of the hazards from smoking, while, on the other, she is nearly mute on the subject of the hazard from asbestos, merely stating (about six times in the book) that the link between asbestos in air and asbestos/lung cancer and mesothelioma is convincingly real. Further, Dr. Whalen cites sources that conclude that the epidemiologic evidence on low occupational exposures provides no evidence that environmental levels of asbestos in water or air pose any significant threat to the general public.

a figure of merit from the life insurance industry. Economists and others have used other explicit approaches. (See Appendix C of the National Research Council's PCB report.[16]) The values derived by various approaches range from 28,000 to 5,000,000 1976-dollars. Risk analysis is derived from cost-benefit analysis and the advantage of the dollar common denominator is that the methods of economics can be used in the analysis; but does not the range argue the futility of the exercise? One is reminded of the futility of trying to put convincing dollar values on ecological goods. How much is a population of snail darters or bog turtles worth? Those species cannot speak for themselves. It would seem more sensible and prudent to try to devise noneconomic value scales, as A. C. Worrell, a forest economist, has tried to do in his thoughtful and thought-provoking book *Unpriced Value.*[24]

In principle, risk assessment yields an objective, reliable estimate of risk. Given such an estimate, the decision-making process, however diffuse, should proceed rationally in a predictable way. Such is not the case, however, for a great variety of reasons: (1) There is more than one kind of rationality, according to Charles Perrow.[25] (2) Most people are not willing to become a statistic for someone else's benefit or for an *estimated* public good. Most people are bright enough to recognize when costs are not internalized, when costs and/or benefits are not likely to be equitably allocated or shared, as pointed out by Comar.[23] (3) To the extent that individuals contribute to a decision they are influenced by their perceptions of risks and benefits rather than those quantities as estimated by experts. Perrow[25] says that people exercise a bounded rationality or a societal rationality rather than an absolute rationality. Fischoff and his colleagues[22,23] have used the techniques of social psychology to attempt to determine what influences our perceptions and our values of risk and benefit from technological activities. Starr[23] attempted to tell us what we thought about risky technology; Fischoff and his colleagues ask us about it. (4) Although science and technology are not suffering a crisis of public confidence, many people do not trust experts, especially those employed by others.

Consider this observation by Perrow: "The new risks have produced a new breed of shamans, called risk assessors. [In] this new alchemy body counting replaces social and cultural values and excludes *us* from participating in decisions about the risks that a few have decided that many cannot do without. The issue is not risk but power."[25] What Perrow is alleging is the corruption of democracy, a failure of the democratic process.

According to David Okrent, a recognized expert,[23] "there are still few risk assessments that (1) provide a detailed statement of the assumptions made in arriving at the conclusion, (2) point out any uncertainties in the results; and (3) have the benefit of a detailed evaluation by a competent independent body."

All three are desirable from the standpoint of the *real utility* of the proposed activity. One's sense of fair play urges that the cost of third-party review should be paid by the proponent party and the third-party experts should be sensitive to the attitudes of the opponent parties.

In June 1979 a party of scholars gathered at Harvard to discuss changing public attitudes to technological progress.[26] The Three Mile Island nuclear accident had occurred earlier in the spring. People have learned that "expertise can be bought" and that we have "government by control of the experts. Technicians often provide sanction for political action, as church officials once did," according to Paul Roszak of the Center for Science and International Affairs at Harvard.

Alexander Morin, Director of the Office of Science and Society of the National Science Foundation, claims that there has been "an enormous shift in the relationship of science and technology to authority: they are now an instrument of state power. If you don't like the authority structure, you do not like the people working for it. People say, 'I want to be involved in the technical decisions that affect my life, and I don't trust the scientists who are speaking for the people I don't trust.'"

In 1970 Chauncey Starr of the Electric Power Research Institute in Palo Alto analyzed the relationships between the economic value of a technology or an activity and its riskiness measured by the number of deaths per unit of activity enacted. One expects a correlation and indeed one was found, albeit a weak one at best. Furthermore the value per unit of acceptable risk tended to be *less* in voluntary activities like flying a light plane than for involuntary activities which entail risk. Starr's measures of economic value (benefit) and risk (number of fatalities per unit of activity) were objective. But what about people's *perceptions* of the value and the risks involved?

## ACCEPTABLE RISK: HOW SAFE IS SAFE ENOUGH?

The psychometric studies of Fischoff and his colleagues[22,23] were designed to measure perceptions and attitudes and to answer that question. In the early rounds of their studies, it emerged that there was no correlation whatsoever between perceived benefits (societal not just economic) and perceived risks. The implications of that finding are startling if Fischoff's group of subjects can be taken as representative. It could mean that in decisions on risky technology we have not been doing what people think should be done. Are we cynical enough to accept that indictment of democracy?

Fischoff and his colleagues decided to refine their measure of perceived risks and they asked their subjects to rate risks for the 30 activities and technologies used in the study according to nine attributes that were thought to influence people's attitudes toward the risk—its relative acceptability:

    (1) the voluntary or involuntary aspect of the risk;

    (2) the immediacy versus delayed experience of the risk;

    (3) knowledge or understanding (personal) of the risk;

    (4) knowledge or understanding on the part of experts about the risk;

    (5) control over the risk (can one by skill or diligence lessen the risk?);

    (6) newness of the risk;

    (7) chronic (one person at a time affected) versus catastrophic aspect of the risk;

    (8) common (a familiar type of risk) or dreadful risk;

    (9) severity of consequences (i.e., whether the risk causes only injury or illness or whether death is the usual result of the risk).

**TABLE 1.**

| Technological Risk Factor | | | Severity Factor | | |
|---|---|---|---|---|---|
| Low Value | | High Value | Low Value | | High Value |
| Not necessarily fatal | *vs.* | Fatal | Known | *vs.* | Unknown |
| | | | Voluntary | *vs.* | Involuntary |
| Chronic | *vs.* | Catastrophic | Controllable | *vs.* | Uncontrollable |
| | | | Old | *vs.* | New |

Now, as you have probably guessed, those attributes tended to be intercorrelated and two factors or scales seem to influence the subject's attitude to a particular risk. Fischoff designated those composite factors *technological risk* and *severity* and they are schematized in TABLE 1. When risk perceptions were weighted by the technological risk factors and the severity factors a correlation between perceived risk and perceived benefits emerged.

People apparently do behave in an understandable manner; democracy does seem to be working to some extent in environmental decision-making. We are accepting large *perceived* risks when we *perceive* large benefits from the activity.

Perrow[25] discusses the important distinction between three rationales:

(1) The absolute rationale; the rationale employed by experts in risk assessment, although the stuffiest expert admits that there is a subjective component.

(2) The bounded rationale; the rationale employed by ordinary persons when they try to reach a rational conclusion with an incomplete understanding of technicalities and with their characteristic prejudices.

(3) The societal rationale; the rationale that accepts the cognitive limits on rational choice. It cannot make the "experts's error" of defining the problem to fit the methodology. One's limited cognitive ability forces social interaction of people with different intuitions and with interdependence in decision-making. Accordingly, argues Perrow, a technological activity which can produce events (like the Three Mile Island accident) that produce confusion, deception, anxiety, and incomprehensible consequences is to be decided against in the societal rationale.

C. W. Churchman in summarizing the 1979 General Motors symposium on societal risk assessment[23] gave sound advice to the assembled experts: "There is [some] reason to be cheerful about the safety profession because it seems to be on the pathway of not treating people as children who should be told how to behave safely, but rather as humans all sharing in the quality of each other's lives, and who realize that some risk, (accepted, even reluctantly) is an essential in the design of a good life."

Slovic *et al.*, at the same General Motors symposium, listed basic premises for the management of hazard in our democratic society:

(1) Both the public and experts are necessary participants.

(2) Assessment is inevitably subjective.

(3) Understanding *judgemental limitation* is crucial to effective decision-making.

The experts should try to understand how the public arrives at its attitudes; the public should try to understand how the experts arrive at theirs.

## STANDARDIZATION OF RISK-ASSESSMENT PROCEDURES

In the course of time analytical processes become explicit, formalized, and eventually standardized. Standardization involves objective evaluation and improvement. Standard procedures carry the mace of authority; they are widely acceptable as applicable to the resolution of disputed issues. Risk- or hazard-assessment procedures are still in the developmental stages.[27]

The National Research Council PCB report[16] gave a provisional scheme for the evaluation of the potential hazards of chemicals to health and the natural ecology; this is shown in FIGURE 2.

Anent the NRC provisional scheme for evaluation of hazard from new chemical products and new uses of manufactured substances, the PCB experiences in the Great Lakes studies[21] are especially noteworthy: "The need for individual consideration of each congener (and not just a vague group of PCBs) is now obvious." Chemical manufacturers have the personal and technical resources to obtain all the information called for in the recommended testing scheme. The public at large bears the cost of regulatory agency budgets and it pays the cost arising from imprudent use of dangerous substances.

## FUTURE PROSPECTS

All of the comprehensive recent studies of the state of the global life-support system,[1,2] while using different methodologies, reach the same general conclusions about the present state of affairs and all indicate similar trends:

(1) Continued economic growth in most areas of the world;
(2) Continued population growth almost everywhere;
(3) Reduced rate of energy *growth*;
(4) An increasingly tight and expensive food situation;
(5) Increasing water problems;
(6) Growing environmental stress.

Predicting the future is an uncertain exercise, even with a solid data base and an elegant computer model like World III, which tries to give appropriate credit for technological innovation; but the prediction leaves little room for confidence or optimism.

What can be done?

The executive committee of the Club of Rome sees global equilibrium, not growth, as the only answer, and if underdeveloped nations are to enjoy a reasonable standard of living, there must inevitably be, under an equilibrium world economy, considerable redistribution of wealth and income.

"Despite the model's material orientation, the conclusions of the study point to the need for fundamental change in the values of society. . . . The achievement of a harmonious state of global economic, social, and ecological

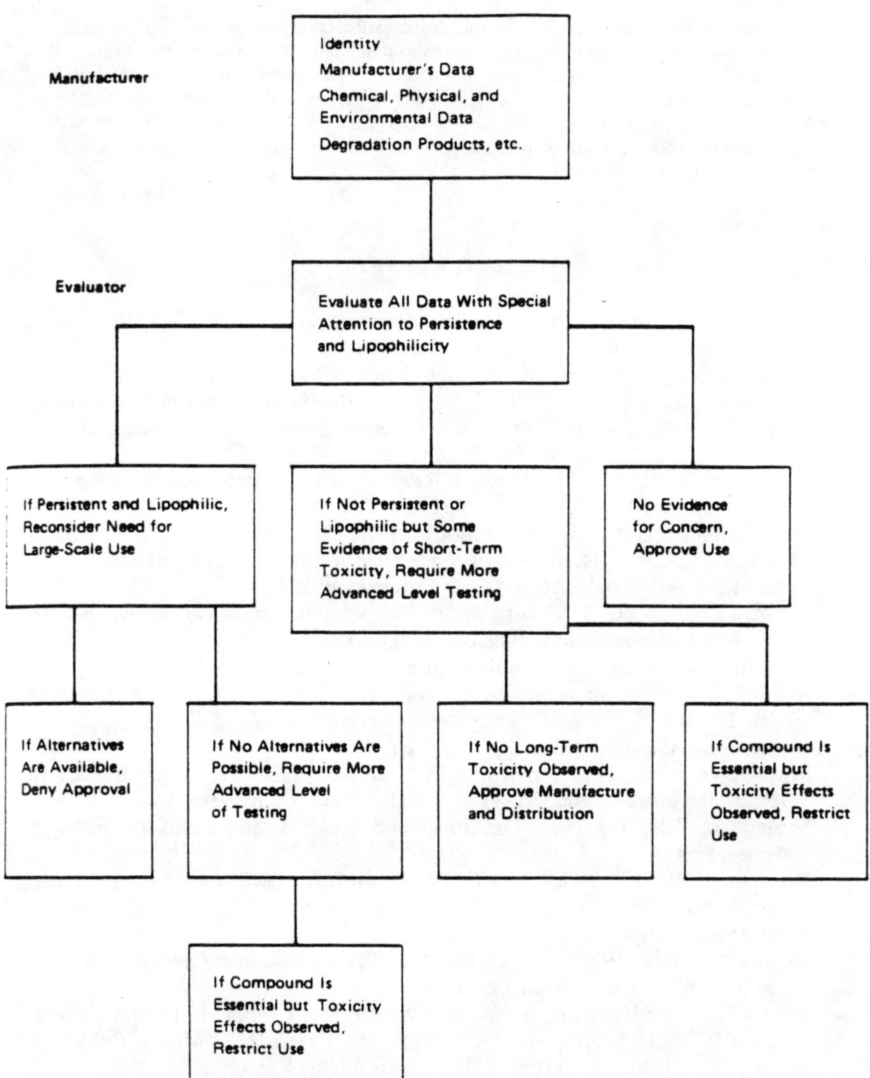


**FIGURE 2.** Evaluation scheme for assessing potential hazards to health and environment. (From the PCB report of the National Research Council.[16])

equilibrium must be a joint venture (between developed and developing nations) based on joint conviction, with benefits for all."

An ancient song from the book of Wisdom says:

Resplendent and unfading is Wisdom, and she is readily perceived by those who love her, and found by those who seek her. She hastens to make herself known in anticipation of men's desires; he who watches for her at dawn shall not be disappointed, for he shall find her sitting by his gate. For taking thought of her is the perfection of prudence, and he who for her sake keeps vigil shall quickly be free from care; Because she makes her own rounds, seeking those worthy of her, and graciously appears to them in the ways, and meets them with all solicitude.

— WISDOM 6:12–16

## NOTES AND REFERENCES


1.  *See* The Global Report to the President — Entering the 21st Century. 1980. Council on Environmental Quality and Department of State. G. U. Barney, Study Director. U.S. Government Printing Office. Washington D.C.

    *See* Commentary Science **208**: 543; and *see also* Global Future: Time to Act Council on Environmental Quality. U. S. Government Printing Office, and commentary by E. Marshall. 1981. Science **211**: 462–3.

2.  MEADOWS, D. H., D. E. MEADOWS, J. RANDERS & W. W. BEHRENS, III. 1972. The Limits to Growth. A Report for the Club of Rome Project on the Predicament of Mankind, 2nd ed. Universe Books. New York, NY.

    MEADOWS, D. L. & D. H. MEADOWS, Eds. 1973. Toward Global Equilibrium: Collected Papers. Wright-Allen Press. Cambridge, MA.

    *See also* Common Crisis: North-South Cooperation for World Recovery. 1981. The Brandt Commission. Pan Books. London.

3.  HUTCHINSON, W. W. 1984. Looking ahead for IUGS. Episodes 7(3): 2.

    ANON. 1983. New international program to study global change. Episodes **6**(4): 29–30.

    ANON. 1984. Global change. Episodes **7**(4): 30.

    MATTHEWS, W. H., F. E. SMITH & E. D. GOLDBERG, Eds. 1971. Man's Impact on Terrestrial and Oceanic Ecosystems. MIT Press. Cambridge, MA.

    VEZIROGLU, T. N., Ed. 1984. The Biosphere: Problems and Solutions. Elsevier. Amsterdam.

4.  WALKER, C. W., L. C. GOULD & E. J. WOODHOUSE. 1982. Too Hot to Handle. Social and Policy Issues in the Management of Nuclear Wastes. Yale University Press. New Haven, CT.

5.  DOMINQUEZ, G. S. 1977. 1983. Guidebook: Toxic Substances Control Act, Vols. I and II. CRC Press. Boca Raton, FL.

    *See also* undated letter prospectus on a conference: Chemical Industry Regulation Briefing, October 15–16, 1984, Washington, D.C. H. W. Mager [Publisher, *Chemical Week*], T. J. Lombardi [Vice President, Executive Enterprise, Inc.], and T. M. Hellman [Conference Chairman, Allied Corporation].

    *See also* Pollution Engineering (February, 1985) concerning a proposed new rule (Federal Register, September 20 and 28, and October 10, 1984) on eight substances, including urethane, that the EPA considers to be possible risks to human health.

6.  NATIONAL ACADEMY OF SCIENCES. 1983. Changing Climate. Carbon Dioxide Assessment Committee. National Academy Press. Washington, D.C.

    *See also* Commentary: $CO_2$ and Changing Climate. 1983. Episodes. **4**: 28.

6a. For ozone layer and greenhouse effect *see:*

    The Changing Atmosphere. Implications for Mankind. A C&EN special issue. Chemical & Engineering News Nov. 24, 1986.

DuPont changes stance; says it is time to limit emissions of CFC's. Research & Development December 1986: 37.

DuPont position statement on the chlorofluorocarbon/ozone/greenhouse issues. September 1986. 4 pp. Available from "Freon" Products Division E. I. DuPont de Nemours & Co. Wilmington, DE 19898.

Atmospheric Ozone 1985: Assessment of our understanding of the processes controlling its presence, distribution, and change. World Meteorological Organization Global Ozone Research and Monitoring Project — Report #16 (3 volumes). Sponsored by NASA, FAA, NOAA, UNEP, WMO, Commission of the European Communities, and Bundesministerium für Forschung und Technologie.

7.  COWGILL, U. M. 1984. Acid precipitation: A review. *In* The Biosphere: Problems and Solutions. T. N. Veziroglu, Ed. Elsevier. Amsterdam.

KRUG, E. C. & C. R. FRINK. 1983. Acid rain on acid soil: A new perspective. Science **221**: 520–525.

BURGESS, R. L., Ed. 1984. Effects of Acidic Deposition on Forest Ecosystems in the N.E. U.S.: An Evaluation of Current Evidence. SUNY College of Environmental Science and Forestry. Syracuse, NY.

GLASS, N. R., D. E. ARNOLD, *et al.* 1982. Effects of Acid Precipitation. Environ. Sci. Technol. **16**(3):162A–169A.

HILEMAN, B. 1981. Acid Precipitation. Environ. Sci. Technol. **15**: 1119–1124.

COWLING, E. B. 1982. Acid precipitation in historical perspective. Environ. Sci. Technol. **16**: 110A–122A.

NATIONAL RESEARCH COUNCIL. 1981. Atmosphere-Biosphere Interactions: Toward a Better Understanding of the Ecological Consequences of Fossil Fuel Combustion. Committee on the Atmosphere and the Biosphere. National Academy of Sciences. Washington, D.C.

8.  HILEMAN, B. 1983. 1982 Stockholm Conference on Acidification of the Environment. Environ. Sci. Technol. **17**: 15A–18A.

DOCHINGER, L. S. & T. A. SELINGA, Eds. 1976. Proceedings of the First International Symposium on Acid Precipitation and the Forest EcoSystem. USDA Forest Service. Gen. Tech. Report NE-23. Upper Darby, PA. — 1985 Second International Conference on Acid Rain, Washington, D.C., March 26–27, 1985. Power. McGraw-Hill. New York, NY.

9.  SCHINDLER, D. W. *et al.* 1980. Can J. Fish. Aquat. Sci. **37**: 313–558.

10.  WELLMAN, *et al.* 1984. Fact Sheet on Acid Rain (revised January 1984). Canadian Embassy. Washington, D.C.

SNOW, 1984. Acid Rain. A Policy for New York State to Reduce Sulfur Dioxide Emissions. Draft Environmental Impact Statement. New York State Department of Environmental Conservation. July 1984.

ULRICH, B., R. MAYER & P. K. KHANNA. 1980. Chemical changes due to acid precipitation in a loess-derived soil in central Europe. Soil Sci. **130**: 193–199.

*See also* remarks of Ulrich at the 1982 Stockholm Conference quoted in Hileman, (Ref. 8): "No forest ecosystem, even forests on calcareous soils, can withstand the present [West German] air pollution without serious damage."

11.  WARD, B. 1984. The intractable politics of acid rain. Pollution Engineering (June, 1984): 20–21.

*See also* the 1985 membership prospectus letter of the Environmental Defense Fund (ACID RAIN Facts the White House Refuses to Face).

*See also* the letter dated March 1985 and attached extract from U.S. Congressional Record of Senator Christopher J. Dodd (Dem., CT). U.S. Senate, Washington, D.C. 20510. Congressional Record **131**: 27.

12.  SINGER, S. F. 1984. Acid rain report. Science **226**: 780.
12a.  NATIONAL ACADEMY OF SCIENCES. 1986. Acid Deposition: Long Term Trends. National Academy Press. Washington, DC.
  SUN, M. 1986. Academy study dispels doubt on acid rain. Science **231**:1500.
  SUN, M. 1986. Acid rain plan draws mixed review. Science **231**:333.
13.  NATIONAL RESEARCH COUNCIL PANEL ON ASBESTOS. 1971. Asbestos. The Need for and Feasibility of Air Pollution Controls. National Academy of Sciences. Washington, D.C.
  BRODEUR, P. 1980. The Asbestos Hazard. New York Academy of Sciences. New York, NY. This book is a popular account of the papers presented in 1978 at an international conference convened by the New York Academy of Sciences. The complete proceedings of this conference, entitled Health Hazards of Asbestos Exposure, are published as Volume 330 of the *Annals of the New York Academy of Sciences* (I. J. Selikoff & E. C. Hammond, Eds.). Because the Brodeur book has no citations of the literature, its value to serious scholars is limited, therefore it should be read in conjunction with the *Annal* it popularizes. An earlier international conference entitled Biological Effects of Asbestos was held in 1964 by the Academy and published as Volume 132, Art. 1 of the *Annals*. The proceedings are also fully integrated into and cited in the 1971 National Research Council report.
13a.  For asbestos damage claims and implications for civil justice reforms *see*:
  HENSLER, D. R., W. L. F. FELSTINER, M. SELVIN & P. A. EBENER. 1985. Asbestos in the Courts. The Challenge of Mass Toxic Torts. The Institute for Civil Justice. Rand Corp. Santa Monica, CA.
  KAKALIK, J. S., P. A. EBENER, W. L. F. FELSTINER, G. W. HAGGSTROM & M. G. SHANLEY. 1984. Variation in Asbestos Litigation Compensation and Expenses. ICJ Rand Corp. Santa Monica, CA.
  KAKALIK, J. S., P. A. EBENER, W. L. S. FELSTINER & M. G. SHANLEY. 1983. ICJ Rand Corp. Santa Monica, CA.
14.  LEVINE, R. J., Ed. 1978. Asbestos: An Information Resource. U.S. Department of Health, Education and Welfare Publication (NIH) 78-1681. National Cancer Institute, National Institutes of Health. Bethesda, MD. Report prepared under contract by Stanford Research Institute, Menlo Park, CA.
15.  SELIKOFF, I. J., E. C. HAMMOND & J. CHUNG. 1968. Asbestos exposure, smoking and neoplasia. J. Am. Med. Assoc. **204**: 106–112.
  ANON. 1985. Tobacco's cost/benefit to the U.S. Economy. Non-smokers' Newsletter **8**: 3. American Lung Association of Connecticut, East Hartford, CT.
  ANON. 1984. Smoking and health. Breathing Views. **14**: 1. Lung Association of Connecticut.
16.  NATIONAL RESEARCH COUNCIL: COMMITTEE ON BIOLOGICAL EFFECTS OF ATMOSPHERE POLLUTANTS. 1971. Asbestos. The need for and feasibility of air pollution controls. National Academy of Sciences. Washington, D.C.
  U.S. ENVIRONMENTAL PROTECTION AGENCY: OFFICE OF AIR QUALITY PLANNING AND STANDARDS. 1974. Background Information on National Emission Standards for Hazardous Air Pollution. Proposed Amendments to Standards for Asbestos and Mercury. USEPA Research Triangle Park, NC.
17.  SUMMERS, A. L. 1919. Asbestos and the Asbestos Industry. The World's Most Wonderful Mineral. I. Pitman & Sons. London.
18.  NATIONAL RESEARCH COUNCIL. 1979. Polychlorobiphenyls. A Report Prepared by the Committee on Assessment of Polychlorobiphenyls in the Environment. National Academy of Sciences. Washington, D.C.
  FARRINGTON, J. W., E. D. GOLDBERG, R. R. RISEBOROUGH, J. H. MARTIN &

V. T. Bowen. 1983. U.S. "Mussel Watch" 1976–78: An overview of trace-metal, DDE, PCB, hydrocarbon. Environ. Sci. Technol. **17**: 490–496.

Gilbertson, M. & L. Reynolds. 1974. A summary of DDE and PCB determinations in Canadian birds. Can. Wildlife Ser. Occ. Pap. #19. Ottawa, Canada.

19. Thomann, R. V. & J. P. St. John. 1979. The fate of PCBs in the Hudson River ecosystem. Ann. NY Acad. Sci. **320**: 610–629.

Nadeau R. J. & R. P. Davies. 1974. Investigation of PCB in the Hudson River. U.S. Environmental Protection Agency. Washington, D.C.

20. Frink, C. R., B. L. Sawney, K. P. Kulp & C. G. Fredette. 1982. Polychlorobiphenyls in Housatonic River Sediments in MA and CT: Determination, Distribution, and Transport. CT Agric. Expt. Stat. Bull. New Haven, CT.

20a. For PCB incineration project at Middletown, CT, *see*:

The Northeast Utilities Program for Burning PCB-Contaminated Transformer Oil at Middletown, CT Station Unit 3. March 1981. Northeast Utilities Co. Environmental Programs Dept. Berlin, CT.

Hall J. M., G. T. Hunt & P. Wolf. 1981. Evaluation of the PCB Thermal Disposal Process in the Unit 3 Boiler at Northeast Utilities Middletown Station. GCA-TR-81-82-G GCA Corp. Bedford, MA.

Anon. n.d. Northeast Utilities Experience in Burning PCB-Contaminated Oil (26 pp). Available from Northeast Utilities Environmental Programs Dept. Berlin, CT.

20b. Whelan, E. M. 1985. Toxic Terror. Jameson Books. Ottawa, IL.

21. Mackay, D., S. Peterson, S. J. Eisenreich & M. S. Simmons, Eds. 1983. Physical Behavior of PCB's in the Great Lakes. Ann Arbor Science Press. Ann Arbor, MI.

22. Fischoff, B., P. Slovic, S. Lichtenstein, S. Read & B. Combs. 1978. How safe is safe enough? A psychometric study of attitudes toward technological risks and benefits. Policy Sci. **9**: 127–152.

According to David Okrent, the art of risk-benefit assessment originated with Starr's 1969 paper in *Science*. Okrent gives a comprehensive list of seminal background papers on risk assessment.

Okrent, D. 1980. Comment on societal risk. Science **208**: 372–375.

Starr, C. 1969. Societal benefit versus risk. Technological risk. Science **165**: 1232–1238.

National Academy of Engineering. 1986. Hazards: Technology and Fairness. NAE Series on Technology and Societal Priorities. National Academy Press. Washington, DC.

23. Selikoff, I. J. & E. C. Hammond. 1979. Public Control of Environmental Health Hazards. Ann. NY Acad. Sci. **329**: 1–405.

Starr, C. 1972. Benefit-cost studies in sociotechnical systems. *In* Perspectives on Benefit-Risk Decision Making. National Academy of Engineering. Washington, D.C.

Starr, C. & C. Whipple. 1980. Risks of risk decisions. Science **208**: 114–119.

Schwing, R. C. & W. A. Albers, Eds. 1980. Societal Risk: How Safe is Safe Enough? Proceedings of a Symposium Sponsored by General Motors. Plenum. New York, NY.

Comar, C. L. 1979. Risk: A pragmatic *de minibus* approach. Science **203**: 319.

Warner, F. & D. H. Slater. The Royal Society Study Group on Risk, 1981. The assessment and perception of risk. Proc. Roy. Soc. London **A367**: 1–206; *see* especially the article by Slovic *et al.*: 17–33.

24. Worrell, A. C. & J. A. Sinden. 1982. Unpriced Values. Wiley. New York, NY. *See also* Benoit, R. J. 1981. Ecologic value scales for small streams in river pollu-

tion control. *In* River Pollution Control. M. J. Stiff, Ed.: 305–318. Ellis Horwood, Ltd. Chichester.
25. PERROW, C. 1984. Normal Accidents: Living with High-Risk Technologies. Basic Books. New York, NY.
26. MARSHAL, E. 1979. Public attitudes to technological progress. Science **205**: 281–285.
27. CAIRNS, J. 1980. Estimating Risk. Bioscience **30**: 101–107.
NATIONAL RESEARCH COUNCIL. 1975. Principles for Evaluating Chemicals in the Environment. National Academy of Sciences Washington, D.C.

# The Meaning of "Nuclear Winter"
## Scientific Evidence and the Human Spirit


H. JACK GEIGER[a]

*President,*[b] *Physicians for Social Responsibility*
*Washington, D.C. 20009*


This subject—the consequences of nuclear war that have come to be known as "nuclear winter"—is at once particularly difficult and particularly important. It is difficult because this is still a young subject, an area of active scientific exploration. New publications are appearing in the scientific literature at a rate of more than one a month, and we can expect new scientific contributions on this subject to continue for several years. There is a very substantial range of scientific opinion, varying findings, and some controversy, as is usually the case with any new area of scientific inquiry.

All this is particularly true with regard to the atmospheric effects of nuclear war, because—fortunately for all of us—there is no "laboratory" of nuclear war and there are no direct experimental data: we must rely upon computer models of the atmosphere and of nuclear weapons' effects. And computer modeling of the atmosphere, in turn—particularly of the global atmosphere as a single system—is relatively new and imperfect.

The topic is important, despite these difficulties, because *nuclear winter, however it is finally defined, significantly changes and **expands** the risks of nuclear war, qualitatively and geographically.* It is also important because the role of physician, as teacher and influencer of public opinion, depends not only on a moral commitment to life, but also on our scientific credibility. In a new and fast-changing scientific field, it is not easy to be accurate about the results of new findings, to avoid both overstatement and understatement. Physicians are neither atmospheric scientists nor ecologists—the two disciplines most relevant to this problem—and so we have a new task. Just as most of us were not physicists or radiobiologists, yet had to learn enough of these disciplines to be able to report the consequences in medical terms of nuclear weapons, the sources and effects of blast, heat and radiation, so must we learn something of the scientific origins of analyses of nuclear winter.

In this paper I propose:

First, to review briefly the history and origins of the basic ideas underlying nuclear winter;


[a] Address for correspondence: H. Jack Geiger, M.D., Arthur C. Logan Professor of Community Medicine, CUNY Medical School/Sophie Davis School of Biomedical Education, City College of New York, New York, New York 10031.
[b] Currently, Immediate Past President.

Second, to present the findings and predictions of the TTAPS (Turco, Toon, Ackerman, Pollock, Sagan)[1] group headed by Carl Sagan, the ecology group headed by Paul Ehrlich, and the data presented in Washington in October–November 1983 and subsequently published in *Science*.

Third, I will briefly review some of the further developments and scientific analyses regarding nuclear winter since the initial Sagan-Ehrlich[2] announcements of 1983, touching on some of the revisions and controversies and trying to indicate the current status of the field.

Finally, I will summarize what, in my view as a physician, can and should be said about nuclear winter.

For most of the nuclear era, discussions of the effects of nuclear weapons have concentrated on blast, heat and radiation. Five years or so ago, damage to the ozone layer of the atmosphere, and the consequent risks of penetrating ultraviolet radiation, was added to the list. Then, only a little over two years ago, the first mention was made of widespread atmospheric consequences of nuclear war. In the Summer 1982 issue of *Ambio*, the international environmental journal of the Royal Swedish Academy of Sciences, Paul Crutzen of Berlin and John Birks of Denver called attention to the probability that nuclear war would have a profound atmospheric impact and create major changes in climate.

The mechanism they described was straightforward. Groundburst nuclear weapons would throw up millions of metric tons of dust, injecting them high into the earth's atmosphere; airburst nuclear weapons, particularly those over cities and forests, would create huge fires which would produce enormous volumes of smoke and soot. The dust particles, and particularly the smoke and soot in the troposphere and stratosphere, would absorb the sun's heat and light. With all that heat absorbed at high altitudes, the top of the atmosphere—normally very cold—would become warm, and the bottom of the atmosphere—at the earth's surface, where we live—would grow cold, in a gigantic temperature inversion. Further, the high-altitude absorption of sunlight would leave the earth's surface in partial or near-total darkness, raising the possibility of the impairment of photosynthesis and the disruption or destruction of the food chain.

It is noteworthy that the U.S. National Academy of Sciences only a few years earlier (1975) had published a study of the world-wide effects of nuclear war that completely overlooked these atmospheric effects.

Building on the Crutzen-Birks work, over the next several years the astronomer Carl Sagan and a group of atmospheric scientists—Turco, Toon, Ackerman and Pollack—constructed a computer model of the global atmosphere and used it to calculate the atmospheric consequences of more than a dozen scenarios of nuclear war, ranging in total yield from only 100 megatons to more than 10,000 megatons, varying from scenarios assuming only airbursts to counterforce scenarios assuming only groundbursts, to mixed scenarios based on a variety of targeting strategies. They adopted as a baseline case a 5000-megaton exchange, almost entirely in the Northern Hemisphere (the U.S., Europe and the Soviet Union) with 20% of the total yield expended in airbursts over cities and industrial targets in the Northern Hemisphere—in sum, given current arsenals, a realistic possibility for a full-scale war. For each scenario, in each computer run (for this is what the process is — setting up a computer model) they focused on six factors:

1. How much dust and smoke were generated;
2. How much sunlight was absorbed by the dust and smoke;
3. How much the temperature changed;
4. How the dust and smoke spread, and how long before it all fell back to the surface;
5. The extent of radioactive fallout over time; and
6. How much ultraviolet light reached the surface after the dust and soot had fallen out.

Their findings were reviewed by a separate group of some 40 biologists, headed by the Stanford University ecologist, Professor Paul R. Ehrlich, who considered the biological consequences and other potential ecological effects.

Their calculations and conclusions were first publicly announced at a meeting in Washington, DC in early November 1983 and subsequently published in two papers in *Science* on December 13, 1983. They have been widely publicized, and I assume their findings are known to most of you. Therefore I will limit myself to a brief summary, in broad general outline, omitting quantitative details except in a few particulars. Their conclusions were:

1. An unbroken pall of darkness would cover the Northern Hemisphere. Within a week after the war, the amount of sunlight at ground level could be reduced to just a few percent of normal. The light would be absorbed primarily by soot and smoke from fires ignited by surface bursts and airbursts; the dust raised by surface bursts, while important, would have less climatic impact since it is typically poorly absorbing.
2. The low light level would disrupt photosynthesis and threaten the food chain. In the early months following a substantial nuclear exchange, the amount of light filtering through the cloud cover might not be adequate to sustain photosynthesis. Even assuming that plants would be otherwise undamaged, which is unrealistic, the lack of light would severely limit growth, and the consequences would cascade through all food chains.
3. *Effects* on the Southern Hemisphere would be greater than previously assumed. Large disturbances in global circulation patterns could greatly accelerate the interhemispheric transport of smoke, dust and radioactivity with consequent darkness and cold over much larger areas. Rapid interhemispheric mixing means that the Southern Hemisphere could be subjected to massive injections of nuclear debris soon after an exchange in the Northern Hemisphere. Previous studies had assumed that Southern Hemisphere effects of a nuclear exchange in the Northern Hemisphere would be minor.
4. A harsh "nuclear winter" would prevail. Nuclear war probably would have a major impact on climate lasting for several years. It would be

manifested by a dramatic drop in land temperatures to subfreezing levels for several months and dramatic changes in local weather and precipitation. Even if the war were to occur in the summer, many areas might be subject to continuous snowfall for months.

5. Subfreezing temperatures would substantially reduce chances for human survival. Except for areas near coastlines, land temperatures would plunge from normal levels to $-15°C$ ($+5°F$) and possibly to $-25°C$ ($-13°F$) with dire consequences for survivors. The impact of dramatically reduced temperatures on plants would depend on the time of year at which they occurred, their duration, and the tolerance limits of the plants. The abrupt onset of cold would be of particular importance, since plants that normally can withstand subfreezing temperatures would have no time to develop tolerance. A spring or summer war would kill or damage virtually all crops in the Northern Hemisphere.

6. Most uncultivated food sources would also be destroyed, as would most farm animals. Many animals that survived would die of thirst, as surface fresh water would be frozen to a depth of one meter or more over the interior of continents. Available food supplies would be rapidly depleted. Most of the human survivors would face starvation.

7. Nontarget areas that import food would be directly affected. Nations that now require large imports of food, including those untouched by nuclear detonations, would suffer the immediate cessation of incoming food supplies. These countries would be forced to rely on their local agricultural and natural ecosystems. This would be especially serious for many less-developed countries, particularly in the tropics.

8. The oceans will not freeze due to their enormous reservoir of heat, and coastal areas will be protected from the temperature drop. However, the enormous land-sea temperature differentials will produce severe and intense coastal storms, and it is unlikely that coastal areas could be a major source of food for survivors of a nuclear war because of the effects of run-off of silt and toxic chemicals from the land, destruction of ships, and concentrations of radionuclides in fish and other marine life.

9. Fire would be a major problem with serious and unanticipated consequences. About one-sixth of the world's urbanized land area, or approximately 240,000 square kilometers, would be partially burned by approximately 1000 of the megatons in the baseline scenario. The remaining 4000 megatons of yield could ignite wildfires and firestorms, and uncontrolled fires could sweep over large areas. For example, multiple airbursts over California in the late summer or early fall could burn off much of the state, leading to catastrophic flooding and erosion during the next rainy season.

10. Urban fires would generate large amounts of deadly toxins. Cities hold large stores of combustible, synthetic materials that would release large quantities of toxic gases (pyrotoxins) as they burn, including carbon monoxide, cyanides, dioxins and furans. Transport by winds to dis-

tant, initially unaffected ecosystems could be an important additional adverse side effect.

11. Ozone depletion would increase exposure to ultraviolet light (UV-B). High-yield explosions would inject nitrogen oxides into the stratosphere, which would result in large reductions in the ozone layer. The dust and soot would absorb the increased UV-B at first, but when the dust and soot cleared some months later, UV-B doses roughly 1.6 times normal would be transmitted to the surface.

12. Tropical forests could disappear. Tropical plants are less able to cope with even short periods of cold and dark than those in temperate zones. The tropical forests, which are the major reservoir of organic diversity, would largely disappear. This would, in turn, lead to the extinction of a majority of the species of plants and animals on earth. Furthermore, the dependence of urban populations in many tropical and developing countries on imported foods would lead to severe effects, even if those areas were not directly affected by the war. Large numbers of people would be forced to leave the cities and attempt to cultivate the remaining areas of forest, accelerating their destruction and the consequent rate of extinction.

13. Even small nuclear exchanges could trigger severe effects. A scenario involving only 100 megatons, if devoted entirely to airbursts over cities, would produce so much smoke and soot in the resulting fires that a two-month interval of subfreezing land temperatures, with a minimum near $-23°C$, could result. In this scenario of only 100 megatons, the subsequent darkness and cold would be almost as severe as that produced in a 5000-megaton scenario of mixed airbursts and groundbursts.

In sum, in the aftermath of a 5000-megaton nuclear exchange, survivors would face extreme cold, water shortages, lack of food and fuel, heavy burdens of radiation and pollutants, diseases, and severe psychological stress, all in twilight or darkness. It is clear that the ecosystem effects alone would be enough to destroy civilization as we know it in at least the northern hemisphere. These long-term effects, when combined with the direct casualties from the blast, suggest that eventually there might be no human survivors in the Northern Hemisphere; human beings, other animals and plants in the Southern Hemisphere would also suffer profound consequences.

The summaries of Sagan-Ehrlich *et al.* went on to say specifically that "the extinction of a significant proportion of the earth's animal and plant species would occur" and that "the possibility of total human extinction cannot be excluded."

This, then, is the image that has come to be known as nuclear winter: a picture of human survivors freezing and starving in darkness, shivering in the howling winds of violent coastal storms, dying of thirst, in a Northern Hemisphere in which civilization had been destroyed and agriculture virtually ended. Further, the tropics would be decimated and the Southern Hemisphere profoundly affected. And all this would occur in *addition* to the effects of blast, heat and radioactivity on which physicians, until now, have concentrated their discussion on medical effects. To these effects, if the nuclear winter scenario is correct, we must now add global atmospheric effects and therefore global environmental effects. And we must further take note of the *interaction* of physical, biological and social effects. With civilizations shattered and the social fabric — the fabric that characterizes us as human — destroyed, survivors will have to turn to natural systems for sustenance, at the very moment when those natural systems have been profoundly disrupted by climatic change.

At the same time that the Sagan-Ehrlich team was at work, similar research was being conducted independently by a Soviet team headed by academician Aleksandrov and G. L. Stenchikov of the Computing Centre of the Soviet Academy of Sciences. Their findings, also announced in the U.S. at the end of October 1983, were strikingly similar to the Sagan-Ehrlich findings. Within a month, preliminary results from another group, headed by M. Mac-Cracken, at the Livermore Laboratory in California, were also similar.

And, as it happened, a new committee of the U.S. National Academy of Sciences was already at work updating that 1975 report on long-term global consequences of nuclear war.[3] Many of the scientists on that committee had been active participants in the Sagan and Ehrlich groups and review committees, and of course the Sagan-Ehrlich publications and data had been circulated among the scientific community to which the NAS committees turn for comment and advice. It was widely expected that this new NAS report would rapidly confirm the Nuclear Winter findings.

But that has not happened. A first draft of the new NAS report drew more than 500 technical and scientific comments, criticisms and proposed revisions. Publication was delayed. A new report is expected. There has been a flurry of new scientific studies and publications, including one major report that I will describe, with results that differ substantially from the Sagan-TTAPS calculations. To make a bad pun about a terribly serious matter, the study of the atmospheric effects of nuclear war is up in the air.

To understand what has happened and is now happening, we must first grasp two facts about nuclear winter calculations.

First: In all the studies everything depends on, or is profoundly affected by, the computer model of the global atmosphere. We have only had such models for a few years. They vary greatly in complexity and sophistication, and they are changing rapidly as atmospheric science and computer science develop and change.

The Sagan-TTAPS calculations, for example, employed a relatively simple computer model of the atmosphere: a one-dimensional radiative-convective model. That means it can consider the temperature effects of heat absorption by smoke only in one dimension, the vertical (that is, altitude). For all horizontal variations (latitude and longitude, or North-South, East-West) it simply "parameterized" or *arbitrarily averaged* the spread of smoke and soot and the resulting temperature changes. The MacCracken-Livermore Laboratory group first used a one-dimensional model as well. But there are also two-dimensional and three-dimensional computer models of the atmosphere that are able to *calculate* rather than assume or average the heat processes in all dimensions. The importance of the model is illustrated by the Livermore Laboratory findings. When a one-dimensional model was used, they found temperature drops of as much as 30°C; when they used a two-dimensional model, their calculations yielded only a 15°C temperature drop.

In addition, models vary not only in their dimensionality but also in their sophistication within dimensions. One model may divide the atmosphere vertically into only two layers, for example, while another may be capable of as many as nine layers. The greater the number of layers, the better is the "resolution" of the model—that is, the better is its capacity to make climatic predictions for relatively small areas.

To add a final complexity, there are now three-dimensional *general circulation* models, which are able to take account of regional variations, seasonal variations, and to make relatively precise large-scale wind calculations. And there is yet another step, now under development: a three-dimensional *interactive* model that is able to calculate dynamic changes—that is, able to take account of smoke-spreading, smoke-transport and smoke-removal processes (such as the removal of smoke and soot by rainfall) *during* the injection of smoke into the atmosphere, during its spread, and during its heating. In sum, the computer model has the most profound effects on the calculation of results.

Secondly, we must understand that the nuclear winter calculations *depend on a whole series of prior assumptions or terms, and these prior assumptions are themselves matters of great scientific uncertainty or dispute.* Let me identify just a few of the questions about which there is uncertainty or disagreement:

1. How many million metric tons (teragrams) of dust will be thrown up by a nuclear explosion of given size?
2. How much of what is injected into the atmosphere will be dust, and how much will be soot?
3. What proportions of cities will burn? What is the fuel loading of different cities, and how will this affect burning? How fast will they burn? How long will they burn?
4. What will be the range and proportion of particle sizes in the soot? How fast will it settle out? How much will be rained out rapidly? How much rain will there be? What will be the effect of conflicting factors, since coastal storms will increase rainfall but temperature inversions will decrease it?
5. What are the characteristics of mass fires? What is the aging pattern of soot in the atmosphere? What are the relevant optical properties of dust and soot?
6. Most important, how uniform, or how patchy, will the smoke and soot distribution be? Will there be holes in the clouds of smoke and soot? Will there be shifting holes, below which temperatures will rise?

These are not mere technical quibbles. Their effects on results can be demonstrated by the major paper I just mentioned—that by Thompson and Schneider of the National Center for Atmospheric Research in Boulder, Colorado.[4]. Using the same nuclear war scenarios as the Sagan-TTAPS authors, and using many of the same prior assumptions, but employing a three-dimensional, nine-layered noninteractive general circulation model which considered the effects of clouds, humidity, seasonal variations in the solar zenith angle, variations in snow and ice cover, and so on, the NCAR group made three 20-day computer model runs for winter, spring and summer wars. Their results differed significantly with regard to the *intensity* of nuclear winter (how cold it will get) as well as its *duration* and its *geographic scope*. In the summertime case, for example, the TTAPS group predicted that the temperature would drop to somewhere between $-15°C$ and $-25°C$ ($+5$ to $-13°F$) in the Northern Hemisphere. But Covey *et al.* found only a 15 to 20°C drop, to temperatures of just above or just below freezing (i.e., to temperatures of about 0°C or 32°F). In the wintertime case, the NCAR group predicted a temperature drop of only about 5°C, or 9°F, below normal prevailing temperatures.

Furthermore, the NCAR group found that many of the temperature drops would be "quick freezes" lasting only one or two days, because of patchiness in the overlying smoke and soot—a picture very different from the Sagan prediction of several months of continuous subfreezing cold. The overall length of time during which such "quick freezes" might occur is uncertain.

And finally, the NCAR group predicted only thin streamers of smoke, with quick freezes beneath them, extending over the tropics and into the Southern Hemisphere, with much of the tropics and the Southern Hemisphere not directly affected.

It should be clear that the NCAR findings still yield a picture of climatic disaster, and they confirm the basic thrust of the Sagan-TTAPS and Aleksandrov findings, but they present a picture that differs significantly from that of unending months of darkness, unrelieved subfreezing cold, widespread and uninterrupted groundwater freezing, and direct destruction of animals and man by exposure, especially over areas outside the band from 50 to 70 degrees north, in the Northern Hemisphere, where effects will be strongest.

We can expect such differences in findings to continue. We can expect continuing vigorous scientific debate. We can expect, I believe, that the National Academy of Sciences report, when it appears, will contain a significant modification of the Sagan findings, and will more closely approximate the NCAR studies. We can expect that Professor Sagan and his colleagues will vigorously defend their findings, even as they press forward with further work. We need to understand that this is not an adversary argument, a controversy between "good guys" and "bad guys" who are seeking either to exaggerate or to minimize nuclear winter findings for political or any other reasons. This is simply the natural process of scientific research and discussion in pursuit of the truth, as best it can be determined.

Given all this, what can we as physicians now say about nuclear winter that is accurate, credible and relevant?

First, and most important, we can point out that it took until the fourth decade of the nuclear era even to discover this effect. In the 37 years following Hiroshima and Nagasaki, while we were told over and over again that all the major effects of nuclear weapons were known and understood, we did not know that there could be major effects on the global atmosphere. What has been discovered is a major new dimension of the consequences of nuclear war, created in part by the insane growth in arsenals and total destructive power.

And we are entitled to wonder what other surprises lie ahead, what other devastating effects of nuclear war are as yet unknown and undescribed, since what we do not know about the effects of nuclear war is surely much greater than what we do know.

Second, we can say that there seems to be no question that, in addition to all the other effects of blast and heat and radiation, there will be a nuclear winter in the Northern Hemisphere, particularly in 50–70 degrees North.

Although the details of intensity and duration of these Northern Hemisphere effects may still be uncertain, we must as physicans point out the consequences of exposure and hunger, the dangers of toxic fog at ground level, the psychological and biological effects of continuing darkness, and the longer-term consequences, both for the Northern Hemisphere and for the areas to which the Northern Hemisphere exports food, of the destruction of agriculture.

Third, we must point out the synergistic interaction of destruction of physical, biological and social systems, and the consequences of simultaneous profound biological damage, damage to physical infrastructures, damage to civilization and social support systems, and damage to the environment.

Fourth, we can say that significant areas of the tropics will be decimated, since even quick freezes will destroy tropical plants and forests, with a huge loss of biomass and of plant and animal species. And we can say that no one can estimate what the long-term consequences of so great an ecological disruption will be. Those who speak of winning, or surviving, or even countenancing a nuclear war are gambling, ignorantly and arrogantly, with the chain of life itself, with the whole intricate web from phytoplankton to man.

Fifth, we should note that Third World nations now have a far larger direct stake in ending the arms race, superpower confrontation, and the risk of nuclear war than they or we have realized until now. Some of them will suffer direct atmospheric effects. Many more will suffer the secondary environmental and social effects, as manifested, for example, by the disruption both of food production and of food transport. We can expect starvation thousands of miles away from the area of a nuclear war.

The message of nuclear winter, then, is that we are one human family, living in one indivisible home, our planet earth. There can be no comfort in the notion that after the superpowers destroy themselves, and much of the Northern Hemisphere, the meeker or the poorer nations will inherit the earth, for the nuclear war will have destroyed that inheritance, the fundamental inheritance that is owed equally to all of the world's children.

What are the things that we *cannot* say, credibly and reliably, about nuclear winter at this point? In my view, they are the following:

First, it is unwise and unsupportable to predict the extinction of the human species as a credible major risk, on the basis of current knowledge. The weight of scientific opinion at present is that extinction is an extremely remote risk. We have a responsibility to be credible and accurate in discussing it.

We also have a responsibility to point out the absurdities of this issue. Is there anyone who believes that human extinction must be proved beyond a shadow of a doubt before we even worry about it? Is there anyone who believes that even a tiny risk of so major an outcome is trivial? Is there anyone who believes that it is acceptable to have a nuclear war as long as it does not result in the extinction of mankind?

Second, I believe we are not justified in speaking of a "threshold" below which global atmospheric changes will not occur—the argument that we should reduce our arsenals to a point below 1500 megatons because such an exchange would not result in nuclear winter. The NCAR data predict quick freezes with even smaller exchanges. There are far better and surer reasons for reducing the world's nuclear arsenals. And is there anyone who believes that even a 1500-megaton nuclear war would be acceptable because it might leave the climate untouched?

Third, as our scientific knowledge increases, we should avoid becoming immersed in mindless, endless arguments about the technical details of temperature drops, particle rainouts, and computer models. To focus on such details, in my view, is to create analogues of the mindless, endless arguments about the minutiae of missile counts, warhead numbers, ranges and accuracies, circular errors probable and the like—all the while distracting attention from the only fundamental fact, the fact that *any* nuclear war would be a catastrophe without precedent in the history of our species.

Nuclear Winter—global atmospheric consequences of nuclear war—is an *additive* reason to oppose nuclear war, not a *unique* reason. We do not need nuclear winter to realize, or teach, that nuclear war is dreadful. Conversely, nuclear winter confirms and enlarges what we, as physicians, have been saying all along about blast and trauma, heat and burns, radiation and disease, and above all the destruction of the social fabric that makes us human.

Let me close with a final, personal word about the meaning of nuclear winter. Lewis Thomas has described the earth and the life it supports as a single system, a thin, richly varied, astonishingly beautiful web of life that is bound to the round warm stone of earth, a web of life "clinging to the round warm stone, turning in the sun."

Against the beauty of that image, perhaps we can recognize that we are *already* in the grip of a nuclear winter, and suffer under it. The nuclear winter that we endure now is a winter of the human spirit—a bleak, cold, contracted willingness to countenance destruction and self-destruction, pain and death, contamination and violence, ripping the web of life in a global madness. Nuclear winter is here. It is our very humanity that is freezing over.

But there is more to humankind than this meanness of spirit, this angry chill in our souls. We are also capable of warmth, generosity, a happy rationality, love, commitment to life, a blossoming of the spirit. Physicians in most corners of the earth strive to represent that alternative. In the face of nuclear winter, in our work here, and when we go home, let us become the first sign of spring.

## REFERENCES

1. EHRLICH, PAUL R. *et al.* 1983. Long-term biological consequences of nuclear war. Science **222**: 1293–1300.

2.  TURCO, R. P. *et al.* 1983. Nuclear winter: Global consequences of multiple nuclear explosions. Science **222**: 1283–1292.
3.  NATIONAL ACADEMY OF SCIENCES-NATIONAL RESEARCH COUNCIL. 1985. The Effects on the Atmosphere of a Major Nuclear Exchange. National Academy Press. Washington, DC.
4.  THOMPSON, S. L. & S. H. SCHNEIDER. 1986. The nuclear winter debate: Comment and correspondence. Foreign Affairs **65**: 171–178.

# Health Aspects of Low-Level Ionizing Radiation

LEONARD R. SOLON

*Director, Bureau for Radiation Control*
*New York City Department of Health*
*New York, New York 10013*

## THE GENERAL PROBLEM

All living organisms — from the first primordial living cells that formed more than several billion years ago to every species that exists on the planet today, including humans — have lived or live in an environment of ionizing radiation. There is little doubt that natural or "background" radiation from cosmic rays and naturally occurring radioactive elements have contributed to the process of Darwinian biological evolution. To what extent this may be so is uncertain and conjectural, but that ionizing radiation can produce perturbations in the genetic molecular structure of reproductive cells is without any scientific doubt. It is also certain that somatic cells in human body organs can be altered by ionizing radiation to produce cancer at sufficiently high doses of ionizing radiation.

The two health effects of primary concern in man, especially at low doses for large human populations, are generally regarded to be cancer production at the somatic level and potentially adverse inherited physiological effects at the genetic level.[1-4] With respect to the production of cancer in man, the information regarding low radiation doses is almost entirely inferential and mathematically deduced from the effects of large radiation doses. The situation with reference to unfavorable genetic mutations produced in man by ionizing radiation is connected by an even more tenuous chain of inference from large doses in animal experiments. In fact, though there is no doubt in the scientific community that man is susceptible to transmittable genetic damage from ionizing radiation doses delivered to the female egg cells or the male sperm cells, such effects have not been demonstrated directly in man even at elevated radiation levels. No effects have been found, for example, from the children of survivors of the Hiroshima and Nagasaki bombings of 1945. It would appear that the estimation of gross inherited genetic risks of ionizing radiation will continue to be based entirely on analysis of laboratory animal data for the foreseeable future. Advanced current research has shown that there are qualifications to this statement, which will be discussed later.

# DEFINITION OF LOW-LEVEL RADIATION EXPOSURE

There is no unanimous consensus as to what constitutes "low" in the designation of low-level radiation. We shall use the term to refer to whole-body yearly doses of 5 rem or cumulative doses of 50 rem. The annual dose of 5 rem per year is, of course, the currently accepted maximum permissible dose for workers in an occupational ionizing radiation environment. For the purposes of orientation it is noted that the average natural or background ionizing radiation dose in the United States from the combination of cosmic radiation and naturally occurring radioactive materials such as uranium, radium, thorium (and their decay daughters), potassium-40, and carbon-14 is about 0.1 rem/year (100 mrem/year). Very approximately the average radiation dose, principally from diagnostic X-radiation, is about the same. Thus the total average radiation dose to Americans is about 0.2 rem/year or a factor of 25 lower than what is characterized as the upper limit of the low-level radiation range for an annual dose.

A caveat is in order, however. Everyone is exposed to background radiation. In the United States the range that is averaged is the range of background between Denver, Colorado (164.6 mrem/year) and Atlantic City, New Jersey (63.7 mrem/year), a factor between two and three.[5] Medical radiation exposure ranges from essentially zero (for large numbers of people) to very high doses (in a small number). It is probably proper to average the dose for genetic effects. It is not altogether clear that the dose for somatic effects deserves the same kind of averaging.


# BASIC STATISTICAL PROBLEM: REQUIRED SAMPLE SIZE

The central difficulty of the epidemiologic problem has its foundation in the very large sample sizes required to observe the impact of ionizing radiation on biological effects. In general the biological effects attributed to low levels of ionizing radiation, particularly cancer production and genetic mutations, also have a substantial incidence in man that is not attributable to radiation. Thus the effect of the augmented risk represented by a low-level dose of ionizing radiation is observed as an increase in incidence of a specific cancer or genetic anomaly. A cancer or genetic change can have, of course, a genesis wholly unrelated to radiation.

Much of the variability in different scientific estimates of low-dose ionizing radiation focuses on the problem in extrapolating reliable data at high doses. In general, a smaller radiation dose yields a smaller excess risk irrespective of the mathematical models employed. The most commonly employed mathematical formulations are the linear-quadratic model of the Committee on Biological Effects of Ionizing Radiation, BEIR-III,[1] the linear and presumably more conservative model used in the earlier BEIR-I report of 1972,[6] or the quadratic model, which was espoused by at least one of the contributors of BEIR-III. Parenthetically it should be mentioned that the chairperson of BEIR-III disassociated himself from one section of the report because of the rejection of the linear model in favor of the linear-quadratic model (Ref. 1, p. 227). The member of the BEIR-III committee favoring the less conservative quadratic model also separated himself from accepting those elements of the report that rely upon the linear-quadratic model (Ref. 1, p. 254). The basic difficulty in estimating dose/response relations is due to the absence of reliable data at low doses.

## REPRESENTATIVE HYPOTHETICAL EXAMPLES[4]

It is constructive to consider a few examples that bring out the essential difficulty in an explicit way. Charles Land of the National Cancer Institute has already done this for us.[4] He starts with the simplest situation, the linear hypothesis, which says that the excess risk of a particular pathology is proportional to the radiation dose received.

Suppose for a hypothetical form of cancer in a particular organ, one requires a large dose of 100-rad in a sample of 1000 persons to see an increase in cancer. Very approximately, the number of persons required to observe an effect at lower doses would increase with the inverse square of dose received. Thus, to see an increase for a population receiving 10 rad would require 100,000 persons and for a population receiving 1-rad would necessitate $\frac{(100 \text{ rad})^2}{1 \text{ rad}} \times$ 1000 persons or a large population sample of 10 million persons.

The essential obstacle to detecting ionizing-radiation-induced physiological effects may be seen from the following specific cases:

### *Breast Cancer from Mamographic Examination*

Suppose a group of comparable women, all aged 35 years, received a tissue dose of 1 rad to each breast as the result of a mammographic examination. If one assumes that no malignancy would occur in the first 10 years as the result of the radiation and one examines the situation in the ensuing 10 years, approximately 1910 cases of breast cancer per million women per year would normally occur in an unirradiated population. From the low-radiation exposure of the particular mammographic examination, one might expect about six additional cases of cancer per million per year. As one might predict, one could not be aware of any radiation effect without a very large population. In fact, with the assumptions made, a population sample of the order of 100 million would be required to be confident that a radiation malignancy increase had indeed occurred.

### *Leukemia from Ionizing Radiation*

If the basic radiation effect on a dose-received basis is of the same order, but the so-called "normal" (i.e., nonradiation-induced) cancer has a much lower incidence, the irradiated population that must be sampled can be materially less. Addressing the normal nonradiation incidence of leukemia in men at aged 35, the rate is only 44 cases per year per million contrasted to the 1910 cases of breast cancer per million per year in the women of the prior example. A 1-rad bone marrow dose to a population of men followed for 15 years after exposure would increase this nonradiation malignancy rate by the order of 3 cases per million per year in an exposed group. Again this latter number is deduced by linear extrapolation from doses several orders of magnitude greater. Although the sample size required to detect a leukemia effect from this low level of ionizing radiation dose would still be very large — between 6 million and 7 million men — it is only one-fifteenth of the 100 million women of the breast cancer hypothetical example.

## ENDPOINTS FOR THE DETECTION OF THE BIOLOGICAL EFFECTS OF LOW-LEVEL RADIATION EXPOSURE[1,2]

There are five general endpoints that are of potential epidemiologic interest in the detection of the biological effects of low-level radiation exposure. There are:

(1) specific cancer or tumor production observed on the large-population sampling level;
(2) genetic changes observed on the large-population sampling level;
(3) genetic changes observed at the molecular or cytologic level;
(4) chromosomal aberrations seen on individual cytologic examination; and
(5) developmental or teratogenic effects in children as the result of prenatal radiation exposure.

With respect to radiation-induced carcinogenesis, believed to be the principal result of low-level radiation exposure to large populations, we have already cited the basic mathematical impediment to meaningful epidemiologic conclusions. In addition to the relatively high normal incidence of cancer unrelated to radiation there is the low average rate per person-rem ($10^{-4}$–$10^{-3}$), i.e., one case of cancer per lifetime for a 1-rem dose to 1000 to 10,000 persons.

Estimated lifetime risks for cancer induction in men and women based on a linear dose-response relationship given by Jacob Fabrikant of the University of California are shown in TABLE 1. They yield an average radiation-induced cancer rate of $2.64 \times 10^{-4}$ per person-rem, with a mortality rate of $1.25 \times 10^{-4}$ per person-rem; i.e., about half of the total number of cases of cancer are fatal.

Another serious disadvantage in epidemiologic investigation is the varying latency periods between radiation exposure and the development of cancer. This may be as short as 2 years, but is more often 10 to 30 years.

With respect to inherited genetic characteristics, the situation is even less favorable than for cancer epidemiology. As mentioned previously, inherited gene mutations have not yet been observed in man, even at the highest doses encountered by humans, either occupationally or as the result of exposure

**TABLE 1.** Lifetime Risk Estimates for Radiation-Induced Cancer per Million Persons Exposed to 0.1 Rem[a]

| | Number of Cases | |
| --- | --- | --- |
| Cancer Type | Fatal | Nonfatal |
| Breast (females) | 5 | 3 |
| Lung | 2 | 0.1 |
| Leukemia | 2 | 0.1 |
| Thyroid | 0.5 | 1.0 |
| Bone | 0.5 | 0.2 |
| All other | 5 | 2 |
| Total (males) | 10 | 12.4 |
| Total (females) | 15 | 15.4 |

[a] From Fabrikant.[7]

to nuclear weapons. However, advanced molecular methods based on the analysis of particular proteins in blood and urine are now being undertaken (Ref. 2, p. 38). These will involve large populations and have latency periods which depend on the time between the irradiation of men and women and their having children.

Of particular promise, both in terms of very much shorter latency periods and the relatively small populations that may be investigated, is the cytologic measurement of chromosome aberrations after irradiation. One study reported in 1979 of nuclear-ship dockyard workers involved only 197 individuals.[8] At present such studies are extremely labor-intensive and unsuitable for large population investigations. Also, the connection between chromosomal aberrations and deleterious physiological effects are not known. However, the basic method is very sensitive and its present use for exposed individuals points to a potential future application with ongoing development.

## RADIATION HORMESIS: POSSIBLE, IMPROBABLE, BUT WORTH FURTHER INVESTIGATION

The principal authoritative scientific reports to which we look for guidance in radiological public health regulatory matters (e.g., BEIR-III or UNSCEAR [United Nations Scientific Committee on the Effects of Ionizing Radiation]-1977)[10] have generally ignored the phenomenon or conjectured phenomenon called radiation hormesis. The term *hormesis* from the Greek "rapid motion" is a word borrowed from pharmacology. It was first employed in the 1940s to characterize the acceleration of fungal growth by low concentrations of oak bark extract that suppressed fungal growth at high concentrations.[11] The designation is now employed generally in pharmacology to refer to substances that may be toxic at high levels, but stimulating or beneficial at low levels.

It has been argued by both T.D. Luckey[12] at the University of Missouri and more recently by Richard J. Hickey and his associates[13] at the University of Pennsylvania that there is a substantial body of epidemiologic evidence to support the position that low levels of radiation exposure, both natural and man-made, are indeed favorable to life-lengthening and cancer reduction.

One of the principal notions of radiation hormesis theory is derived from evolutionary biology: Since we have emerged, in the evolutionary trajectory, from an environment of low-level ionizing radiation, it can be reasonbly anticipated that levels ecologically compatible with the background radiation can be salutary to human life.

That the prestigious group of experts in radiation biology and radiologic health responsible for BEIR-III, for example, have not elected to address the radiation hormesis hypothesis suggests that the peer-review threshold of scientific acceptability may not have been reached by its proponents. However, and I speak as a firm skeptic and nonbeliever in telepathy, psychokinesis, and radiation hormesis, I believe that the latter, at least, warrants far more complete scientific investigation. The truth of whether radiation has potential favorable implications for public health should be either confirmed or refuted by objective scientific scrutiny once and for all.

## CURRENT STATUS

The penultimate overall conclusion that emerges at present is that there is no identifiable irradiated population that can be epidemiologically investigated in order to define with reasonable accuracy the low-level radiation-induced risk of cancer.

However, there are fairly large occupational groups that lend themselves to prospective study, which may furnish direction for future follow-up and investigation (Ref. 2, p. 109).

(i) There are twenty thousand nuclear-power plant workers with an average annual dose of 0.5 rem, one-tenth of the current permissible occupational radiation exposure. In this group one would anticipate a nonradiation-induced cancer incidence of 2730, with a projected excess incidence of between 26 and 45 cases of cancer — an increase of between about 1 and 2 percent.

(ii) A second group comprises 3000 workers in Department of Energy facilities, each of whom has received at least one annual dose of 5 rad or more. If the average dose of this group was 15 rem, one would anticipate an excess incidence of 6 to 11 cases of cancer over a normal expectation of about 1000.

(iii) The third and largest occupational group is made up of 170,000 radiologic and nuclear medicine technologists (about 2/3 women and 1/3 men). Over a 15-year period the estimated average cumulative dose of this group was 8 rads. On the basis of linear extrapolation, one would anticipate an excess cancer incidence of between 99 and 277 cases of cancer as contrasted with a normal nonradiation cancer expectation of 26,000. This is, of course, a small mathematical increase of only 0.4 percent to 1.1 percent, which would not be statistically significant at any confidence level.

In addition, one environmental as contrasted to occupational group warrants further follow-up investigation. Between 1951 and 1963, nuclear weapons tests were conducted above ground at the Nevada Test Site in Mercury, Nevada (Ref. 2, p. 110). Residents of states near the site numbering about 172,000 men and women are estimated to have received an average dose from weapons fallout of 0.3 rem prior to 1958. After 1958 tests were conducted principally underground, so fallout dose essentially vanished. It is anticipated that most excess cases of radiation-induced cancer, estimated to be between 6 and 16 cases, will appear between 1980 and 1990. Of course, simple radiation-induced cancer production should not be detectable in this population since the normal cancer incidence will be of the order of 51,000.

In connection with this group it was asserted in 1979 that there was a demonstrated increase in childhood leukemia from Nevada test fallout in 17 counties of southern Utah between 1951 and 1958. This contention has very recently been rejected in a recent paper reporting investigation by epidemiologists from the National Cancer Institute.[9]

I mentioned earlier that gross inherited genetic risks were not demonstrated directly from the children of survivors of the Japanese detonations. However, a careful examination of chromosomal aberrations and cytologic changes done by Schull, Otake and Neel in behalf of the Atomic Bomb Casualty Commission (ABCC) and its successor agency (since 1975), the Radiation Effects Research Foundation (RERF), points to strong confirmation of theoretically expected genetic effects.[18] These investigators arrive at an average doubling dose of 156 rem for genetic mutations. On the basis of the linear hypothesis, this would make low-level radiation effects on the reproductive cells of sufficient concern to warrant strenuous continued efforts in the reduction of superfluous ionizing radiation used in the healing arts.

Some inferences that have been drawn from the work of Schull, Otake and Neel have been challenged in part by Hamilton, Nagy, and Robinson of Brookhaven National Laboratory.[19] At this point the matter must be viewed as unresolved and requiring more investigation.

## POSSIBLE PROGRAMS FOR THE CITY OF NEW YORK RELATING TO RADIATION EPIDEMIOLOGY

The following suggestions are offered for consideration in New York City:

(1) The City of New York can take the initiative in establishing a radiation-worker exposure registry as a prototype for a national radiation-worker registry.

(2) A broad-spectrum study using traditional epidemiologic methods as well as the most advanced biochemical and cytologic tests for genetic and chromosomal anomalies can be undertaken in New York for a selected group of individuals occupationally connected with radiation. The group could include dentists, radiologists, nuclear medicine technologists, X-ray technicians, certain categories of nurses, and particle-accelerator research physicists. It is hoped that research funding may be obtained from the appropriate federal agencies.

(3) The FDA X-Ray Record Card (HHS Publication No. FDA 80-8024) should be made generally available for use through schools and health facilities for all individuals. Its mandated employment could be incorporated in the City of New York Health Code.

# SOME CONCLUDING OBSERVATIONS

In addition to these suggestions for our own jurisdiction, the City of New York and its good neighbors in the metropolitan area, there are some general observations which I would like to offer.

## *Another BEIR Report Needed*

Among the authoritative and frequently cited inventories of scientific information on this subject are the BEIR reports of the National Academy of Sciences. BEIR is acronymic for Biological Effects of Ionizing Radiation. Three such reports have been issued. The first, in 1972, and referenced as BEIR-I or more often as BEIR-1972, has the title *The Effects on Populations of Exposure to Low Levels of Ionizing Radiation.*[6] The second, published in 1977 as BEIR-II, endeavored to apply the scientific findings of BEIR-1972 to practical problems of radiation regulation and standard setting. It has the title *Considerations of Health Benefit–Cost Analysis for Activities Involving Radiation Exposure and Alternatives.*[14] The most recent BEIR report, to which we have made prior reference, was published in 1980 after some protracted polemics involving strenuous technical disagreements among the BEIR-III committee members. This dispute received considerable exposure not only within the scientific community, but even in the media.

BEIR-III has the same title as the first report and materially updates the scientific subject matter of BEIR-1972. For different and nearly opposite reasons, strong exception to the linear-quadratic cancer-risk dose-response hypothesis advanced in the report was taken by Edward P. Radford, Professor of Environmental Epidemiology of the Graduate School of Public Health, University of Pittsburgh and by Harald H. Rossi, Professor of Radiology of the Columbia University College of Physicians and Surgeons.

As mentioned earlier, Dr. Radford, Chairman of the BEIR-III Committee and also Chairman of the Subcommittee on Somatic Effects, defended the linear hypothesis of the earlier BEIR-1972 report. On the other hand Dr. Rossi argued that both the earlier linear model and the linear-quadratic model adopted in BEIR-III are inapplicable to most biologic response situations actually encountered. He contended that the positions adopted in the main BEIR-III report tend to seriously overstate the radiation-induced cancer risk of low linear energy transfer (LET) (e.g., gamma radiation) while underestimating the hazards of high-LET radiation (e.g., neutrons or alpha particles).

Now a very substantial fraction of the scientific edifice upon which rests today's perception of the effects of low-level radiation exposure for human populations is the intensively studied survivors of the atom bomb explosions in Hiroshima and Nagasaki on August 6 and August 9, 1945, respectively. Let me conjecture that these tragic events in recent history contribute, in a scientific epidemiologic sense, more than half of our real effective data-points in the dose-biological response relationship for human beings.

As we know, there has been a recent re-evaluation of the original Oak Ridge

T-65 (for Tentative 1965) dosimetry, which calls into serious question the dose estimates received by the Hiroshima-Nagasaki survivors.[15-17] In inadequate and still arguable summary, it would appear that the high-LET neutron dose from the uranium-235 gun-type Hiroshima weapon was overestimated by an order of magnitude and the low-LET gamma dose was underestimated importantly. Furthermore, the plutonium-239 implosion weapon employed at Nagasaki yielded a gamma-ray dose profile to the survivors which was seriously overestimated.

I should mention parenthetically, and not altogether positively, that the dose re-evaluation would tend to support Radford's dissent rather than Rossi's opinion, but this is only ancillary to the main issue. What is really important is that the BEIR-III and other authoritative expert reports rely to a decisive extent on dosimetry that is questionable.

I am constrained to conclude that it is time for new objective scrutiny of all the data, i.e., a new BEIR report which would take into full account the changes in the Hiroshima-Nagasaki dosimetry as well as other developments in radiobiology since 1976, when the BEIR-III committee first began its work.

### The Need for Reduced Invective and More Illumination in the Discussion of Low-Level Radiation

As a public health scientist whose principal career interest has been in radiologic health research and local (i.e., New York City) regulatory jurisdiction, I have been a witness to or participant in numerous official, scientific, and public-interest forums where the level of polarization, and indeed invective, has reached incredible proportions.

The subject of the effects of low-level radiation effects is one about which the discussion tends to get particularly heated. Environmentalists, or as they are referred to by some representatives of the nuclear industry with the pejorative adjective "so-called" environmentalists, are characterized as totally uninformed about modern scientific principles and are accused of exploiting the public's unrealistic and unwarranted fears of ionizing radiation.

On the other hand, our friends among good-quality environmental and public-interest groups have retaliated, or sometimes initiated, in kind. Advanced-technology advocates, especially in the nuclear-power field, have been depicted as totally indifferent to human life and sound ecology in the pursuit of large energy objectives and larger corporate earnings.

I would like to propose that the time has come for a major reduction in invective irrespective of the conviction with which we hold our scientific, environmental, technological, or public health opinions. Democratic dialogue, with respectful attention to each other's point of view, caveats, and reservations, is no guarantor of sound public policy in the area of nuclear energy. However, I do not know a better way.

### One Last Word

It has been often asserted that excessive attention has been directed to the effects of low-level ionizing radiation, particularly connected with nuclear-power generation, in view of the larger risks we encounter in our daily lives from numerous other sources.[20] Thus we tend to identify this nuclear fission, not with useful electric power, but rather as an exceptionally destructive weapon of war.

There is a certain usefulness to this conception inasmuch as it serves to remind us of the need to control these weapons. We live in a world where the aggregate number of large strategic nuclear warheads is estimated at 17,000, with an average yield of about 700 kilotons or about forty times that of each of the Hiroshima-Nagasaki devices of World War II.[21]

The world nuclear arsenal also contains about 30,000 tactical warheads of various yields.[21] In round figures the number of nuclear warheads approaches a horrendous 50,000, the detonation of any one of which—either intentionally or by accident—could develop into a weapon exchange resulting in the extinction of our civilization and maybe our species.

Our concerns about the biological effects of ionizing radiation, then, work as a dialectic catalyst to motivate all of us, both within and without the environmental scientific community, to recognize the need for reducing and ultimately eliminating these weapons.

## SUMMARY

Biological effects of small multiples of the natural ionizing radiation environment are addressed and attention paid to potential public health problems in nuclear technology. Implications for the employment of radiation in the healing arts are discussed. Formidable mathematical obstacles are noted due to the fact that health effects at low-level radiation doses are generally inferred from observations at much larger doses. Practical cost-effective decision-making consistent with sound radiologic health must take into account the breadth of current scientific uncertainties.

## ACKNOWLEDGMENTS

I want to thank Dr. Helen Q. Woodard of the Biophysics Laboratory of the Memorial Sloan-Kettering Cancer Center for a scholarly and helpful reading of an early draft of this paper. Dr. Woodard, of course, is not responsible for any of its inadequacies, nor does she necessarily share the opinions offered.

## REFERENCES

1.   —.1980. BEIR III. The Effects on Populations of Exposure to Low-Levels of Ionizing Radiation: 1980. Committee on the Biological Effects of Ionizing Radiation, Division of Medical Sciences, Assembly of Life Sciences, National Research Council. National Academy Press, Washington, DC.
2.   —. 1980. NUREG/CR-1728. The Feasibility of Epidemiologic Investigation of

the Health Effects of Low-Level Ionizing Radiation (Final Report). N.A. Dreyer, H.I. Kohn, R.W. Clapp, S.J. Covins, Jr., F.H. Fahey, E.R. Friedlander & J.E. Laughlin, Eds. U.S. Nuclear Regulatory Commission.

3.   —1981. GAO REPORT EMD-81. Problems in Assessing the Cancer Risks of Low-Level Ionizing Radiation Exposure, (Vols. 1 and 2). Report to the Congress of the United States by the Comptroller General. U.S. General Accounting Office.

4.   LAND, C.E. 1980. Estimating cancer risks from low doses of ionizing radiation. Science **209**: 1197–1203.

5.   OAKLEY, D.T. 1972. ORP/SID 72-1 Natural Radiation Exposure in the United States. U.S. Environmental Protection Agency.

6.   —1972. BEIR I. The Effects on Populations of Exposure to Low-Levels of Radiation. Committee on the Biological Effects of Ionizing Radiation, Division of Medical Sciences, National Academy of Sciences, National Research Council, Washington, DC.

7.   FABRIKANT, J.I. 1982. Carcinogenic effects of low-level radiation. Health Physics Society Newsletter **10**: 1ff.

8.   EVANS, H.J., K.E. BUCKTON, G.E. HAMILTON & A. COROTHERS, 1979. Radiation-induced chromosome aberrations in nuclear-dockyard workers. Nature **277**: 531–534.

9.   LAND, C.E., F.W. MCKAY, & S.G. MACHADO, 1984. Childhood leukemia and fall-out from the Nevada nuclear tests. Science **223**: 139–144.

10.  UNSCEAR. 1977. Sources and Effects of Ionizing Radiation. United Nations Scientific Committee on the Effects of Scientific Committee of the Effects of Ionizing Radiation, 1977 Report to the General Assembly, with annexes. United Nations, New York, NY.

11.  SAGAN, L. A. 1981. Some thoughts on dose-response, hormesis, and all that. Nuclear News. October: 80–82.

12.  LUCKEY, T.D. 1980. Hormesis with Ionizing Radiation. CRC Press. Boca Raton, FL.

13.  HICKEY, R. J., E.J. BOWERS & R.C. CLELLAND. 1983. Radiation hormesis, public health and public policy: A commentary. Health Physics **44**:207–219.

14.  —1977. Beir II. Considerations of Health Benefit—Cost Analysis for Activities Involving Ionizing Radiation Exposure and Alternatives. Committee on the Biological Effects of Ionizing Radiation, Assembly of Life Licenses, National Research Council, National Academy of Sciences. U.S. Environmental Protection Agency EPA 520/4-77-003. Washington, DC.

15.  B. M. SCHWARZSCHILD. 1981. Studies revise dose estimates of A-bomb survivors. Physics Today. September: 17–20.

16.  LOEWE, W.E. & E. MENDELSOHN. 1981. Revised dose estimates at Hiroshima and Nagasaki. Health Physics **41**: 663–666.

17.  JONES, T.D. 1982. Neutrons in Hiroshima? Health Physics **42**: 537–538.

18.  SCHULL, W.J., M. OTAKE & J.V. NEEL. 1981. Genetic effects of the atomic bombs: A reappraisal. Science **213**: 1220–1227.

19.  HAMILTON, L.D., J. NAGY & C.V. ROBINSON. 1983. Response to "genetic effects of the atomic bombs: A reappraisal." Health Physics **44**: 435–437.

20.  UPTON, A. C. 1982. The biological effects of low-level ionizing radiation. Scientific American **246**: 41–47.

21.  TURCO, R. P., O.B. TOON, T.P. ACKERMAN, J.B. POLLACK & C. SAGAN. 1983. Nuclear winter: Global consequences of multiple nuclear explosions. Science **222**: 1283–1292.

22.  EHRLICH, P.R. *et al.* 1983. Long-term biological consequences of nuclear wars. Science **222**: 1293–1300.

# Magnetic Field Exposure Related to Cancer Subtypes

NANCY WERTHEIMER[a]

*Department of Preventive Medicine and Biometrics*
*University of Colorado Medical Center*
*Denver, Colorado 80262*

ED LEEPER

*Monitor Industries*
*Salina Star Route*
*Boulder, Colorado 80302*

In studies of both children[1] and adults[2] cancer was reported to be associated with high-current power-line configurations (HCCs) near the home. Such power lines can expose residents of the home to fairly continuous 60-Hz alternating magnetic fields (AMFs) ranging from about 0.001 to 0.01 gauss.

Fulton *et al.*[3] failed to find any association between HCCs and childhood leukemia in Rhode Island, but their controls may have had an urban bias. A rough correction for case-control differences in their data produced a positive HCC-cancer association.[4] More recently, a positive association has been reported in Sweden[5]: Children with cancer were found to live near large power lines and/or at homes with high 50-Hz magnetic fields (measured at the front door) more often than control children. In addition, a number of epidemiologic studies have reported that persons exposed to electromagnetic fields by their occupations may show increased rates of certain kinds of cancer.[6-11]

Although these epidemiologic studies are quite consistent in reporting associations between cancer and exposure to sources of AMF, the interpretation of such associations is still decidedly controversial. Only further research can determine whether the observed associations result from (1) increased initiation of cancer, (2) promotion of the carcinogenic process, or (3) unrecognized factors in the different studies that connect cancer and AMF exposure artifactually rather than causally.

In the study relating adult cancer to HCCs,[2] several patterns were seen suggesting that cancer promotion might underlie the association between HCCs and cancer: (1) The effect of HCC exposure seemed to be reversible. (That is, the cancer incidence was no longer excessive in persons who had lived at an HCC home if they had been gone from that home for 3 years or more.) (2) A latency period of 7 years or less was typically observed between first occupation of an HCC home and cancer manifestation; this is a shorter latency

[a] Address for correspondence: Nancy Wertheimer, Ph.D., 1330 5th Street, Boulder, Colorado 80302.

than has usually been seen for most known carcinogens. (3) A wider range of cancer subtypes seemed to be associated with HCC exposure than would be expected from past experience with factors believed to initiate cancer. In this report we will examine the relationship between cancer subtypes and AMF exposure more thoroughly.

## METHODS

### AMF Exposure from Residential Power Lines

Addresses of 1179 adults with cancer (the "cancer cases") were assembled from Colorado Death Certificate and Cancer Registry records, covering Denver and its suburbs and two smaller towns near Denver. Each address of a person with cancer (a "cancer address") was matched with a control address: (1) Control addresses for the small-town cases were addresses of persons dying from non-cancer causes, matched to the cancer case for town, age, sex, year of death, year when the subject lived in the house, and socioeconomic level of census tract. (2) Control addresses for cases in Denver and the Denver suburbs were neighborhood addresses, chosen randomly from those addresses listed in the city directory within two blocks of the case address. Each cancer case was compared with its matched control (using power-line configurations as a rough index of historical AMF exposure) to see whether the power lines near the case or the control home showed the greater potential for exposing the house occupants to AMFs. The procedure used in assembling these case-control pairs, as well as those used in determining the relative AMF exposure to be expected at the case and control homes, are described in detail elsewhere.[2]

### Occupational Exposure to AMFs

Besides comparing people with high or low potential AMF exposure from power lines near the home, we wished to compare people with high or low potential exposure at their work place. To do this we analyzed data from three published sources on occupational mortality. For the United States data[12] we have calculated the SMRs (standard mortality ratios) for total cancer and for each subtype for men aged 20–64 years employed in those occupations where high AMF exposure (in the 60-Hz range) might be expected: power station operators; electricians; electrical engineers; stationary engineers; welders and flame-cutters; electrical and electronic technicians; linemen and servicemen; and employees in the primary nonferrous metal industry, where electrolytic separation of metal from ores involves high DC (direct current) magnetic fields which will have an AMF component (ripple) due to the rectifying process.

For the data from Washington state[13] PMRs (proportional mortality ratios) were calculated for a similar group of occupations (for men aged 20–64): electricians; electrical engineers; stationary engineers; welders and flame-cutters; electrical and electronic technicians; linemen and servicemen; aluminum workers; and motion-picture projectionists (who get high AMF exposures from arc lamp equipment and electrical control panels).

For the California data[14] PMRs were again calculated for men aged 20–64 employed in the following occupations: electricians, electrical engineers; stationary engineers; welders and flamecutters; motormen (street, subway, and elevated railway); linemen and servicemen; and motion-picture projectionists.

The occupations were chosen from the categories provided by each data source before the data were analyzed, on the basis of their presumed exposure to occupational sources of low frequency AMFs. It should be recognized that these occupational listings are crude. For instance, some of the occupational categories provided include both "exposed" and "unexposed" occupations (e.g., welders and flame-cutters, where electric welders would be exposed, while gas welders would not). Furthermore, some of the occupations may be associated with carcinogens whose effects would be confounded with any effects of AMF exposure from the occupation.

## RESULTS

In the data relating adult cancer to HCCs, we identified a Denver/exposed subgroup. This group is composed of all residents of the city of Denver, and also all suburban males aged 20–69 (because they would be likely to work, or recently to have worked, in Denver). These Denver/exposed cases are distinguished from the non-Denver/exposed cases (females and retired males [aged 70+] residing in the suburbs, and all residents of the two smaller towns sampled). TABLE 1 shows that those cases least apt to be exposed to Denver's environment show fairly strong HCC-cancer associations for cancer of each system except the respiratory system. The group more highly exposed to Denver's environment, on the other hand, generally shows little or no evidence of such an association. This difference cannot be explained on the basis of the different control procedures used in the small towns and the Denver area (including suburbs), since the suburban women and retired males (who would presumably be relatively unexposed to Denver's urban/industrial environment) showed HCC-cancer associations fully as strong as those seen in the small-town residents.

In TABLE 2 two types of cancer are differentiated on the basis of their incidence-age patterns: (1) *Type A cancer* is defined as cancer that peaks in incidence before extreme old age and shows a declining incidence late in life. (Lung cancer shows this pattern and is included as Type A cancer even though the early peaking has been shown to be a cohort effect.[15] For our purposes here it does not matter whether or not the early peak is due to a cohort effect.) (2) *Type B cancer* is defined as cancer that continues to increase in incidence throughout life.

The breast cancer curve suggests a Type A pattern for premenopausal cancer, overlapping with a postmenopausal Type B curve. Consequently, we have analyzed breast cancer as a premenopausal Type A cancer when diag-

**TABLE 1.** Number of Case-Control Pairs That Showed Higher Potential 60-Hz Magnetic Field Exposure in the Case as Compared to the Control Pair-Member

| | Non-Denver/Exposed Higher Exposure in: | | | Denver/Exposed Higher Exposure in: | | | Total Higher Exposure in: | | |
|---|---|---|---|---|---|---|---|---|---|
| | Case | Control | C-ratio[1] | Case | Control | C-ratio | Case | Control | C-ratio |
| Nervous system | 16 | 7 | 229[d] | 9 | 4 | 225 | 25 | 11 | 227[c] |
| Lymphatic system: | | | | | | | | | |
| Leukemia | 17 | 14 | | 6 | 9 | | 23 | 23 | 100 |
| Hodgkins' lymphoma | 6 | 4 | | 6 | 4 | | 12 | 8 | 150 |
| Non-Hodgkins' lymphoma | 13 | 6 | | 13 | 12 | | 26 | 18 | 144 |
| Multiple myeloma | 5 | 2 | | 3 | 1 | | 8 | 3 | 267 |
| Total | 41 | 26 | 158[d] | 28 | 26 | 108 | 69 | 52 | 133 |
| Reproductive system | | | | | | | | | |
| Ovary | 14 | 11 | | 4 | 3 | | 18 | 14 | 129 |
| Uterus | 11 | 3 | | 7 | 3 | | 18 | 6 | 300[c] |
| Breast | 51 | 27 | | 36 | 26 | | 87 | 53 | 164[b] |
| Prostate | 18 | 10 | | 12 | 9 | | 30 | 19 | 158 |
| Total | 94 | 51 | 184[a] | 59 | 41 | 144[d] | 153 | 92 | 166[a] |
| Urinary system | | | | | | | | | |
| Kidney | 8 | 2 | | 2 | 4 | | 10 | 6 | 167 |
| Bladder | 15 | 6 | | 9 | 8 | | 24 | 14 | 171 |
| Total | 23 | 8 | 288[b] | 11 | 12 | 92 | 34 | 20 | 170[d] |
| Digestive system | | | | | | | | | |
| Mouth | 3 | 1 | | 10 | 7 | | 13 | 8 | 163 |
| Pancreas | 20 | 13 | | 10 | 5 | | 30 | 18 | 167 |
| Stomach | 8 | 6 | | 6 | 10 | | 14 | 16 | 88 |
| Colon/rectum | 40 | 24 | | 21 | 31 | | 61 | 55 | 111 |
| Total | 71 | 44 | 157[c] | 47 | 53 | 89 | 118 | 97 | 122 |
| Respiratory system: | | | | | | | | | |
| Bronchus/lung | 35 | 30 | 117 | 17 | 14 | 121 | 52 | 44 | 118 |
| Other and unknown | 33 | 27 | 122 | 25 | 24 | 104 | 58 | 51 | 114 |

NOTE: The C-ratio is calculated by dividing the number of matched pairs where the case address shows more potential AMF exposure than the control address by the number of pairs where the reverse is true, and multiplying by 100. Equivalent pairs are dropped from consideration. Superscripts indicate the significance level by sign test: [a] $\leqslant$ 0.001; [b] $\leqslant$ 0.01; [c] $\leqslant$ 0.05; [d] $\leqslant$ 0.1.

**TABLE 2.** Association of Potential AMF Exposure from Residential Power Lines with Cancer by Type of Cancer

| Type of Cancer | Non-Denver/Exposed Age (yr) at Diagnosis: | | | | Denver/Exposed Age (yr) at Diagnosis: | | | |
| --- | --- | --- | --- | --- | --- | --- | --- | --- |
| | 20–69 | | 70+ | | 20–69 | | 70+ | |
| *Type A Cancer[a]:* | | | | | | | | |
| Nervous system | [b]14 | 5 | 2 | 2 | 7 | 3 | 2 | 1 |
| Hodgkins' lymphoma | 6 | 3 | 0 | 1 | 5 | 3 | 1 | 1 |
| Non-Hodgkins' lymphoma | 10 | 3 | 3 | 3 | 12 | 8 | 1 | 4 |
| Multiple myeloma | 3 | 0 | 2 | 2 | 3 | 1 | 0 | 0 |
| Ovary | 12 | 6 | 2 | 5 | 3 | 2 | 1 | 1 |
| Uterus | 7 | 1 | 4 | 2 | 6 | 3 | 1 | 0 |
| Premenopausal breast cancer (aged < 55) | 27 | 6 | — | — | 16 | 9 | — | — |
| Mouth | 2 | 1 | 1 | 0 | 10 | 5 | 0 | 2 |
| Kidney | 4 | 2 | 4 | 0 | 1 | 4 | 1 | 0 |
| Bronchus/lung | 22 | 20 | 13 | 10 | 15 | 11 | 2 | 3 |
| Total | 107 | 47 | 31 | 25 | 78 | 49 | 9 | 12 |
| C-ratio[c] | 228 | | 124 | | 159 | | 75 | |
| | $p < 0.0001$ | | NS | | $p < 0.01$ | | NS | |
| *Type B Cancer[d]* | | | | | | | | |
| Adult leukemia | 4 | 9 | 13 | 5 | 2 | 7 | 4 | 2 |
| Postmenopausal breast cancer (aged 55+) | 14 | 10 | 10 | 11 | 10 | 13 | 10 | 4 |
| Prostate | 8 | 2 | 10 | 8 | 4 | 7 | 8 | 2 |
| Pancreas | 12 | 5 | 8 | 8 | 6 | 1 | 4 | 4 |
| Stomach | 4 | 4 | 4 | 2 | 5 | 6 | 1 | 4 |
| Colon/rectum | 16 | 10 | 24 | 14 | 14 | 22 | 7 | 9 |
| Bladder | 4 | 5 | 11 | 1 | 6 | 6 | 3 | 2 |
| Total | 62 | 45 | 80 | 49 | 47 | 62 | 37 | 27 |
| C-ratio | 138 | | 163 | | 76 | | 137 | |
| | $p = 0.12$ | | $p < 0.01$ | | NS | | NS | |

[a] Type A cancer is that in which incidence peaks before extreme old age (incidence data from Young *et al.*[25]).

[b] Each pair of numbers represents (1) the number of pairs where the case had a higher exposure than the control; and (2) the number of pairs where the reverse was true. Pairs with equivalent exposure are dropped from the analysis.

[c] The C-ratio is calculated by dividing the number of matched pairs where the case address shows higher potential AMF exposure than the control address by the number of pairs where the reverse is true, and multiplying by 100.

[d] Type B cancer is that in which incidence increases into extreme old age.

nosed before age 55, and as a postmenopausal Type B cancer when diagnosis occurred later.

TABLE 2 shows that for Type A cancer as a group, both the Denver/exposed and the non-Denver/exposed samples show a significant association with HCCs — but only for cases diagnosed before age 70. For Type B cancer, on the other hand, the association is less strong (especially in the Denver/exposed group) and what association there is is most marked in persons aged 70 or older.

TABLE 3 presents data for two sources of AMF exposure: occupational exposure and exposure due to residential wiring configurations. These two sources

**TABLE 3.** Cancer Subtype Related to 60-Hz Magnetic Field Exposure as Indicated by Occupation and by Power-Line Configurations at the Residence

| | Occupational Exposure Data: (males, aged 20–64) | | | | Residential Power-Line Exposure (adults, both sexes) C-ratios |
| --- | --- | --- | --- | --- | --- |
| | U.S. 1950 SMRs | Washington 1950–1979 PMRs | California 1959–1961 PMRs | Total Observed/ Expected | |
| Nervous system | 146[a] | 142[a] | 96 | 135[a] (196) | 227[c] (36) |
| Urinary system | | | | | |
| Kidney | 118 | 129 | 136 | 125[c] (103) | 167 (16) |
| Bladder | 145[b] | 126 | 167[c] | 142[a] (117) | 171 (38) |
| Total | 133[b] | 127[c] | 150[c] | 134[a] (220) | 170[d] (54) |
| Lymphatic system | | | | | |
| Leukemia | 112 | 124[d] | 111 | 116[d] (171) | 100 (46) |
| Lymphoma | 109 | 111 | 118 | 111[d] (248) | 159[d] (75) |
| Total | 110 | 116[d] | 115 | 113[c] (419) | 133 (121) |
| Digestive system | | | | | |
| Mouth | 87 | 75 | 91 | 84[d] (96) | 163 (21) |
| Stomach | 104 | 92 | 108 | 102 (276) | 88 (30) |
| Pancreas | — | 136[b] | 56[c] | 110 (108) | 167 (48) |
| Colon/rectum | 114 | 90 | 116 | 106 (388) | 111 (116) |
| Total | 106 | 99 | 97 | 102 (868) | 122 (215) |
| Reproductive system | | | | | |
| Prostate | 123 | 78 | 117 | 106 (104) | 158 (49) |
| Respiratory system | | | | | |
| Bronchus/lung | 121[a] | 108 | 108 | 113[a] (847) | 118 (96) |
| Other and Unknown | 110[d] | 98 | 95 | 104 (721) | 109 (94) |
| Total cancer (excluding cancer of female reproductive organs) | 114[a] | 107[c] | 105 | 110[a] (3375) | 131[a] (665) |

NOTE: For the occupational data the values presented are observed/expected ratios multiplied by 100 (SMRs for the U.S. data; PMRs for the Washington and California data). Values presented for residential power-line exposure are C-ratios (the number of matched pairs where the case address shows more potential AMF exposure than the control address, divided by the number of pairs where the reverse is true, multiplied by 100). Superscripts indicate the significance level by $\chi^2$ or sign test: [a] $\leqslant 0.001$; [b] $\leqslant 0.01$; [c] $\leqslant 0.05$; [d] $\leqslant 0.10$.

The numbers in parentheses indicate the number of cancer cases used to calculate the ratios presented.

of AMF exposure show fairly similar associations with the various cancer subtypes: For both the combined occupational data and the power-line data, cancer of the nervous system shows the strongest association, followed by cancer of the urinary system; cancer of the lymphatic system shows an intermediate degree of association; and digestive cancer shows a low degree of association. Only cancer of the reproductive system (prostate) and cancer of the respiratory system differ by more than one rank order for the two sources of exposure.

## DISCUSSION

### *Urbanicity Patterns*

The HCC-cancer associations shown in TABLE 1 are generally stronger for the non-Denver/exposed sample than for the Denver/exposed sample. The lack of association between HCCs and cancer in the Denver/exposed sample may simply mean that there are urban factors that cause and/or promote cancer, which are encountered so often in Denver that they tend to obscure any cancer increase due to power lines near the home. (These factors may include the ubiquitous sources of AMF exposure commonly found in cities.)

### *Type A and Type B Cancer*

In TABLE 2 we saw that the age at which the strongest HCC-cancer association is found varies from one subtype to another in a way that seems related to the incidence age pattern (Type A or Type B) characteristic of each subtype.

These patterns are broadly consistent with the hypothesis that AMF exposure accelerates the development of cancer. For Type A cancer, latent cancer existing in the more highly AMF-exposed population living at HCCs is hypothetically accelerated and should show a peak incidence relatively early in life; similar latent cancer in the less-exposed population, being unaccelerated, should peak later. This delayed peak in the unaccelerated cancer would occur at an age when the accelerated cancer of the HCC dwellers had already entered the declining phase of its incidence curve, so the accelerated and the unaccelerated cancer would tend to balance each other out in old age (FIG. 1, top).

Type B cancer would be expected to show a different pattern, because the incidence of this type of cancer does not decline in old age. For both the Type A and the Type B cancer the incidence curve is hypothesized to be shifted to earlier ages in the HCC-exposed group. But for the Type B cancer the unaccelerated group cannot balance out the accelerated (HCC) group in old age, because the accelerated group will include numerous cases of cancer that have been accelerated into the life-span — cancer that would never have become manifest before death occurred from some other cause, if it had not been accelerated (FIG. 1, bottom).

This theoretical formulation is weakened, however, because it does not explain the lack of HCC-cancer association generally seen for the Type B cancer

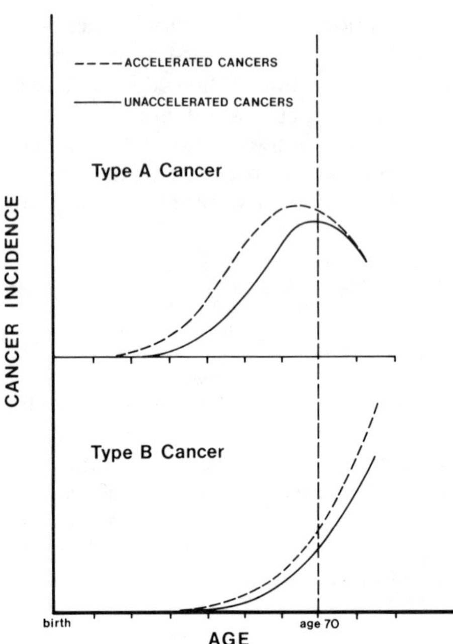



**FIGURE** 1. The expected effect of acceleration in time on Type A and Type B cancer. Accelerated Type A cancer (disease that shows a peak and then declines in incidence in old age) would be expected to have a higher incidence than unaccelerated Type A cancer at younger ages, but this might be balanced out at older ages by the delayed peak occurring in unaccelerated cases of Type A cancer. Accelerated Type B cancer (disease that continues to increase in incidence throughout the lifespan) should show increased incidence at all ages.

before age 70. However, since this lack of association was seen largely in the Denver/exposed population, we suspect that it may result from major urban-industrial impacts on some of these forms of Type B cancer. Such impacts might overwhelm any effect due to HCCs at the residence, particularly during the years of employment when people spend less time at home. After age 70 retirement may reduce the impact of such nonresidential factors and emphasize cancer-related factors at the residence, such as HCCs hypothetically represent.

## The Nature of the AMF-Cancer Association

The essential question is whether the epidemiologic association seen between AMF exposure and cancer is a direct one, or whether it is an artifact of some unassessed third variable. One would like to see this question answered by experiment, but the investigator is hampered because so little is yet known about electric- and magnetic-field dosimetry. Experiments have shown that exposure to fields or currents at low intensity sometimes produce effects that disappear as the intensity is increased.[16-20] This finding of intensity "windows" must cause us to question the traditional experimental assumption that, by applying large doses to a small experimental population, we can detect any effects that might be produced by low doses applied to a large population.

Given the experimental difficulties, it is especially important to note whether the epidemiologic evidence supports the hypothesis of a direct link between AMF exposure and cancer. Earlier[2] we pointed out that a number of patterns in our data were consistent with a direct link: (1) The HCC-cancer association appeared to be dose-related (over the dosage range explored); (2) it did not seem to be an artifact of age, urbanicity, neighborhood, or socioeconomic conditions; (3) it held regardless of which of three control procedures were used (one for children[1] and two for adults[2]); (4) it was most evident where the competing carcinogenic influences that might be expected from an urban/industrial environment were minimized; and (5) the patterns of latency and onset-age observed were consistent with a hypothesis of cancer promotion. This latter point—that the pattern is consistent with cancer promotion—is further supported by the cancer-subtype data in this report, which suggest movement of the characteristic incidence-age curves to a younger age among the AMF-exposed.

A further check on artifact is the confirmation coming from the occupational data in TABLE 3. Although either the residential or the occupational associations, or both, could result from artifact, it is somewhat unlikely that the same subtypes would be affected artifactually for the two different sources of exposure.

### *Leukemia*

Since several recent communications[6-9] have suggested an association between leukemia and occupational exposure to electric and magnetic fields, a comment seems called for on the minimal associations we see with leukemia in TABLE 3.

The occupational data we present are reasonably consistent with the above-mentioned reports: Although the associations with leukemia presented are not very strong, they are consistently positive. (Furthermore, the other reports sometimes had access to more refined occupational categories than we could obtain from our sources, and they included certain electronics workers whom we have excluded because we wished to assess only exposure to extremely low frequency electric and magnetic fields. The other studies also included males over age 64, which we excluded since information over age 64 was not available for the United States occupational data we used.

Our data on residential exposure show no overall association between adult leukemia and AMF exposure, but we do find an association of HCCs with leukemia in two age groups: in children[1] and in the very old (see TABLE 2). It is only in the middle age ranges that our limited data showed no positive association.

This age pattern is consistent with our hypothesis of cancer promotion. Leukemia is a cancer subtype that shows a Type A curve in childhood, followed after many years of minimal incidence by a Type B curve in old age. It seems possible that in children we see an HCC-promoted increase in the youthful form of leukemia, whereas in old age we see an HCC-promoted increase in the old-age form of the disease. In the intermediate age ranges we would expect to see a rather unpredictable combination of (1) late-occurring cases of the youthful form, and (2) early-occurring cases of the old-age form.

The late-occurring cases of the youthful form would be unaccelerated cases, expected to occur largely in persons with low AMF exposure; the early-occurring cases of the old-age form would be accelerated cases, expected to occur largely in persons with high AMF exposure. The exact balance between the two components would probably vary under different conditions; specifically, it might be different for occupational than for residential exposure.

In summary, our data on cancer subtypes suggests that increased cancer rates in AMF-exposed persons are not to be expected "across the board"; they may be demonstrable only in those age ranges where accelerated manifestation would produce a concentration of cases.

### Nervous System Cancer

The strong association with nervous system cancer is of particular interest. Lin *et al.*[11] have recently reported an excess of brain cancer in electrical and electronics workers for still another population than the three studied here (workers in Maryland). They report not only that brain cancer is increased in such workers, but also that it occurs at an earlier age than for other occupations, a finding consistent with our data in TABLE 2.

At present the most consistently demonstrated experimental effects of low-frequency electric and magnetic fields seem to be those involving the nervous system.[17,21-24] The findings of this report may provide further evidence of nervous system susceptibility, in that cancer of the nervous system seems to be especially strongly associated with both residential and occupational sources of exposure to AMFs.

## SUMMARY

Certain subtypes of cancer (notably nervous system cancer) showed an association with two indices of exposure to 60-Hz alternating magnetic fields (AMFs): Subtype similarities were seen in those people potentially exposed to AMFs by their occupations, and in those potentially exposed by high-current power lines near their homes. The incidence-age patterns observed in exposed and nonexposed groups suggest that prolonged AMF exposure may act as a cancer promoter.

## REFERENCES

1.  WERTHEIMER, N. & E. LEEPER. 1979. Electrical wiring configurations and childhood cancer. Am. J. Epidemiol. **109**: 273–284.
2.  WERTHEIMER, N. & E. LEEPER. 1982. Adult cancer related to electrical wires near the home. Int. J. Epidemiol. **11**: 345–355.

3.  FULTON, J. P., S. COBB, L. PREBLE, *et al.* 1980. Electrical wiring configurations and childhood leukemia in Rhode Island. Am. J. Epidemiol. **111:** 292–296.
4.  WERTHEIMER, N. & E. LEEPER. 1980. Re: "Electrical wiring configurations and childhood leukemia in Rhode Island". Am. J. Epidemiol. **111:** 461–462.
5.  TOMENIUS, L. 1986. 50-Hz electromagnetic environment and the incidence of childhood tumors in Stockholm county. Bioelectromagnetics **7:** 191–207.
6.  MILHAM, S., JR. 1982. Mortality in workers exposed to electrical and magnetic fields. N. Engl. J. Med. **307:** 249.
7.  WRIGHT, W. E., J. M. PETERS & T. M. MACK. 1982. Leukaemia in workers exposed to electrical and magnetic fields. Lancet **ii:** 1160–1161.
8.  COLEMAN, M., J. BELL & R. SKEET. 1983. Leukaemia incidence in electrical workers. Lancet **i:** 982–983.
9.  McDOWALL, M. E. 1983. Leukaemia mortality in electrical workers in England and Wales. Lancet **i:** 246.
10.  SWERDLOW, A. J. 1983. Epidemiology of eye cancer in adults in England and Wales, 1962–1977. Am. J. Epidemiol. **118:** 294–300.
11.  LIN, R. S., P. C. DISHINGER, J. CONDE & K. P. FARRELL. 1985. Occupational exposure to electromagnetic fields and the occurrence of brain tumors. J. Occup. Med. **27:** 413–419.
12.  GURALNICK, L. 1963. Mortality by Occupation and Cause of Death among Men 20 to 64 Years of Age: United States, 1950. Vital Statistics Special Reports 53. U.S. Department of Health, Education and Welfare. National Vital Statistics Division. U.S. Government Printing Office. Washington, DC.
13.  MILHAM, S., JR. 1983. Occupational Mortality in Washington State, 1950–1979. HEW Publication No. (NIOSH) 83-116. U.S. Government Printing Office. Washington, DC.
14.  PETERSEN, G. R. & S. MILHAM, JR. 1980. Occupational Mortality in the State of California. HEW Publication No. (NIOSH) 80-104. U.S. Department of Health, Education and Welfare. Cincinnati, Ohio.
15.  DORN, H. F. & S. J. CUTLER. 1959. Morbidity from cancer in the United States. Public Health Monograph No. 56. U.S. Government Printing Office. Washington DC.
16.  BECKER, R. O. & D. G. MURRAY. 1970. The electrical control system regulating fracture healing in amphibians. Clin. Orthoped. Rel. Res. **73:** 169–198.
17.  ADEY, W. R. 1980. Frequency and power windowing in tissue interactions with weak electromagnetic fields. Proc. IEEE **68:** 119–125.
18.  AARHOLT, E., E. A. FLINN & C. W. SMITH. 1981. Effects of low-frequency magnetic fields on bacterial growth rate. Phys. Med. Biol. **26:** 613–621.
19.  BLACKMAN, C. F., S. G. BENANE, L. S. KINNEY, *et al.* 1982. Effects of ELF fields on calcium-ion efflux from brain tissue *in vitro*. Radiation Res. **92:** 510–520.
20.  LYMANGROVER, J. R., S. T. HSIEH & W. P. DUNLAP. 1983. 60-Hz Electric Field Effects on Adrenocortical Function, *in Vitro*. Project resumé, Contractor's review, November 1983. U.S. Department of Energy, Division of Electric Energy Systems, Washington DC.
21.  SISKEN, B. F. & S. D. SMITH. 1975. The effects of minute direct electrical currents on cultured chick embryo trigeminal ganglia. J. Embryol. Exp. Morphol. **33:** 29–41.
22.  HANSSON, H. 1981. Lamellar bodies in Purkinje nerve cells experimentally induced by electric field. Brain Res. **216:** 187–191.
23.  JAFFE, R. A. 1979. Neurophysiology. *In* Biological Effects of High Strength Electric Fields on Small Laboratory Animals. :195–227. Prepared by R.D. Phillips *et al.* Pacific Northwest Laboratory, Richland, WA.

24.  DELGADO, J. M. R., J. LEAL, J. L. MONEAGUDO & M. G. GARCIA. 1982. Embryological changes induced by weak, extremely low frequency electromagnetic fields. J. Anat. **134:** 533–551.
25.  YOUNG, J. L., JR, C. L. PERCY & J. A. ARDYCE. 1981. Surveillance, Epidemiology, and End Results: Incidence and Mortality Data, 1973–77. National Cancer Institute Monograph 57. U.S. Government Printing Office. Washington, DC.

# Influence of Power Frequency Electric and Magnetic Fields on Human Health[a]

SOL M. MICHAELSON

*Department of Radiation Biology and Biophysics*
*University of Rochester*
*School of Medicine and Dentistry*
*Rochester, New York 14642*

Recently, public concern has been expressed about possible risks to human health, function, and wellbeing arising from exposure to the electric and magnetic fields associated with overhead transmission lines. Discussion of these issues in the light of available information and concepts is necessary to provide perspective and understanding in this area. In any review of human surveys concerned with possible extremely-low-frequency (ELF) field effects, one has to be aware of the critical importance of how epidemiologic studies are designed and executed and how results are interpreted. Reports of clinical effects of electric and/or magnetic fields among human populations present serious difficulties which stem from the many publications in which pertinent material is not presented, and the data are variable or internally inconsistent. One has to beware of "spurious correlations." Correlations in themselves do not necessarily establish cause-effect relationships. An outstanding problem of epidemiologic studies of high-voltage transmission exposure is related to exposure assessment. Typical of the situations that exist in the literature are studies that provide no documentation of the basic element of exposure such as voltage. Often only broad categorizations such as "exposed" or "nonexposed" are provided. Most of the reports on humans demonstrate a lack of adherence to rigorous study design, analyses, and discussion. Control groups are frequently lacking. Population selection criteria and methods are often not described. Control of such confounding variables as age, occupation, health status and habits, and economic status is often not considered. Applied statistical methods may not be described or, largely because of study design, do not measure the strength of association via relative risk or odds ratio. It is imperative that we recognize the fact that practical decisions must be made on the basis of *what is known* rather than what is postulated or assumed without any rational basis.

Several studies have been performed to evaluate whether exposure to power frequency (50 or 60 Hz) electric or magnetic fields constitutes a health hazard. The study populations have included electric utility personnel, workers in other industries with electrical environments, and persons living in varying proximities to distribution lines. Only those reports of the last five years are the subject of this review. Previous studies have been reviewed by Michaelson[1] and Mehn.[2]

## OCCUPATIONAL EXPOSURES

A study was reported by Stopps and Janischewsky[3] in which clinical examinations were performed on 30 high-voltage maintenance men from Ontario Hydro together with 30 other employees matched for age and educational level but not exposed to electric fields. The investigation included electroencephalogram (EEG), electrocardiogram (ECG), blood biochemistry including tests of liver function and serum electrolytes; and complete physical and psychological assessment with psychometric and personality tests. At the time of examination the subjects were not identified as being *exposed* or *control.* The exposed group consisted of: (*a*) 19 linemen with an exposure experience calculated to be 7 kV/man-hours per day up to 8000 kV/m hours over ten years, and (*b*) 11 substation workers with average calculated exposure of 13 kV/m hours per day up to 36,000 kV/m hours over ten years.

It was concluded from this study that extra-high voltage (EHV) work does not cause chronic ill health in substation staff in Ontario.

In 1974 the Swedish State Power Board started physiological, psychological, and physical investigations on the effects of electric fields (EF) on personnel in Swedish EHV substations. In this study, 53 workers with long-term (more than five years) exposure to the EF of 400-kV substations were examined and compared with a matched control group of 53 nonexposed workers from the same power companies.[4] Matching considered age, geographical location, and length of employment. It was determined that some substations ("higher") had large areas with field intensities up to 10–15 kV/m, whereas others ("low-built") may have substantial areas with field intensities of about 20 kV/m. The aim of the study was to investigate whether there were any persistent, chronic health effects in the exposed group as a consequence of exposure. Examination included study of the nervous system for neurologic symptoms, blood pressure, ECG, and complete blood count. The results indicated no differences between the exposed and control groups as a consequence of long-term exposure of the EF. The groups differed, however, in that the exposed group had (1) consistently better results in psychological performance tests; (2) fewer number of children, especially boys; and (3) somewhat higher education. The differences in test results were apparently a reflection of the higher educational level among the exposed individuals. The difference in number of children was also thought to be related to factors other than exposure since this difference was found to be present 10–15 years prior to employment in 400-kV substations.

Attempts were also made to elucidate any "acute" effects of exposure, i.e., exposure to the EF at some time during the 14 days prior to examination. When this questionable criterion for "acute" exposure was used, the performance in the psychological tests was found to be lower among the "acute" exposed as compared with the "nonacute" exposed. No other results indicated the presence of an "acute" effect. It should nevertheless be pointed out that any "acute" effect might be ascribed to such other aspects of work in substations as stress, awareness of risks of accidents, and discomfort caused by, and fear of, spark discharges or shock.

Differences in the ratio of male and female offspring between exposed and nonexposed populations were mentioned by both Knave *et al.*[4] and Roberge,[5] although in neither case did the authors ascribe any significance to them. The trends were in opposite directions in the two surveys and in the case of Knave *et al.*[4] were known to have begun before the subjects started work in the substation.

The general conclusions to be drawn from all of the these studies are reassuring, notwithstanding the occasional suggestion still being made that various health effects may occur. In Sweden[6] preliminary studies revealed eight cases of congenital deformity among 113 children of substation workers exposed to 500 kV at the time of conception. Only 2.6 percent of nonexposed workers' children showed such deformities. In addition, chromosome analyses on eight of the workers showed a trend toward increased frequency of breakage.

Nordström *et al.*[7] analyzed the pregnancy outcome of male substation workers and found an increased frequency of congenital malformations. Exposure data, however, go back 40 years in time and part of the control group was the exposed workers at a younger age. The authors emphasize that their results must be interpreted with caution and that further research is needed to clarify any health hazards. Chromosome aberrations were also studied in 20 workers and compared to 20 selected referents for clinical standard material. More chromatid gaps and chromosome breaks were noted among the substation workers. Again, the authors point to the necessity for caution in any interpretations, mainly because of the limited number of persons examined and the special character of the control material. Chromosome analyses from an earlier study were used for five control subjects. Thus, different technicians could have been involved, possibly introducing a score bias. In addition no information is given on electric and magnetic field exposure, nor are any statistical analyses provided. Consequently no inferences can properly be drawn from this report.

Bauchinger *et al.*[8] repeated the chromosome studies on German substation workers. Chromosome analyses were done in lymphocytes of 32 workers occupationally exposed for more than 20 years in 380-kV switchyards. As compared with a control group of 22 workers of similar age and occupation but without field exposure, there was no increase in structural chromosome changes. In German substation workers, furthermore, no increase in abnormal pregnancy outcome could be demonstrated in epidemiologic studies.[9]

In another study, Nordström *et al.*[7] addressed the potential relation that may exist between pregnancy outcome and the occupation of the father exposed to different electrical voltages. This is essentially a descriptive study of pregnancy outcome in a group of individuals exposed to different electrical fields. In their analysis, the authors subdivided the group into exposed and nonexposed, and compared the two groups. Exposure information was obtained from questionnaires distributed to workers. According to their responses, the workers were grouped into three subgroups (less than 70 kV, 130 to 200 kV, and 400 kV). They were also grouped into occupational groups (high-voltage switchyard, construction or repair of switchyard, other tasks).

Questions on the following were asked: miscarriages, birth defects, still-births, fertility problems. The authors related these problems to many variables without really controlling for them. A decrease in the number of normal pregnancies, especially as a result of birth defects, was indicated for fathers who worked in contact with high electrical voltage.

In their discussion, the authors attribute these changes to the fact that "The employees' bodies and/or objects in the switchyards are electrically charged and then discharged . . ." They stress also that the presence of chromosomal aberration in these workers could be associated with the observed problems.

There are a great number of weaknesses in this study. The epidemiologic design and statistical analysis are very poor. The number of subjects on which the authors based their analysis was very small. Exposure was measured according to a questionnaire sent by mail to the subjects. It is not certain that the questionnaire discriminates in a valid manner between the "exposed" and "nonexposed" men. The regrouping of the exposure by different voltages and by occupational groups is not clear. The observation of problems of pregnancies by occupational groups cannot be attributed to the exposures to different voltages since within each occupational group there are exposed and nonexposed persons.

The main problem as to retrospective cohort studies relates to the insufficiency in documentation of the different exposure factors and relevant confounders. For instance, in the Swedish study on fertility and reproduction outcome among substation workers[7] subjective exposure data go back 40 years. In this type of epidemiologic study there is a definite need for knowledge of individual exposure data. Before we have that, no reliable conclusions can be drawn.

Milham[10] in a "letter to the editor" suggested that of 11 occupations with potential exposure to electrical and magnetic fields 10 have a raised proportionate mortality ratio (PMR) for leukemia. This excess is statistically significant for electricians, power station operators, and aluminum workers.

Milham's observations were extracted from a study of much greater scope in which the author examined the total mortality among men between 1950 and 1979 (438,000 deaths) in the state of Washington. The observed illnesses are grouped according to leukemias and acute leukemias. He found 136 deaths due to cancer among the 438,000 deaths of white adult males (20 years and older) working in 11 industries in the state of Washington. He related the 11 industries to "exposure to electric or magnetic fields," and computed the proportionate mortality rate as a ratio of actual deaths per industry to expected deaths per industry from cancer. He stated that he was unaware of obvious leukemogenic agents in the occupations, and therefore claimed that his findings suggested that "electrical and magnetic fields may cause leukemia."

Milham considers no other possible causes of cancer than electromagnetic fields in reaching his conclusion, and offers no explanation for ignoring other possible causes except a lack of awareness of "obvious leukemogenic exposures" in the listed industries. He also ignores the possibility of nonoccupational factors, and does not provide any information on ages at death. The numbers of deaths over a 20-year period (136, or about 7 per year in the state) seem low as a basis for such a sweeping conclusion. Questions can also be raised regarding the accuracy of the diagnosis of leukemia on the death certificate, particularly over the time period from 1950 to 1979, when there were improvements in medical care and consequently in methods of diagnosis. At best, Milham's views are highly speculative.

The PMR is not the best analytical technique in studies of mortality. The PMR for a cause of death may be raised simply because there is a decrease in other causes of death among a group of workers. The author does not indicate the constraint he utilized for each occupational group. What constitutes, for example, the occupation "electrical engineer" or "electrical technician?" Does this include the persons really exposed to magnetic and electrical fields? The author does not tell us whether the occupational classes correspond to the last occupation or to the principal occupation of the individual during his life.

In his conclusion, Milham indicates that his results may signify that electrical and magnetic fields cause leukemia. This remark clearly goes beyond the strength the study possesses. A letter to the editor does not ordinarily become an object of peer scientific review. It is necessary then to be judicious in the interpretation of these results.

Wright *et al.*[11] listed the proportional incidence ratios (PIRs) for leukemia, acute leukemia, and acute myeloid leukemia among white males in Los Angeles from 1972 to 1979 for 11 electrical occupations (occupation at time of diagnosis). The occupational categories were the same as those used by Milham.[10] The number of cases was zero in two categories, one in three categories, and six or less in all other categories except one (electricians).

The particular type of analysis used, proportional incidence rate (PIR), was obtained by comparing the proportion of exposed workers among the total of other cases of cancer reported in this register.

The authors suggest there is a link between the incidence of leukemia and work in certain occupations where there may be exposure to electric and magnetic fields. This is particularly the case for power and telephone linemen. In their discussion, Wright and coworkers note that it is not the electric or the magnetic field itself which is responsible for this excess, but that it may be related to another unmeasured factor that was common in the exposed occupational groups.

This report provides no data except a table listing cases of leukemia for the eleven occupations over a period of 7 years. The occupations are related to electrical systems only, and the absence of other data and lack of analysis suggest that the findings are entirely speculative.

There are several weak points in this study. The first involves the small numbers. The excess of acute myeloid leukemia in power linemen and telephone linemen is based on observation of only two cases. There are also reservations about the statistical analysis used. PIR is a recent method which, although theoretically correct, has not yet stood the test of time. In general, this study is weak. The conclusion of the authors ("The hypothesis that electric and magnetic fields or other shared exposures cause leukemia deserves further study") tells us very little.

In a third report, McDowall[12] examined the alleged link between mortality from leukemia and exposure to magnetic and electric fields in the workplace. The author produced in reality two types of studies — one study of mortality (PMR) and a case study.

Somewhat in the fashion of Milham, McDowall computed proportionate mortality rates (PMR) for cancer deaths in ten electrical industries for males aged 14–74 for the years 1972–74. The industrial categories were, however, somewhat different from Milham's categorization. And, unlike Milham's calculations, those of McDowall do not find a significantly high PMR for the 85 deaths in the ten industries. Some occupations were included erroneously in his occupational categories, and the data are therefore biased.

McDowall also computed the relative risk of death by cancer for a different category of electrical industries, using the number of deaths from cancer (537) in males in England and Wales in 1973 as "cases." He selected 1074 controls from remaining deaths for that year. He matched his controls with the cases within 5-year age groups. He did not describe how these data were used to compute relative risk (RR). He stated that his data for 1973 generally support those of Milham[10] and Wright *et al.*[11] (1982), but noted that other causative factors cannot be ignored.

In the study of mortality, McDowall used as a population men 15 to 74 years old who died of leukemia between 1970 and 1972. In his case study, he compared the occupations of 537 men who died from acute myeloid leukemia after the age of 15 with those of 1074 controls (men who died of all causes except leukemia).

For measurement of exposure, the author utilized the occupational classification. The diseases concerned were: all the leukemias, lymphoid leukemias, and myeloid leukemias.

In the study of PMR the author observed 85 cases of leukemia. The PMR is not in excess for all the occupations. It is in excess for four of them: electrical engineers (including three subclassificaitons) and telegraph radio operators. In his study McDowall observed 36 cases of acute myeloid leukemia. The relative risk is significantly elevated for all occupations in the electrical industry.

In his discussion, McDowall concluded that these observations serve to reinforce the previous observations made in the American studies. He stressed (by contrast) that there are some differences between the studies. While the Americans found an excess for all the workers exposed to magnetic and electric fields, McDowall found some excess for the professional subclassification only. He attributes this difference to differences in mobility of workers in various occupations.

McDowall recognizes that many factors in an industrial environment may cause death by cancer and he does not insist that electromagnetic fields may be among those factors. He does not provide sufficient information to clarify his tabular data, nor does he discuss his results in any detail. Any conclusions based on his findings would therefore be speculative. It is not known whether McDowall speaks here of the principal occupation or the last reported occupation. For each specific occupation, the numbers are small.

In another letter to the editor, Coleman *et al.*[12a] analyzed the proportional registration ratio (PRR) of the incidence of cancer in workers in ten electrical occupations in southeast England for the period 1961–1979. The total registration number (cases reported by hospitals) was 113, and was six or less in six of the ten categories. The PRR was less than expected in two categories (radio/radar mechanics and professional electrical engineers). The investigators note the small number of cases used in the analysis, and differences in listing occupation by hospitals (cases) and by government census (controls). They further note that the literature does not support a conclusion that electromagnetic fields and death by cancer are related.

Coleman *et al.* generally followed the methods used by others, but point out several limitations of methods employed. They do not conclude clearly that a relationship exists between electromagnetic fields and cancer. The obvious limitations of the study and inconsistencies between their tabular data and those of others suggest that no conclusions should be drawn from the report.

## RESIDENTIAL EXPOSURE

Wertheimer and Leeper[13] have suggested a link between the development of cancer in individuals less than 20 years of age and the configuration of electric currents around the homes where these persons lived in Denver, Colorado. They related the records of death due to childhood cancer (up to age 19) in the greater Denver area to the presence of electric power lines.

The cases consisted of 344 children who died from cancer before the age of 19 during years 1976–1977. The same number of children not diagnosed as having cancer were matched to cases according to date of birth. The residences of the cases and the controls were classified into two groups: (*a*) HCC or high-current configuration and (*b*) LCC or low-current configuration, starting from the distribution of electric cables around the houses. The cancer was classified as leukemia, lymphoma, cancer of the nervous system, and others.

The analysis indicated that more children resided in homes where the configuration of electric cables suggested an elevated exposure (HCC). In addition, the authors mentioned that there is a dose-response relation between the onset of cancer and the levels of current. In their discussion, they attribute these results to the alternating magnetic fields that accompany the electric current.

This report includes oversimplified explanations, questionable assumptions, debatable interpretations of data, and seemingly unjustified conclusions. It is additionally troublesome because the body of the report presents interpretations that appear unequivocal, but are not supported in the authors' final discussion.

Wertheimer and Leeper did not measure the intensity of the electromagnetic fields they were interested in. They estimated the exposures by looking at wires outside the houses. On this basis they made distinctions between high and low electromagnetic fields, leading to the conclusion that high-exposure homes produced more cancer victims.

In their discussion, the authors go much too far in the interpretation of their results; their commentaries concerning the mechanisms of action are more personal hypothesis than observed facts. The greatest weakness of the study resides, however, in the fact that the exposures were not codified according to a blind study. The person who estimated the exposure knew the group to which the subject belonged. This weakness became at the outset a significant bias.

The oversimplified explanation of the nature of electric power systems, which apparently is believed by the investigators to justify their not measuring magnetic field intensities, is the most troublesome aspect of the report, and entirely erodes confidence that views expressed by the investigators should be accepted. Other apparent deficiencies might be satisfactorily explained in a more detailed report, but the lack of measured magnetic field data is paramount and cannot be overcome.

In describing their reason for being interested in studying the relationship between power lines and cancer, the investigators cite several references, implying that laboratory experiments support the hypothesis that low-level ELF electromagnetic fields produce cancer. The investigators also comment upon reported measurements of magnetic field intensities from appliances, and caution that the data were obtained "as close as possible" to the appliances, and therefore are not pertinent to electromagnetic exposure of humans. The view also is expressed that the duration of exposure may be more important than exposure levels.

Specific deficiencies in the Wertheimer and Leeper[13] report include the following:

*Field procedure:* The birth and death address of each "case" was visited (or only the birth or the death address, if the information was limited). A map of the immediate vicinity was drawn, and the measured distances between the residence and visible power lines and substations (in a few cases) were noted. The same was done for the controls. Primary distribution lines were classified as "large-high-current configurations (HCCs) gauge" or "thin," low-current configurations (LCCs) depending on whether they were larger than secondary wires.

*Data selection:* Death certificates were obtained for persons up to the age of 19 in the Denver area (not defined) who died of cancer and had addresses in the area between the years 1946–1973. Birth addresses were obtained from birth certificates, and death addresses were those listed in city directories for the two years before the diagnosis of cancer. If no birth address could be found, only the death address was used; if no death address could be found, only the birth address was used. A total of 344 cases was studied. Of these, 72 birth addresses (21 percent) could not be found; death addresses could not be found for 16 (5 percent). Assumptions therefore had to be made for 26 percent of the addresses. The investigators do not mention intermediate addresses or time of residency at any address.

Controls were selected from the next listed birth in the Denver area birth certificate listing. Siblings were skipped. Presumably the listing was chronological.

It appears that possible intermediate addresses of cases between birth and death were not determined or traced. There is no reason to believe that intermediate addresses were not possible for at least some cases. The same holds true for controls. A complete residence history is essential, and if obtained, should be reported. In the absence of this information, the report is limited and not definitive.

No justification is provided for assuming the same birth and death addresses for cases when one or the other was not available. Since 26 percent of the cases were not traceable with regard to both addresses, and the time of residency is neither stated nor considered, the data are unreliable to a significant degree.

*Proximity to HCCs and LCCs:* Many more cases than controls are reported as being near HCCs. Both the number of cases and controls near LCCs are substantially higher than those near HCCs.

The summary of magnetic field intensities can be misinterpreted and mislead some readers. The locations of the power systems and measurement points are not identified. The measurement methods are not described, and the important characteristics of the power system are not given. The currents conducted in any of those systems depend upon their electrical design characteristics, their cable geometries, the type of connection to and the nature of system loads, the impedance characteristics of the entire system network, temporal and spatial load variatons, interconnections with other electric networks, and many other factors. The data have very limited practical application in the absence of adequate information about these factors.

The investigators state that line imbalance on a line is greatest at locations where current is highest. The statement is not supported, and is not universally true. Utility operators prefer to maintain balance on their systems, connect loads in ways which help in maintaining balance, and invariably will resort to more complex engineering to maintain balance on high-current branches than on low-current branches. It is not uncommon, in fact, to operate some relatively low current lines unbalanced. A fundamental understanding of line balance for understanding power line magnetic fields, and other factors are also important, including the geometric configuration of even a well-balanced line.

The applicability of findings in this report is questionable, especially since the investigators themselves note that "current flow most probably has altered since subjects' residency" and state that "it was rarely feasible to measure fields." Current flow was nevertheless presumed *not to have changed* over the years in categorizing about 25 percent of the cases studied. While it is recognized that measuring field intensities in residences is an imposition on occupants, the measurements are not time-consuming and are imperative if a study of this nature is to be of value.

The investigators classify power systems solely on the basis of limited physical characteristics and spatial distribution of residences. The classifications are therefore *qualitative*, and have extremely limited usefulness.

The investigators note that more cases than controls lived near *presumed* HCCs for stable addresses and moved residence (the importance of intermediate addresses was noted earlier). The total numbers of cases and of controls are different and unexplained. It is not clear whether a particular address was listed more than once if it was in proximity to more than one type of wire configuration, but that seems to be the case. If so, the table of measured magnetic field data could be especially misleading. The magnetic field intensity at an address could be produced by multiple sources and be much higher than that inferred from the reported data. Furthermore, the field intensity at a presumed LCC address conceivably could be higher than at a presumed HCC address under some circumstances.

*Literature citations:* The authors imply that the literature on physiological effects of ELF electromagnetic fields is limited. A reader could be left with that impression from the few references cited by the investigators. Many more are readily available. More importantly, the cited references have questionable applicability to the reported study.

*Type of neighborhood:* The investigators suggest that cases at HCCs outnumber controls at HCCs regardless of neighborhood. The neighborhoods, however, are insufficiently identified and described in terms of too wide a range of possible ambient environmental conditions for suggestion to be meaningful. The apparent failure to match cases and controls by general neighborhood dilutes the credibility of the data and their interpretations, even if a neighborhood match had been made.

Environmental characterization of neighborhoods is important, and especially so for urban and suburban neighborhoods. Many types of pollutants have been identified in air, water and soil in urban-suburban areas, and at least some are either known or suspected carcinogenic agents. None of the common pollutants or those that might be unique to particular neighborhoods are addressed by the investigators, with one exception (heavy traffic). The presence of certain types of industries could be a contributing factor as, for example, the production of compounds that create hazardous waste as a byproduct.

*Use of suburban controls:* The investigators note that an excess of suburban controls was used in the study, but they were insignificant to the analysis. No numbers are given. The number or proportion of suburban controls used in the study is not identified, but is said to be insignificant. It is unclear whether any cases were located in suburbs. Ambient environmental conditions, which characterize neighborhoods, may identify significant differences between neighborhoods of cases and controls, and which can be important in health studies, are not reported or discussed, thereby limiting the credibility of data interpretations.

*Traffic routes:* A reference is cited that supposedly relates cancer to heavy traffic routes. A "mild" excess of cases of cancer on heavy traffic routes is suggested by the investigators. Heavy traffic is defined as a daily traffic count of 5000 vehicles or more, and is based on highway department maps. The investigators comment that the HCC-case excess was apparent in all neighborhoods, and a significant excess occurred only when the nearby traffic and HCCs coincided, not just when heavy traffic was evident.

*Substation proximity:* The investigators note that none of the 702 control addresses (including those not included in the study) were within 150 meters of a substation, but 6 of 491 cases' (1.5 percent) addresses were within 150 meters of a substation and within 40 meters of a large primary. They suggest that only one in 1000 homes in Denver is within 150 meters of a substation. They also note that each of the six cases resided at that address no more than three years prior to illness.

Since only 6 of 272 cases resided relatively close to substations, the information does not appear to be particularly significant. It may be that the investigators regard substations as especially high magnetic field sources of the same type as very long power lines. Substations may produce locally high magnetic field intensities, depending upon their capacities and loading of supply lines.

*Blind study:* The investigators regard their study as exploratory, and therefore did not classify addresses in the blind. They suggest that there was no bias, as demonstrated by a check with an assistant. The assistant classified 70 case addresses and 70 control addresses blind, and agreed with the investigators' classification in 91 percent of the instances, and in 5 of 12 diageements upgraded the classification to HCC. It is not stated if the agreement was more pronounced for cases or for controls, and which were upgraded. Another blind study was conducted in Colorado Springs and Pueblo. In that study 32 percent of 65 cases were HCC, and only 18 percent of the controls were HCC.

*Type of cancer:* The authors list data according to four types of cancer for birth addresses and for death addresses. The cases near HCCs exceed the controls near HCCs in every category for both birth and death addresses. Both cases and controls at LCCs are greater than at HCCs in both address categories. The investigators suggest that the data may show that the HCC-cancer relationship *may not be a causal one*, alternative explanations being artifactual or that there was an unknown effect on the body's ability to resist cancer.

*Socioeconomic class:* A reference is cited as a source of the statement that cancer is more prevalent in higher socioeconomic groups. The investigators believe that the trend in their data is insignificant. Socioeconomic class is based on the father's occupation at the time of the case or control's birth, and was determined according to classes given in a referenced document.

Identifying social class by occupation of the father at the time of the birth of a child may be inappropriate. Doing so presumes no upward mobility. It is not apparent whether the cited reference supports this method of defining social class. The investigators apparently do not question the fact that their data seem too inconsistent with the conclusions of the reference, and may be inconsistent due solely to methodology.

*Sex differences:* The investigators state there is a prevalence of cancer among males in contrast to females (no age information or reference provided). Males constituted 57 percent of the cases and 49 percent of the controls. There was an excess of HCC for both males and females over controls, with the trend stronger for males: 51 percent HCC for male cases at birth, death, or both addresses and 45 percent for females; and 28 percent HCC for male controls at birth, death or both addresses, and the same for female controls. A significant male case excess also was noted when the birth address was LCC and the death address was HCC, and when cancer developed after 9 years of age at a primary HCC address. The investigators state that significant male case excess occurred only for two conditions: when birth addresses had a lower current configuration than the death address, and when cancer developed 1 year or more postnatally at a stable address near primary wires.

The analysis of sex differences is especially tenuous. The references used to categorize the data are inappropriate. The cited work refers to direct-current (static) electromagnetic fields and fields of much higher intensity than those produced by distribution lines. No information or any other justification is presented for the conjecture, suppositions, and speculations about aborted male fetuses and exposure initiated postnatally. The scenario presented by the investigators simply cannot be taken seriously. The analysis of sex factors is based on inappropriate reference material, unsupported suppositions, and unjustified speculations, and is therefore not credible.

*Assessment:* The investigators' discussion is relatively brief. They note that it is not clear how HCCs might affect the development of cancer in children, and suggest the following:

(*a*) An unknown factor may affect the findings;
(*b*) There may be a direct cause, but the evidence is ambiguous and probably negative;
(*c*) Electromagnetic fields may change the nature of some ambient carcinogen; or
(*d*) Electromagnetic fields may alter human cell mechanisms.

The reported association between leukemia in the young and residential exposure to electric currents was investigated by Fulton *et al.*[14] in Rhode Island to confirm or invalidate the results of the earlier study conducted by Wertheimer and Leeper.[13]

The cases consisted of 119 persons up to 20 years of age who developed leukemia between 1964 and 1978. The controls consisted of 240 persons matched according to birth date.

As in the study by Wertheimer and Leeper[13] exposure was measured by considering the configuration of electric cables near the homes of the cases and the controls. In their analysis, the authors controlled for age at the onset of the leukemia and socioeconomic level.

Leukemia in persons up to 20 years of age reported by hospital authorities during the years 1964–78 was studied. Addresses at birth, intermediate years, and time of treatment were obtained. There were 119 patients with a total of 209 addresses. Controls were selected by a stratified random sampling of birth certificates, and two controls were matched to each case by year of birth only (240 total). Only the birth addresses of controls were obtained.

The power lines within 50 yards of each case address were noted by visual inspection, as were those near control addresses at birth. The investigators note that some power lines may be buried, and therefore were not recorded.

The investigators conclude that no relationship exists between the incidence of cancer in children and power line proximity in Rhode Island. They suggest that the Wertheimer-Leeper finding, if valid, may be insignificant, or may be due to some interacting variable unique to Denver. They further suggest that measuring fields in residences may be important.

This report by Fulton and coworkers, as in the Wertheimer-Leeper study, has several flaws which diminish the strength of the study. There are insufficient health, medical and environmental data to support the negative conclusion (that there is no effect) of the authors. The investigators depend upon the electromagnetic categorizations espoused by Wertheimer and Leeper, and these are clearly open to question.

It is not possible to estimate magnetic field intensities produced by power lines without measured data. There are too many factors that affect power line currents, and it is unlikely that all those factors could be identified reliably at a reasonable cost, even with substantial assistance from utility operators.

Epidemiologically, there is an important weakness in the methodology of the study by Fulton *et al.* The cases are of children with leukemia; the controls are in reality the addresses of birth. This method of selecting controls is inappropriate. As a consequence, the authors did not measure the same things in the cases as in the controls. This method of procedure would tend to invalidate the results of the study. This report by Fulton *et al.* does not convincingly reject the hypothesis raised in the Wertheimer and Leeper report.[13]

Wertheimer and Leeper conducted another survey[15] of cancer incidence among adults in two small cities in Colorado (Boulder and Longmont), in suburban Denver, and in central Denver. An attempt was made to relate the incidence of cancer to the presence of presumed high-current-carrying power lines.

The study was patterned generally along the lines used by the investigators in their earlier study.[13] There were some differences in detail, but the same presumptions about current levels and magnetic field intensities formed the basis of the analysis. Without the benefits of measured data, power lines were categorized qualitatively according to expected current levels. Six categories were used rather than the four used in the previous study.

The cases consisted of 1179 adults from four different geographic regions, who were deceased or who had developed cancer during the years 1967 to 1975, and the controls comprised 1179 adults deceased or living who did not have cancer. The authors say they controlled for sex, age, year of death, urbanicity, neighborhood, and socioeconomic status. The measurement of exposure was based on the configuration of the distribution of the electric field near the homes of the subjects. In the estimation of exposure, the authors considered the place of residence for the 10 years that preceded the diagnosis of cancer.

The cases all had cancer. According to Wertheimer and Leeper the adults who developed cancer lived for longer periods in the houses considered to have high electric currents (HCC). They suggest a dose-response relation between the level of current and the incidence of cancer. The types of cancer for which a significant excess was observed included cancer of the nervous system, the uterus, the breast, and lymphomas.

In their discussion, Wertheimer and Leeper[15] attribute the cause of a marked excess of cancer to the magnetic fields that accompany the alternating current.

This study has numerous weaknesses. The methodology is awkward and nebulous. The weakest point is the fact that the exposure was not coded blind. This weakness alone causes significant bias at the outset. The method allows considerable imprecision; the authors do not tell us that their evidence includes people who changed address; the data were not chosen in a uniform manner in the four regions; and the proportion of the deceased varies greatly between the towns and the age group, which ought to be the subject of a separate analysis.

As in their earlier study, Wertheimer and Leeper go much too far in the interpretation of their results. The engineering basis of the study is inadequate.

The critique for their study of cancer in children[13] generally applies. In addition, there are other concerns. The statement is made that currents in single-phase lines are especially low, and therefore that magnetic fields produced by such lines are also very low. The statement is not supported, and is of questionable validity. Single-phase currents can be highly unbalanced, with high resultant fields.

Discussion is presented as to why a causal relationship should be expected between cancer and high-current conductors. None is supported by data or clinical references. In several cases the discussion contradicts statements made in the 1979 study. For example, a socioeconomic difference is noted, and by means of a series of assumptions, the difference is connected to a presumed length of exposure. The discussion leads the investigators to conclude that the latent period for the cancer is quite long. The opposite conclusion was reached in the 1979 study from a different set of presumptions.

This study, like the previous one, should be regarded as speculative, and does not provide any useful information about any possible relationship between power systems and cancer. Usually there is a specific association between a causal agent and a type of cancer. Without being confirmed by other studies, the outcome of this research cannot be accepted as conclusive.

Tomenius *et al.*[16] attempted to show a relationship between transmission line magnetic fields and cancer in Stockholm, Sweden. Visual inspections were made at the listed birth residences and at the addresses listed at the time of diagnosis of cancer. Electrical systems and equipment that were visible and within 150 meters of the residents were noted, but only one of five possibilities was recorded, in the following order of priority: (1) 200-kV transmission lines, (2) substations, (3) transformers, (4) electric railways, and (5) subways.

These investigators note that many buried electrical systems are used in Sweden, but no attempt was made to identify their locations. Magnetic field intensities were measured at entrances of each residence. Measuring equipment is identified, and calibration method briefly described.

A total of 400 cases was identified as occurring in persons born and resident in the same church district of Stockholm County at the time of diagnosis. It is not reported whether addresses changed. The remaining 316 cases moved from their church districts, but remained in the county. Nothing is said of intermediate addresses. The total number of case addresses is given as 1162 rather than 1132 as in the text, and only 1129 in an accompanying table.

Controls were matched to the cases by age, sex and church district of birth. The total number of control dwellings is given as 1015 in the text, and 969 in the table. The text notes that some residences were demolished, and others were vacant.

Only 119 of the 1129 case addresses have visible electrical systems within 150 meters of the dwelling. Almost one-third (32) were near 200-kV transmission lines, and 36 were near electric railways. Those two categories were also the most prevalent for the controls (45 and 37). Only 77 of the 969 control addresses were near visible electrical systems.

The range of measured magnetic field intensities was $1 \times 10^{-3}$ to $1 \times 10^{-6}$ gauss, with an arithmetic mean of 0.69 mgauss. The investigators state that very low values were expected because most electrical systems are buried cable, and substantial field cancellation was therefore expected. The highest fields were measured near 200-kV transmission lines (2.2 mgauss, mean value).

All measured data are listed in a table under columns for the various types of observed electric systems or equipment, arranged in rows of 1-mgauss increments up to 20 mgauss. The number of case and control addresses are listed for each increment. Tomenius and coworkers conclude that the study neither confirmed the existence of a causal relationship between magnetic fields nor provided data that would prove that no such relationship exists.

The reported study is incomplete in too many respects to be conclusive or to provide much useful data. The measuring equipment is not described in sufficient detail to ascertain that measured intensities existed only at 50 Hz. A filter was used between the sensing coil (which was quite large) and the amplifier, but evidently provided a flat response above 30 Hz. The band width of the oscilloscope used to "register" the intensity is not given, but probably is relatively wide. Even though there are possible sources of error, the range of measured intensities seems reasonable, with high values being rare. For example, 1653 of 2098 measurements were less than 0.001 gauss (79 percent). The percentages were nearly the same for cases (80 percent) and controls (78 percent).

The mean arithmetic values of field intensity between cases and controls were substantially different at case addresses for only three types of systems. They were:

| System | Case | Controls |
|---|---|---|
| 200-kV transmission | 1.8 mG | 3.3 mG |
| Subway | 0.8 | 0.5 |
| Substations | 0.8 | 0.5 |

The intensities were very similar for all other systems, and were practically the same for the total number of cases (0.69 mG) and controls (0.68 mG). It may be particularly significant that the number of cases near 200-kV transmission lines was much higher than the number of controls (32 versus 13), but the magnetic field intensities for cases were significantly lower. At substations the numbers were 7 and 5, and at subways the numbers were 14 and 16, respectively, for cases and controls.

The data show no consistent trends about the possible relationship between magnetic field intensity and cancer.

## CRITIQUE OF EPIDEMIOLOGIC ASSOCIATION STUDIES

Associational studies such as those done without specific hypotheses are relatively crude and do not take into account important potential confounding variables. As such, they cannot be used to prove a causal association between a population and a type of cancer.

Many of the associations seen in any one surveillance study will be chance associations. Any time many statistical comparisons are made, a certain number will achieve statistical significance by chance alone.[17]

The strengths and limitations complement each other to some extent. For example, the quality-of-occupation and cause-of-death information obtained from death certificates is known to be mediocre.[18-25]

Investigators inevitably express their own biases and prior knowledge. While results may point to an occupational or residential cancer problem, seeing it as a problem may be influenced by previous reports.

Numerous factors need to be taken into account in proposing a hypothesis that a cancer problem exists in a given population. These factors include the following:

(1) The magnitude, level of significance, and consistency of the association of the population with a type or types of cancer.
(2) The number of cancer sites with which the population is associated, as well as the relationship among those sites. For example, if there is an increased amount of cancer incidence in several sites in the respiratory and upper alimentary tracts, the possibility of an inhaled carcinogen is suggested.
(3) The possibility of confounding factors or biases explaining the association(s).
(4) The likelihood of exposure to known or suspected carcinogenic agents.

As already noted, most surveillance studies are relatively crude and do not take into account potential confounding variables such as cigarette smoking. It is important to consider the possibility of confounding factors explaining cancer associations with selected populations identified from surveillance studies. Any attempt to rigorously test hypotheses generated from these data must carefully control for confounding.

Alternatively, at least two possible biases exist that could lead to an artifactual association of cancer types with certain populations. The accuracy of cause-of-death certification for specific forms of cancer on death certificates may range from 60 to 90 percent on the average.[23-25].

The second possible bias that could lead to an artifactual association of cancer types with selected populations would occur if the death or incidence rate of one or more common diseases in the occupation of interest were lower than average. In proportionate analyses, this would result in an artifactual increase in the proportion of the other diseases.

Most surveillance studies have not been able to control for confounding, and there are many possible biases. Dubrow and Wegman[17] have identified the 178 strongest and most consistent occupation-cancer associations from twelve major surveillance studies. These include leukemia among lawyers and judges; lymphatic leukemia among farm owners, and leukemia among lawyers, judges, mechanics and repairmen. They also found the strongest and most consistent brain and nervous system cancer among students, bank officers, lawyers and judges, teachers, engineers, accountants, and auditors. In regard to the report by Milham,[10] it should be noted that the author himself,[26] in a study of leukemia death certificates in Washington and Oregon, was able to confirm and extend a previously reported[27] association of leukemia with farming occupations, which tends to weaken his association of disease with magnetic field/electrical workers.

Occupational mortality studies suggest that observed differences in the mortality pattern within or between occupations primarily reflect variations in specific exposures within or between occupations and only secondarily reflect behavior patterns associated with individuals. This assumption does not actually hold true, since it is known that certain types of individuals select their vocation or are prescreened by physical examinations as a condition of entry.[28] Disease experiences of persons in the years prior to choosing a vocation play a major role among preselection factors. Similarly, the disease experiences and psychological factors associated with a primary job may aid in determining transfers of certain individuals from one class of occupation to another. This selection leads to the problem of interpreting final jobs, since associations with the final job may lead to false conclusions if the job was entered relatively late in life.

In determining *proportional mortality rate* the number of deaths ascribed to a particular disease is expressed as a proportion of all deaths within a specified population. Similarly, in clinical studies the incidence of a disease is sometimes reported as the number of patients with that disease as a proportion of all patients seen at the same institution.

Such proportional rates do not, of course, express the risk of members of the population contracting or dying from the disease. Comparison of such rates between areas or between population subgroups may suggest that a difference exists that is worth investigating. But until rates can be computed against a population base, it will not be known whether the difference relates to differences in the sizes of the numerators or the denominators of the compared rates.[29]

The major flaw of the proportionate morality ratio (PMR) is that it says nothing about total force of mortality for a given occupation.[30] All occupations have a total PMR or 100. Also, since the cause-of-death specific PMRs must sum to 100, a very high or low PMR in a common cause-of-death group will affect the other PMRs for that occupation. The method has had numerous critics. A large excess or deficit from one cause of death or several causes will decrease or increase the proportions dying from other causes. Thus, the PMR indicates *only* the importance of a specific cause of death relative to other causes in the same occupation, and does not measure the risk of death or the overall mortality.

## CONCLUSION

There is no doubt that the works examined so far are not very probing. The numbers are small, the epidemiologic methodologies are often weak, and the statistical analyses utilized are not solid. The possibility of a link between leukemia and cancer and exposure to electrical and magnetic fields has been raised and only responsible research can refute or confirm these reports.

It has been suggested these reports establish a causal link between exposure and cancer or leukemia. This link is by no means valid. To affirm the existence of a causal link, studies must be carried out that are much more sophisticated and of a far greater scope. It has also been suggested the reported association between leukemia and cancer and exposure to electric or magnetic fields is perhaps confounded by factors other than the fields per se.

Wright *et al.*[11] have noted that: "The precise cause of the excess of leukemia [they] have observed is not clear. The occupation grouped as sharing exposure to electric and magnetic fields undoubtedly shares other exposure. While significant exposure to ionizing radiation is probably not present in most of these jobs, metal fumes, solvents (including benzene), fluxes, chlorinated biphenyls, synthetic waxes, epoxy resins and chlorinated naphthalenes are other exposures that may be shared."

In further studies it will be necessary to consider very closely these other occupational exposures if one wants to clarify the link between leukemia and exposure to electric and/or magnetic fields.

Reports of effects of electric and/or magnetic fields among humans must be put in perspective. There are numerous problems in designing epidemiologic or incidence studies. It is essential that the multiple environmental factors which may interact among themselves and with personal characteristics of an individual be evaluated. There is always the danger that real factors may be overlooked, leading to a false association with factors of initial interest. The validity of application of the epidemiologic method to the study of the health impact of a factor such as a transmission line or substation environment is largely determined by the ascertainability and definition of an effect as well as the intensity of the exposure.

One problem is to select paired populations that are not systematically loaded with some other bias. The control or comparison group should be comparable with the case (i.e., the exposed group) in all relevant characteristics except for the exposure itself. The report of the study should provide the data needed to assess whether the study has met these requirements, and should note potential biases in selection of subjects, in measurements, and in followup that may affect the inferences being derived regarding the effects of the fields. The sample must also be large enough to make it possible to detect an increased risk.

An outstanding problem of epidemiologic studies of high-voltage transmission exposure is related to exposure assessment. Typical of the situations that exist in the literature are studies that provide no documentation of the basic element of exposures such as voltage. Often only broad categorizations such as "exposed" or "nonexposed" are provided. Most of the reports on humans demonstrate a lack of adherence to rigorous study design, analyses, and discussion. Control groups are freuqently lacking. Population selection criteria and methods are usually not described. Control of confounding variables, such as age, occupation, health status and habits, and economic status, is often not considered. Applied statistical methods may not be described or, largely because of study design, do not measure the strength of association vis-á-vis relative risk or odds ratio.

There is considerable difficulty in establishing the presence of and quantifying the frequency or severity of "subjective" complaints. Individuals suffering from a variety of chronic diseases or acute ailments may exhibit the same dysfunctions as those reported to be a result of exposure to electric or magnetic fields. Psychosocial interactions can also influence physiological reactions. Thus, it is extremely important to rule out other factors in attempting to relate electric or magnetic field exposure to clinical conditions.


## SUMMARY

Several reports have appeared in the last ten years in which the authors suggest an association between cancer/leukemia and exposure to electric and magnetic fields in the workplace or in the vicinity of distribution lines from overhead transmission systems. Several of these reports are reviewed and critiqued. The reports of clinical effects of electric and/or magnetic fields among human populations present serious difficulties which stem from the many publications in which pertinent material is not presented, and the data are variable or internally inconsistent. The reports presented so far are not very probing. The numbers of subjects are small, the epidemiologic methodologies are often weak, measurements of intensity of field are absent, and the statistical analyses utilized are not always appropriate. Although suggestive associations of electric/magnetic fields and cancer or leukemia have been made, no one has established a causal relationship between these fields and cancer or leukemia. All the reports on human exposure have numerous deficiencies which include: lack of or imprecise measurements of electric or magnetic field intensities, questionable subject identification, lack of statistical significance, confounding with uncontrolled variables such as socioeconomic differences, smoking, X rays, drugs, population mobility, and the unreliability of occupational classification. Nevertheless the possibility of a link between leukemia and cancer and exposure to electric and magnetic fields has been raised and only responsible research can refute or confirm these reports.

# REFERENCES


1. MICHAELSON, S. M. 1979. Human responses to power-frequency exposures. *In* Biological Effects of Extremely Low Frequency Electromagnetic Fields. R. D. Phillips, *et al.*, Eds.: 1–20. Annual Hanford Life Sciences Symposium, October 16–19, 1978, Richland, WA. U.S. Department of Energy Symposium Series. Washington, DC.

2. MEHN, W. H. 1979. The human considerations in bioeffects of electric fields. *In* Biological Effects of Extremely Low Frequency Electromagnetic Fields. R. D. Phillips, *et al.*, Eds.: 21–27. Annual Hanford Life Sciences Symposium October 16–19, 1978, Richland, WA. U.S. Department of Energy Symposium Series. Washington, DC.

3. STOPPS, G. J. & W. JANISCHEWSKYJ. 1980. Epidemiological Study of Workers Maintaining H.V. Equipment and Transmission Lines in Ontario, Canadian Electrical Association. Montreal, Quebec.

4. KNAVE, B., G. FRANCESCO, S. BERGSTROM, E. BIRKE, A. IREGREN, B. KOLMODIN-HEDMAN & A. WENNBERG. 1979. Long-term exposure to electric fields. Scand J. Work Environ. Health **5**: 115–125.

5. ROBERGE, P. F. 1976. Study of the State of Health of Electrical Maintenance Workers on Hydro Quebec's 735 kV Power Transmission System. Health Department, Hydro-Quebec, Montreal, Quebec.

6. NORDSTROM, S. & E. BIRKE. 1979. Studies of Possible Genetic Hazards in Workers at Vattenfall/Sydkraft Exposed to 400 kV. A Summary of Preliminary Results. Umeå Universitet. Umeå, Sweden.

7. NORDSTROM, S., E. BIRKE, L. GUSTAVSSON & I. NORDENSSON. 1981. Reproductive Hazards among Workers at High Voltage Substations. Report from Umeå University, Sweden. (Later published in: Bioelectromagnetics **4**: 91–101, 1983).

8. BAUCHINGER, M., R. HAUF, E. SCHMID & J. DRISP. 1981. Analysis of structural chromosome changes and SCE after occupational long-term exposure to electric and magnetic fields from 380 kV systems. Radiat. Environ. Biophys. **19**: 235–238.

9. HAUF, R. 1981. Studies of the Effects of Technical Energy Fields on Man. Special Report No. 9. Publ. Forschungsstelle fur Elecktropathologie. Freiburg im Breisgau, FDR.

10. MILHAM, S. 1982. Mortality from leukemia in workers exposed to electrical and magnetic fields (letter). N. Engl. J. Med. **307**: 49.

11. WRIGHT, W. E., J. M. PETERS & T. M. MACK. 1982. Leukemia in workers exposed to electrical and magnetic fields (letter). Lancet **11**: 1160–1161.

12. McDOWALL, M. E. 1983. Leukemia mortality in electrical workers in England and Wales (letter). Lancet **1**: 246.

12a. COLEMAN, M., J. BELL & R. SKEET. 1983. Leukemia incidence in electrical workers. Lancet **1**: 982–983.

13. WERTHEIMER, N. & E. LEEPER. 1979. Electrical wiring configurations and childhood cancer. Am. J. Epidemiol. **109**: 273–284.

14. FULTON, J. P., S. COBBS, L. PREBLE, T. LEONE & E. FORMAN. 1980. Electrical wiring configurations and childhood leukemia in Rhode Island. Am. J. Epidemiol **111**: 292–296.

15. WERTHEIMER, N. & E. LEEPER. 1982. Adult cancer related to electrical wires near the home. Int. J. Epidemiol. **11**: 345–355.

16. TOMENIUS, L., 1986. 50 Hz electromagnetic environment and the incidence of childhood tumors in Stockholm County. Bioelectromagnetics **7**:191–208.

17. DUBROW, R. & D. H. WEGMAN. 1983. Setting priorities for occupational cancer

research and control: Synthesis of the results of occupational disease surveillance studies. J. Natl. Cancer Inst. **71**: 1123–1142.

18.  MILHAM, S. 1976. NIOSH Research Report: Occupational Mortality in Washington State 1950–1971. Vols. 1–3. Cincinnati: National Institute for Occupational Safety and Health. (DHEW Publication No. (NIOSH) 76–175).

19.  BUECHLEY, R., J. E. DUNN, G. LINDEN & L. BRESHLOW. 1956. Death certificate statement of occupation: Its usefulness in comparing mortalities. Public Health Rep. **71**: 1105–1111.

20.  ANDERSON, M. R. 1972. Some sources of error in British occupational mortality data. Br. J. Ind. Med. **29**: 245–254.

21.  WEGMAN, D. H. & J. M. PETERS. 1978. Oat cell lung cancer in selected occupations: A case-control study. J. Occup. Med. **20**:793–796.

22.  WIGLE, D., Y. MAO, G. HOWE & J. LINDSAY. 1980. Comparison of occupation on survey and death records. Am. J. Epidemiol. **112**: 443–444.

23.  GITTELSOHN, A. & J. SENNING. 1979. Studies on the reliability of vital and health records. I. Comparison of cause of death and hospital record diagnoses. Am. J. Public Health **69**: 680–689.

24.  ENGEL, L. W., J. A. STRAUCHEN, L. CHIAZZE & M. HEID. 1980. Accuracy of death certification in an autopsied population with specific attention to malignant neoplasms and vascular diseases. Am J. Epidemiol. **111**: 99–112.

25.  PERCY, C., E. STANEK & L. GLOECKLER. 1981. Accuracy of cancer death certificates and its effect on cancer mortality statistics. Am. J. Publ. Health **71**:242–250.

26.  MILHAM, S. 1971. Leukemia and multiple myeloma in farmers. Am. J. Epidemiol. **94**: 307–310.

27.  FASAL, E., E. W. JACKSON & M. R. KLAUBER. 1968. Leukemia and lymphoma mortality and farm residence. Am. J. Epidemiol. **87**: 267–274.

28.  PETERSEN, G. R. & S. MILHAM. 1980. Occupational mortality in the State of California. DHEW Publication No. (NIOSH) 80–104.

29.  MacMAHON, B. & T. F. PUGH. 1970. Epidemiology: Principles and Methods. Little, Brown. Boston, MA.

30.  MILHAM, S. 1983. Occupational Mortality in Washington State, 1950–1979. (DHHS Publication No. (NIOSH) 83–116).

# Fire and Brimstone

JAMES P. LODGE

*Consultant in Atmospheric Chemistry*
*385 Broadway*
*Boulder, Colorado 80303*

In the prehistoric world, on a date unrecorded (by definition, since it was prehistoric), but nevertheless clearly real, a caveman, whom we might as well call Ug, encountered a fire. He was certainly not unique in this; Ug and his ancestors and fire had already shared the Earth for as long as it is reasonable to call hominid life forms ancestors of Ug.

The novel thing that Ug did on that unrecorded day was not, then, his encounter with fire, but his response to it. He picked up one or more burning brands and took them home. He and his successors learned how to feed a fire so as to keep it going, and the whole list of the do's and don'ts concerning fire, such as keeping it out of the bed-straw, and not sealing the door of the cave too tightly if you have a fire inside. An extremely remote descendant even learned how to *make* fire.

At a single stroke, Ug founded civilization and invented air pollution. On balance, had Ug looked ahead to all that he would cause through his innovation, he might well have thrown the fire back out of the cave and gone and bashed a saber-toothed tiger, an act entailing fewer consequences. But Ug was not strong on this sort of forethought, which makes it ironic that he comes down to us in Greek mythology through the figure of Prometheus, who stole fire from the gods and gave it to man, and whose name, translated, means "forethought."

In other writings[1] I have used this same concept to trace the history of air pollution from that humble beginning. Here, however, I would like to raise a different question. What was the origin of that fire? There are probably only two viable alternatives, lightning and volcanism, and both had, or rapidly acquired profound religious significance. Which was the immediate cause of the fire that involved Ug cannot be determined, the more so as we have no clear notion of just where Ug lived. However, by the beginnings of historic times, the deification of both lightning and volcanos was well underway. The Greeks clearly had deities for both, and fairly clear references to both occur in the Hebrew scriptures. For example, Mt. Sinai was clearly a volcano, and thus a fitting abode for deity. It is also altogether possible that the mountain to which the prophet Elijah fled when King Ahab tried to kill him was a volcano; the sequence of wind, earthquake and fire would certainly support such an idea. In the New Testament, much of the array of horrible things foretold for God's opponents, including the lake of burning sulfur, bespeaks the great power of the volcanic image.

Between eruptions, a volcano tends to draw its lava back down below the top of the mountain, giving a vision of complex subterranean plumbing, which still fumes and sputters on occasion. In the Graeco-Roman mythology as well as the early Hebrew thought, the descriptions of the abode of the dead came quickly to resemble exactly what is seen in such a circumstance — the endless caves through which Orpheus sought Eurydice, and the Hebrew view of Sheol, attest to this. It was apparently only during the intertestamental period that the Hebrews took the idea seriously that good people might be with God in Heaven (i.e., above the Earth) and that only the evil might descend to what most of them now called by the Greek name of Hades. That this was viewed as a place of fire and vile fumes is testified by the use in the New Testament of the term Gehenna as a synonym, or probably more properly a circumlocution, for Hell. Gehenna was an actual piece of neighboring geography, Ben Hinnom, a ravine just outside Jerusalem. Some commentaries naively attribute this to the fact that this ravine was cursed as a site of some pagan human sacrifice during the reign of some of the bad kings, but it is much more likely that this name arose from the fact that, being cursed, the area was used as the garbage dump of Jerusalem. It is almost self-evident that the area caught fire, with a resulting emission of smoke and vile odors, already clearly associated with Hell as the inside of a volcano. Clearly this imagery is the source of the previously mentioned "burning lake of sulfur" from the Apocalypse, and numerous other citations that could be made.

Thus, at an early age, the principal characteristics of a volcano — fire (smoke) and a sulfurous odor — became fixed in the collective consciousness as the attributes of Hell. I venture that nobody reading the title of this presentation doubted that I would allude to Hell, and I have obviously not disappointed anyone.

Although, as noted, Ug the caveman was certainly the inventor of air pollution, his invention did not cause serious problems in his own day, with the exception of a few truly painful experiences, such as the death of some entire families from carbon monoxide when the oxygen supply to a fire was unduly restricted. The chief reason was that Ug simply did not live long enough to die of lung cancer. He would probably have been delighted to do so. Instead, he died at what we would call a very young age from disease or the saber-tooth or food poisoning or starvation. (I should probably state at this point that my use of the masculine pronoun is generic, with no intent to exclude a woman from the possible role as a bringer of fire.) In all likelihood it was a millennium after the writing of the last of the Scriptures before air pollution, in anything like its modern sense, become important. The location of the problem is now clearly known — England.

Up until the time of the Norman conquest, or thereabouts, the British had done virtually all of their heating with charcoal, and in the process had largely deforested England. One of the first priorities of the early Norman kings was the reforestation of tracts of land, to this day called the "New Forest." These were declared royal preserves to keep out the charcoal burners and let the trees grow to maturity. The result was a hideous increase in the cost of charcoal, and the beginning of English experiments with another energy source, known as sea-coal because it was brought down to the London area by ship. Today we call it bituminous coal, and England was blessed with an abundance of the stuff. If was far cheaper than charcoal, because the ships that brought it actually used it as ballast; by and large, they had been carrying goods from London up the coast and coming back empty. It gave a hotter fire than charcoal, and it was cheaper. There was one problem. Burning only charcoal, the British had never invented the chimney.

Beginning less than a century after the arrival of the Normans, various laws were passed to try to ban sea-coal from the city. There is even a legend of one artificer being put to death for his persistent use of coal, but this appears to be a misreading of the court records of the day; his property was confiscated, and his lime-kiln was leveled. But overall the ban on coal burning was precisely as effective as Prohibition was in the United States.

One of the more interesting pieces of legislation attempted occurred during the reign of Elizabeth I. It forebade the burning of sea-coal in London "whilst Parliament was in session." The announced rationale for this was concern over the delicate health of the Virgin Queen, a claim almost as tenuous as Elizabeth's claim to virginity. She may have been a small woman, but she had the constitution of a stevedore. The transparent fact was that the Members themselves found the environment uncomfortable.

However, things simply got worse, and by the time of the Restoration it must have been quite awful. In 1661 the agile civil servant and diarist John Evelyn prepared a pamphlet addressed to the king. (That he was agile is shown by the fact that he successfully served Charles I, Oliver Cromwell, and Charles II in succession, without any loss of position, property, or perquisites.) This tract[2] was called "Fumifugium," denounced the current state of the air in London as totally injurious to health and happiness, and proposed a cure, which was, in modern terms, zoning the dirty industries out of London, and replacing them with parks. But there is here a clue to Evelyn's thinking: At one point he gives a lengthy list of major sources of smoke, states that these are far more important than home fires, and then writes:

> And that this is not the least hyperbolic, let the best of judges decide it, which I take to be our senses: whilst these are belching it forth their sooty jaws, the City of London resembles rather the face of Mt. Aetna, the court of Vulcan, Stromboli, or the Suburbs of Hell, than an assembly of rational creatures, and the imperial seat of our incomparable Monarch. For when in all other places the air is most serene and pure, it is here eclipsed with such a cloud of sulfur, as the Sun itself, which gives day to all the world besides, is hardly able to penetrate and impart it here; and the weary traveler, at many miles distance, sooner smells, than sees the City to which he repairs.

Now, it is obvious that Evelyn has no epidemiologic data. More or less contemporary with him there were a few studies that showed some sort of urban factor in health, but that did not distinguish between the effects of crowding, poverty, and all of the other urban factors besides pollution. Evelyn simply relied on sensory data and came up with a simile: London looks and smells like Hell or a volcano. Note the coupling of the two terms. It looks and smells like something horrible, the most horrible thing he can think of, and this must be the direct cause of all of the ill-health he observes.

A bit more than a century later, Evelyn's tract was reprinted with a brief introduction. In it, the editor remarks that while "[Evelyn] complains that the gardens about London would no longer bear fruit . . . the complaint at this time would be, not that the trees were without fruit, but that they would not bear any leaves." Obviously, the situation had gotten far worse, and apparently continued to do so until nearly the end of the nineteenth century. It is ironic that the most notorious episode of pollution in London occurred in 1952, during a long period of gradual lessening of the average amount of pollution.

One important point is not immediately obvious from these brief excerpts. Neither Evelyn nor his anonymous editor, who wrote just before the American Revolution, uses the word *pollution* in connection with this subject. In fact, the word was not applied until the latter part of the nineteenth century. Prior to that time it was an essentially theological term, restricted to describing very grave states of moral transgression, usually sexual, and not infrequently homosexual. Thus, when the King James version of the Bible has the Prophet Amos laying on the people the curse that they "shalt die in a polluted land," he was not consigning them to England specifically, but to lands that were unclean — essentially non-Kosher. That this term would ever come to represent smoky air implies a powerful moral judgment on those responsible for causing the air to be dirty.

It is tempting to continue with the devious history of air pollution and attempts at its control. However, both I and others have done so in other writings. Instead, let us halt the historical survey at this point and summarize it. First, however, let me make it clear that I do not present John Evelyn as the author of the association between polluted air and perdition; he is simply a striking example. On the contrary, there was a very general perception that dirty air was bad for man and beast, as well as tree and building. It was further perceived that this badness was a direct result of the presence of sulfurous vapors and smoke — fire and brimstone, if you will — and that the reason that this was bad was that this mixture was diabolical. To inflict this mixture on someone was to do the devil's work for him, and hence it is almost surprising that the name of the moral sin of pollution was not applied to it at an earlier time, instead of waiting almost until the time of the American Civil War.

If you doubt this, look at some more contemporary evidence. Probably the first specialty magazine given over to the promotion of air pollution control was "Smokeless Air," published in England. The first head of the Los Angeles Air Pollution Control District announced as his priorities for Los Angeles the control of smoke and sulfur dioxide. These were in fact controlled, and of course the smog got worse, and he left, to make a bad pun, under a cloud. The first priority of the ancestor of EPA, saddled with producing those marvelous compendia called "Criteria Documents," was to produce documents on sulfur dioxide and particulate matter.[3] The largest single effort in the second round of revision of Criteria Documents and Standards has been on a *combined* Criteria Document on Particulate Matter and Sulfur Compounds.[4] Despite the fact that that document reveals comparatively low toxicity for both, the EPA is now working very hard to justify stiffer standards for them. The fact that health effects have not been found at low levels admittedly does not prove, in a philosophical sense, that no effects exist. Nevertheless, I would maintain that the same belief that impelled Evelyn to name them as the cause of all ills of London and other large cities is still at work. There just has to be something bad about them; they are hellish. And this from people who are nominally, in many cases, neither believing Jews nor Christians. At this point it is too deeply embedded in our culture, like the Protestant work ethic, for anyone to escape.

As a result, resources that could well have gone into the study of other contaminants have been instead directed to a desperate search to find effects of these two pollutants, and the present flap over acid rain, justified though it may well be, probably owes at least as much to what must well be felt by many as real relief — at last something truly bad has been found to justify controlling the sulfur oxides! Early in the study of acid rain, it was suggested[5] that perhaps one reason it seemed to be getting worse was progressive control of coarse particles that are predominantly alkaline, and that this might be expected to neutralize some of the acid. That went over with a dull thud, and has hardly been heard from since, despite the fact that it is a reasonable hypothesis and relatively easy to test. In fact, comparison of a very good data set from the 1950s with contemporary measurements[6] shows that the important difference is not an increase in sulfate and/or nitrate, but a decrease in alkaline earth ions. These largely arrived as particles far too large to inhale; scientifically speaking, they posed no threat to health. Yet to allow them to be emitted seems unthinkable. They are part of the hellish complex, and they simply have to be bad.

Let me state clearly at this point that I am not in favor of dirty air; all too often I have been stereotyped as wearing a black hat; simply because I number some industries among my consulting clients. There are perfectly good reasons why at least some interests may wish to enforce extremely stringent standards for both smoke (or generalized particulate matter) and/or sulfur dioxide. For example, it is clear that some types of vegetation are adversely affected by levels of sulfur dioxide far below those that have any demonstrable effect on health, quite aside from any considerations of acid rain. There are perfectly good economic reasons, in the Mountain West, for preserving visibility to protect tourism, and this means control of both fine particles that are emitted directly and of those gases that form secondary fine particles. However, to promote standards to protect these interests is one thing, and to promote them in the name of health is another. I am continually reminded of an old friend who had numerous dislikes among food. These included rare beef, milk, and several other items that I personally enjoy on occasion. However, he would never admit that he was simply exercising preferences; he always explained that rare beef, milk, etc. were all carcinogenic. He chain-smoked unfiltered cigarettes. Had he been religiously inclined, I have the feeling he would have made some theological argument, no better founded than the theory that particulate matter and sulfur compounds simply have to be bad for you.

I have at home a copy of a U.S. Department of Agriculture Yearbook[7] dating from around the beginning of World War II. The theme for that year was the effect of trace substances on vegetation and health. In numerous respects, it is one of the most up-to-date references available. I find that frightening. Some 40 years have passed, during which we have expended enormous sums to control generalized particulate matter and a few gaseous pollutants that have been virtually studied to death. At the same time the truly critical analysis of the inhalation toxicity of boron, selenium, vanadium, and a host of other individual airborne species has received no emphasis whatever, and their respective ecological impacts have received even less study. In truth, it is not even clear that we have made any significant headway in methodology for assessing ecosystem impacts; we have not even settled the question of whether a term like "climax forest" describes anything that truly exists. In many cases we even lack analytical data to pair off with such knowledge as we have of toxicity or ecosystem impact to judge whether or not there is likely to be any sort of problem.

Parenthetically, in closing, I strongly suspect that our bad theology—the association of fire and brimstone with the devil and all his angels—has distorted our thinking in other ways as well. A couple of months ago a particularly stupid upset in a chemical plant in India resulted in 2000 deaths. There is no question that this is a dreadful thing to have happen. However, I have the impression that what made it newsworthy was the fact that these were not casualties to fire. An explosion could well have killed off as many people, and I strongly suspect it would have stayed in the news for only about 24 hours; it is highly dubious that it would have attracted the attention of Melvin Belli. It is as though we expect to be done in by fire, but get terribly upset when something else is responsible. Can it be that this is also a factor in our lack of concern about nuclear winter? Do we really expect to be done in by the aftermath of fire, and consequently have little interest in the details?

It is clear that, given the public interest, many of our unanswered questions could be answered. It is also clear that, given enough push by public opinion, many of our current environmental problems could be solved. Solving problems, in this sense, costs money, and money is allocated on the basis of perceived priorities.

I once had a sort of vision. The year was a time far in the future—far enough to balance our beginning far in the past. The event written about by science fiction writers finally occurred: Earth was visited by intelligent beings from another world. They came down in the flying saucers, but I was unable to answer the crucial questions that they had come to ask. Was there, or had there ever been, intelligent life on Earth?

## REFERENCES

1.  LODGE, J. P., JR. 1980. An anecdotal history of air pollution. *In* Advances in Environmental Science Technology, Vol. 10. Wiley. New York, NY.
2.  EVELYN, J. Fumafugium. Reprinted (1969) *in* The Smoke of London, Two Prophecies. Maxwell Reprint. Elmsford, NY.

3. National Air Pollution Control Administration. 1969. Air Quality Criteria for Particulate Matter. NAPCA Publication No. AP-049. U.S. Government Printing Office. Washington, DC.

4. Environmental Protection Agency. 1982. Air Quality Criteria for Particulate Matter and Sulfur Oxides. EPA-600/8-82-029. Environmental Criteria and Assessment Office, EPA. Research Triangle Park, NC.

5. Likens, G. E. & F. H. Bormann. Acid rain: A serious regional environmental problem. Science **184**: 1176.

6. Stensland, G. J. & R. G. Semonin. 1982. Another interpretation of the pH trend in the United States. Bull. Am. Meteorol. Soc. **63**:

7. U.S. Department of Agriculture. 1938. Soils and Men, Yearbook of Agriculture. U.S. Government Printing Office. Washington, DC.

# Both Sides Now
## The Chemistry of Clouds[a]

STEPHEN E. SCHWARTZ

*Environmental Chemistry Division*
*Brookhaven National Laboratory*
*Upton, New York 11973*

## INTRODUCTION

The past several years have seen considerable research effort directed toward examination of the chemistry of clouds. Much of the motivation for undertaking such studies has arisen from concern over the so-called "acid rain" phenomenon and the desire to understand the role of clouds in the transport, transformation, and delivery to the earth's surface of pollutants involved in acid deposition. Cloud chemistry is thus of interest in this context not only because clouds are the source of precipitation, but also because of the possibility that clouds may represent an important medium for chemical reaction in the atmosphere.

This article is an account of research on cloud chemistry that has been conducted over the past several years, with examples taken principally from work carried out by myself and my colleagues at Brookhaven National Laboratory. In carrying out this research, we have followed a two-sided approach:

(1) Laboratory studies of the thermodynamics and kinetics of processes involved in cloudwater acidification, and

(2) Field measurements directed to determination of cloud composition and elucidation of the mechanisms by which this composition is established. The title of this article refers to the two "sides" of our approach to the study of this chemistry and is taken from the popular song, "Both Sides Now" by Joni Mitchell[1] whose refrain begins "I've looked at clouds from both sides now."

## BACKGROUND

Since the impetus for research into cloud chemistry is the presumed role of clouds in the acid deposition process, it seems worthwhile briefly to review

[a] The research was performed under the auspices of the United States Department of Energy under Contract No. DE-AC02-76CH00016. Portions of this work have received support from the Electric Power Research Institute (RP2023-1), the Federal Aviation Administration (High Altitude Pollution Program), and the National Acid Precipitation Assessment Program through the Processing of Emissions by Clouds and Precipitation (PRECP) project funded by the United States Department of Energy.

this process in order to place the cloud chemistry research in a context. Much more exhaustive accounts of the present understanding of the acid deposition process have been given in the report of a National Academy of Sciences panel chaired by Calvert,[2] in a compilation of review papers edited by Altshuller,[3] and in the report of the U.S.-Canada Work Group on Transboundary Air Pollution.[4] For a historical account of acid precipitation studies see Cowling.[5] A brief but insightful description of the overall process of acid deposition and the implications of acid deposition research on control strategies has been given by Gordon.[6]

A schematic representation of the subprocesses involved in the overall acid deposition process is shown in FIGURE 1. In brief these subprocesses are:

• Emission of acid precursors, mostly $SO_2$ and NO, into the atmosphere. These species are emitted largely as byproducts of combustion reactions: $SO_2$ from oxidation of sulfur in fuel; NO from oxidation of nitrogen in fuel or from partial oxidation of atmospheric $N_2$. Noncombustion emissions of $SO_2$ (e.g., from metal smelting) are also substantial in some regions. Natural emissions of $SO_2$ and $NO_x$ ($NO_x$ refers to the sum of NO plus $NO_2$) are substantially less than anthropogenic emissions, except in regions of very low industrial activity.[7]

• Transport of acidic and related pollutants in the atmosphere. Transport processes include motion with the mean wind, vertical motion, motion brought about by clear-air convective processes or associated with storms, and mixing processes.

• Transformations—processes altering the chemical or physical state of acidic pollutants and related substances. These processes include both chemical reactions (e.g., oxidation of $SO_2$ and nitrogen oxides to higher oxidation state acids) and physical processes (e.g., aerosol formation, dissolution of gaseous species by atmospheric liquid water, and coagulation of aerosol particles on cloud droplets).

• Deposition—the delivery of acidic pollutant to the earth's surface. This deposition is conveniently distinguished into two categories: wet deposition, the delivery of dissolved materials in precipitation, principally rain and snow; and dry deposition, the loss of gaseous or aerosol materials to the surface of the earth in the absence of precipitation. For gases the dry deposition process consists of turbulent transport followed by molecular diffusion across a laminar sublayer. For materials present in aerosol particles dry deposition processes include inertial impact and gravitational settling in addition to Brownian diffusion. It might be observed that the distinction between wet and dry deposition is somewhat arbitrary; Lovett *et al.*[8] have noted that impaction of cloud droplets to mountainous terrain represents a major fraction of both the water budget and pollutant deposition at such sites.

The foregoing subprocesses represent the phenomena that must be included in the description of the overall acid deposition process. It is toward elucidation of these phenomena that much of the current research into acid deposition is directed, the overall objective being an improved description of the relation between emissions of these materials and the location and amount of deposition attributable to these sources.

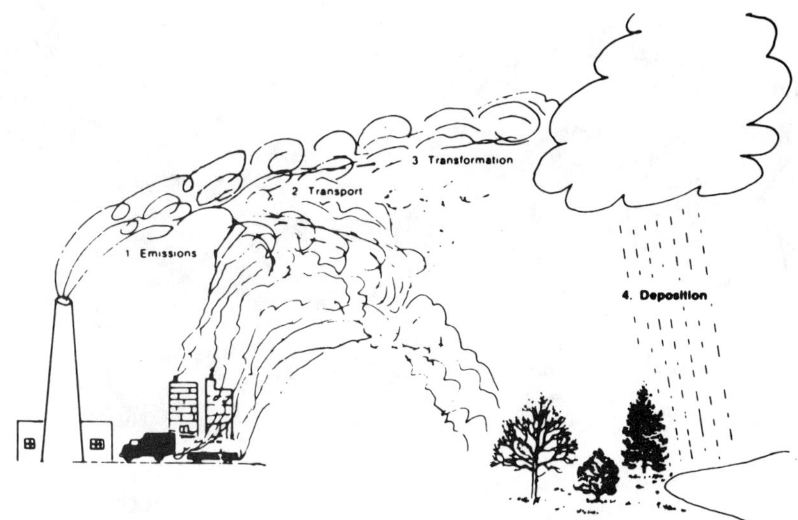


**FIGURE 1.** Schematic diagram of components of overall process of acid deposition.

It may be noted that while deposition of acidic pollutants represents the end of the process of concern to atmospheric science, it represents the beginning of a second major focus of acid precipitation research, namely research into the effects of deposition of acidic pollutants on biological systems, including human health effects, and, as well, effects on man-made materials. Detailed accounts of current understanding of such effects are given in a compilation of review papers edited by Linthurst,[9] in a report of a United States-Canada work group on trans-boundary air pollution[10] and in a National Research Council Report.[11] A brief resumé of health and environmental consequences of acid fog has been given by Hoffmann.[12]

As may be inferred, there is considerable ongoing research into all aspects of acid deposition. Attention is called to listings of such ongoing research in the public[13] and private[14] sectors.

## *Precipitation Composition*

It is useful briefly to consider precipitation composition as a basis for comparison of cloud composition and also insofar as precipitation represents a major deposition mode of acidic pollutants. Considerable effort has been devoted over the past several years to documentation of the concentrations of major ionic species in precipitation. It should be noted that while the concentration of ionic species in precipitation is highly variable from storm to storm and even within storm events,[15] nonetheless a rather smooth geographic pattern of concentrations is observed for sufficiently long averaging periods, e.g., seasonal or yearly. Examples of such patterns in North America are shown

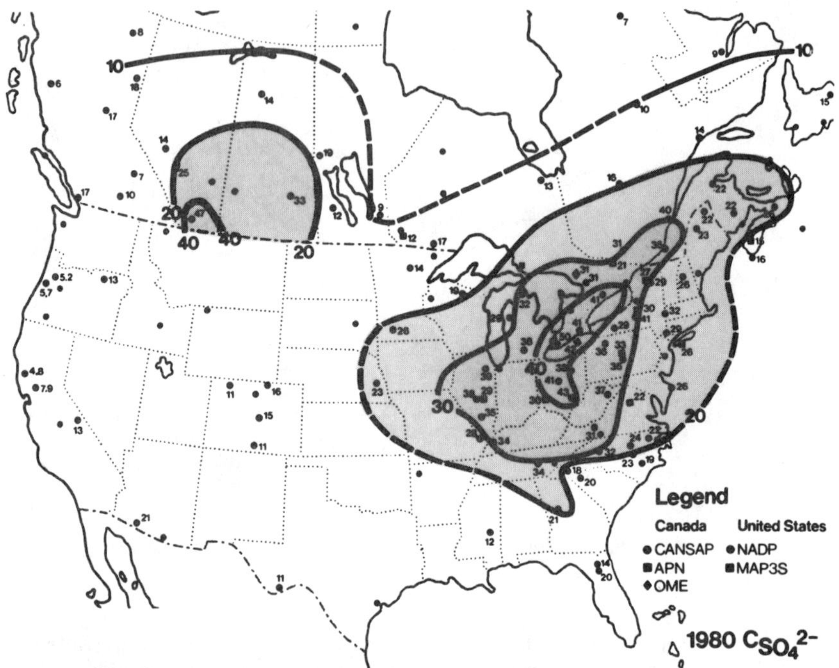



**FIGURE 2.** Mean concentration (μmol/liter) of sulfate ion in precipitation in North America for 1980, weighted by precipitation amount. (From the United States—Canada Memorandum of Intent on Transboundary Air Pollution. Reprinted by permission.)

in FIGURES 2–5.[10] Concentrations are peaked in the region of the United States east of the Mississippi River and north of the Ohio River corresponding to the region of maximum emission of sulfur and nitrogen oxide precursors. It may be noted that the concentrations of sulfate and nitrate in the regions in which acidic precipitation has its greatest impact exceed those in regions less affected by an order of magnitude or more. It is interesting to note that on a molar basis the deposition of sulfate and nitrate in this region is roughly equal; this equality is roughly matched in emissions of the precursor oxides. Total sulfate plus nitrate concentration (equivalents per liter) is roughly equalled by total hydrogen ion plus ammonium ion; other cations, usually present to a much lesser extent, include calcium and magnesium. Thus precipitation in the northeast United States is seen to consist of a dilute solution of sulfuric and nitric acid, partially neutralized by atmospheric ammonia. The extent of neutralization may be inferred from the ratios of the ionic concentrations to be ~25% in much of the northeastern United States. This is further illustrated in FIGURES 6 and 7,[16] which are frequency distributions of the fractional acidity, $[H^+]/([NO_3^-] + 2[SO_4^=])$, in precipitation samples collected by event. It is seen that the peaks of the distributions represent a 30% neutralization, although substantial variance exists about such a value. We shall return to

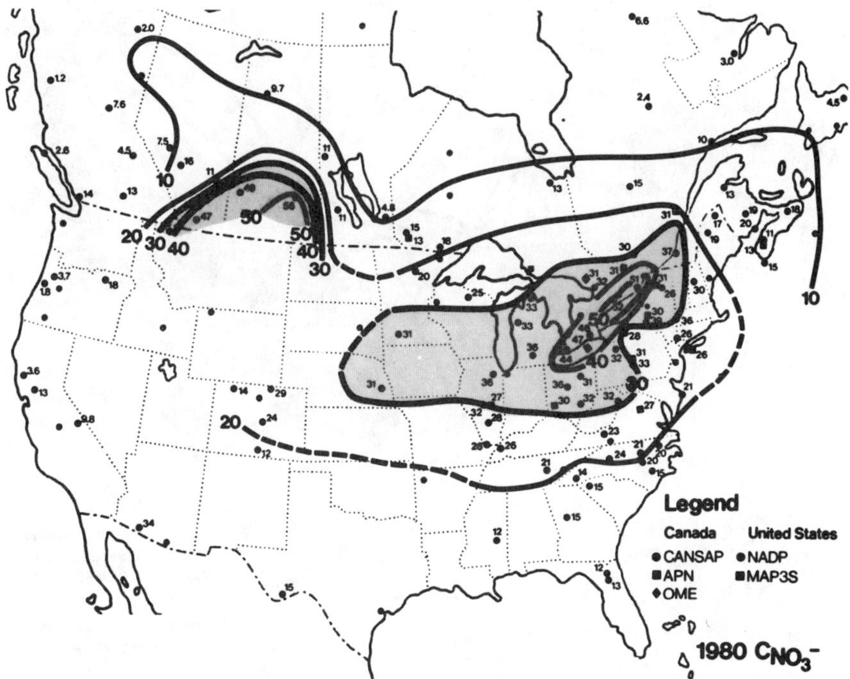



**FIGURE 3.** Mean concentration (μmol/liter) of nitrate ion in precipitation in North America for 1980, weighted by precipitation amount. (From the United States—Canada Memorandum of Intent on Transboundary Air Pollution. Reprinted by permission.)

consideration of the fractional acidity of rainwater as a point of comparison for cloudwater samples.

One further point of interest in consideration of rain composition is the absolute concentration in comparison to emissions of sulfur and nitrogen oxides. Emissions of sulfur and nitrogen oxides for the region of the United States east of the Mississippi River and north of (and including) Tennessee and North Carolina are given in TABLE 1. In order to compare emissions with wet deposition, an average composition of precipitation is evaluated assuming that the entire emissions of these species are dissolved in an annual rainfall amount of 100 cm. It is seen that the spatial-average annual deposition of $SO_4^=$, $NO_3^-$, and $H^+$ evaluated in this way substantially exceeds (by a factor of perhaps 4 or 5) the magnitude of wet deposition of these species indicated in FIGURES 2–4. The fate of the fraction of sulfur and nitrogen oxides not deposited in precipitation in this region is not known in detail, but undoubtedly consists primarily of dry deposition and of deposition outside this region.

### Transformation Pathways

Returning to consideration of processes responsible for acid deposition,

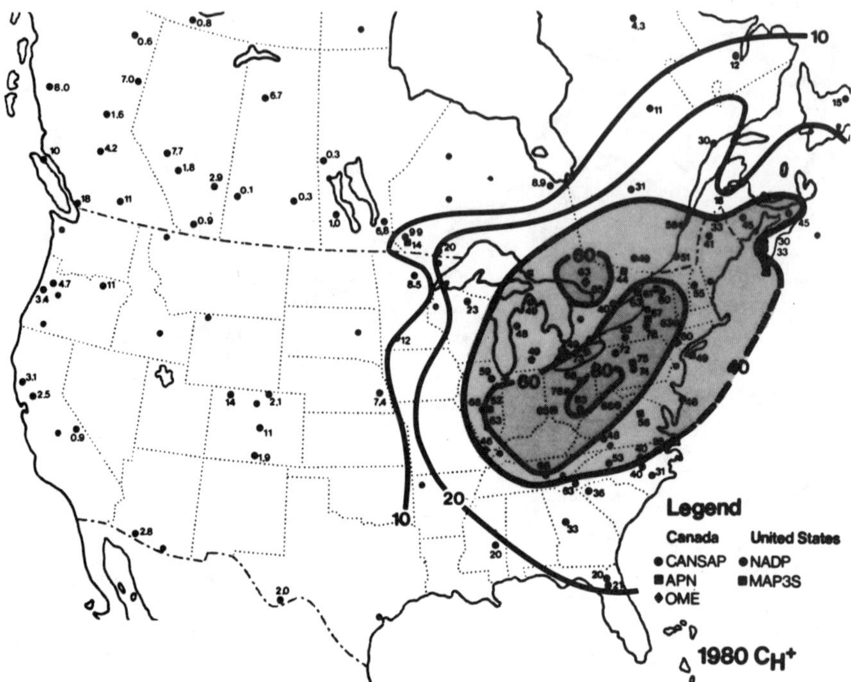



**FIGURE 4.** Mean concentration (μmol/liter) of hydrogen ion in precipitation in North America for 1980, weighted by precipitation amount. (From the United States—Canada Memorandum of Intent on Transboundary Air Pollution. Reprinted by permission.)

we first observe that the overall process of transformation of sulfur dioxide and nitrogen oxides to sulfuric and nitric acid present in and delivered in precipitation consists of two components: a chemical process, namely the oxidation of the relatively insoluble, lower-oxidation-state oxides initially present in the gas phase to the highly soluble acids; and the physical process of uptake of the gaseous species by cloud- or rainwater. It is thus useful conceptually to distinguish two pathways for this process. These are indicated in FIGURE 8 and are distinguished by the phase in which oxidation takes place, namely the gas phase or aqueous-solution phase. The overall process consists of going from the initially emitted gaseous oxides (upper left on the figure) to the dissolved acids (lower right). We denote the gas-phase path as that consisting of gas-phase oxidation followed by uptake of the high-oxidation-state oxide or acid by cloud- or rainwater, i.e., right and then down on the figure. The aqueous-phase path consists first of dissolution of the oxide followed by aqueous-phase oxidation. Much of our research and that of a number of other groups concerned with the mechanism of acid deposition has been directed to elucidation of the contributions of these two pathways to the overall process. One point that should perhaps be explicitly mentioned here is that in

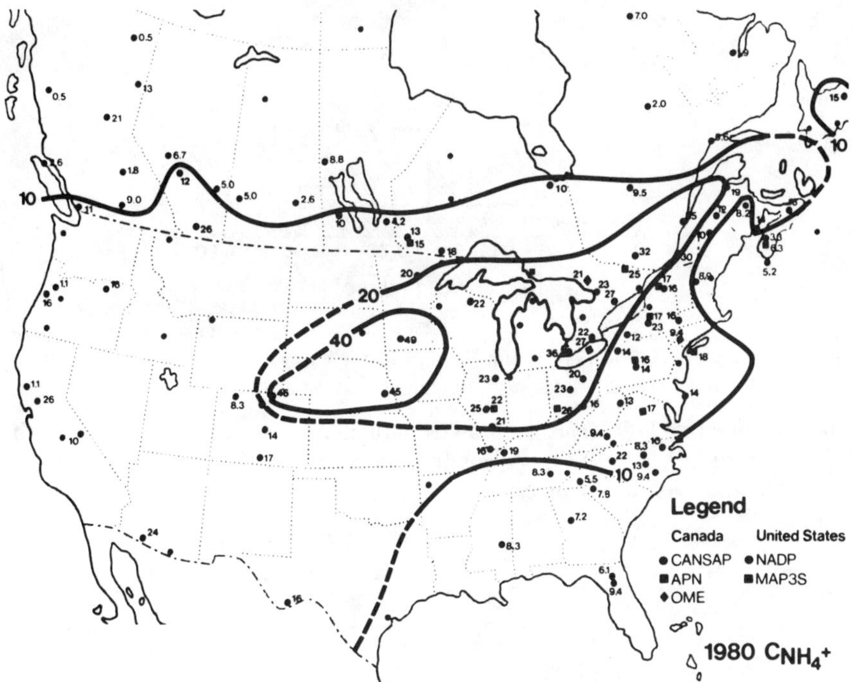


**FIGURE 5.**  Mean concentration (μmol/liter) of ammonium ion in precipitation in North America for 1980, weighted by precipitation amount. (Form the United States — Canada Memorandum of Intent on Transboundary Air Pollution. Reprinted by permission.)

the presence of atmospheric $O_2$, and all the more so in the presence of oxidant species such as $O_3$ and $H_2O_2$, there is strong thermochemical driving force for oxidation of $SO_2$ and $NO_x$ to the high-oxidation-state acids. The question that must be addressed is therefore not one of chemical equilibrium, but rather of the mechanism and rate whereby the initially emitted oxides are transformed to the acids that are deposited in precipitation.

Of the two pathways for this oxidation that have been outlined here, the gas-phase pathway is much the better understood, the various gas-phase reactions having received much study in a number of laboratories. For a recent review of present understanding see Calvert.[17] A brief synopsis of important gas-phase oxidation reactions of $SO_2$ and nitrogen oxides is given here.

First, with respect to $SO_2$, the principal gas-phase oxidation reaction appears to be initiated by the addition of free radical OH,

$$OH + SO_2 \rightarrow HO-S\begin{smallmatrix}\nearrow O\\ \searrow O\end{smallmatrix} \qquad\qquad [1]$$

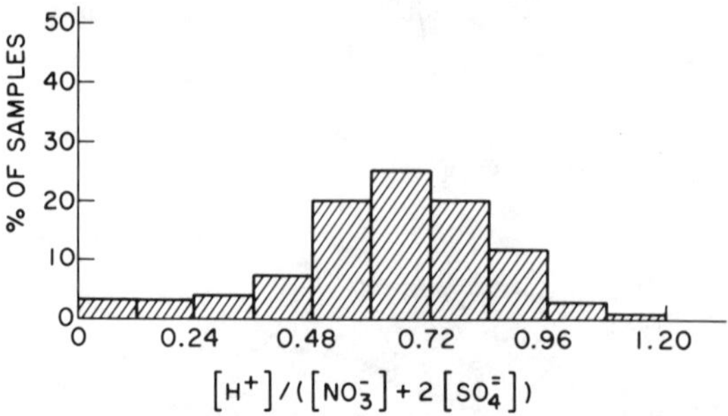

FIGURE 6. Fractional acidity in precipitation samples by event at Brookhaven National Laboratory for the years 1977 to 1981. (From Daum.[16] Reprinted by permission.)

The intermediate $HOSO_2$ is itself a free radical and is presumed to lead with high yield to $SO_3$ or $H_2SO_4$. A possible reaction sequence, based on recent work of Stockwell and Calvert[18] is:

$$HO-S{<}^{O}_{O} + H_2O \xrightarrow{?} \quad [2]$$

$$+ O_2 \xrightarrow{?} \quad + HO_2 \quad [3]$$

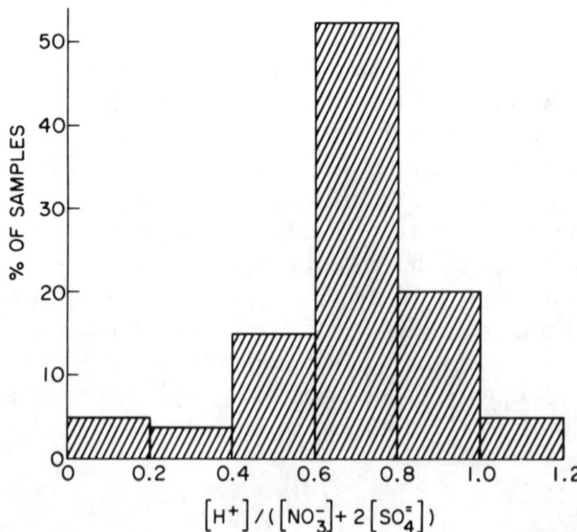

FIGURE 7. Fractional acidity in precipitation samples by event at Lewes, Deleware, summer 1980. (From Daum.[16] Reprinted by permission.)

**TABLE 1.** Emissions and Maximum Wet Deposition in the Northeastern United States[a]

|  | $SO_2$ | $NO_x$ | Acid Equivalents[b] |
|---|---|---|---|
| Emissions[c] (mol/yr) | $2.4 \times 10^{11}$ | $2.0 \times 10^{11}$ | $6.8 \times 10^{11}$ |
| Area (km²) | | $1.4 \times 10^6$ | |
| Precipitation amount (cm/yr) | | 100 | |
| Precipitation volume (liter/yr) | | $1.4 \times 10^{15}$ | |
| Average concentration[d] (µmol/liter) | 170 | 140 | 480 |

[a] East of Mississippi River and north of (and including) Tennessee and North Carolina.
[b] $NO_x + 2 \times SO_2$.
[c] Reference 7.
[d] Under assumption that all emissions are deposited in precipitation in the region.

The gaseous sulfuric acid product is unstable to formation of an aerosol and apparently readily either adds to existing aerosol particles or nucleates with water vapor to form new aerosol particles. On the basis of the rate constant $k_1$ for reaction [1] and estimates of representative average OH radical concentration $[(1-2) \times 10^6 \text{ cm}^{-3}]$, the average rate of $SO_2$ oxidation by this path is perhaps 0.5 to 1% hr$^{-1}$ (Ref. 17), and a rate of this magnitude is consistent with inferences based on field measurements of $SO_2$ oxidation in various point-source plumes.[19]

Gas-phase atmospheric oxidation of nitrogen oxides is similarly rather well understood. We have noted above that the principal nitrogen oxide emitted as a combustion byproduct is nitric oxide, NO, with nitrogen dioxide, $NO_2$, present as a relatively minor component. The gas-phase oxidation of NO to $NO_2$ is strongly coupled to ozone chemistry, which has been the object of much

PATHWAYS FOR FORMATION OF ATMOSPHERIC SULFATE AND NITRATE

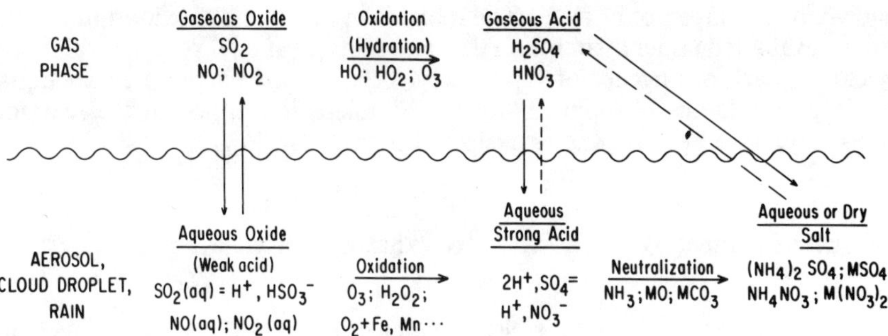


**FIGURE 8.** Schematic representation of pathways for atmospheric formation of sulfate and nitrate. (From Schwartz.[41])

study as an environmental concern in its own right.[20] Without reviewing that chemistry, suffice it to say that the gas-phase oxidation of NO to $NO_2$ is a fairly rapid process, taking place on a time scale of one to several hours, depending on the amount of $O_3$ present and/or on the extent of photochemical activity in the system. The extent of conversion of NO to $NO_2$ depends as well on insolation, since photolysis of $NO_2$ sets up a steady-state NO concentration. However, except in areas rather strongly influenced by local source emissions, $NO_2$ is the dominant of the two species.[21]

Gas-phase oxidation of $NO_2$ to nitric acid can take place by OH addition, similar to the situation described above for $SO_2$,

$$OH + NO_2 \rightarrow HONO_2 \tag{4}$$

In this case the reaction appears to be a simple addition reaction without complication of subsequent steps. The rate of $NO_2$ oxidation by reaction [4] for estimated representative OH concentrations is perhaps 5 to 10% $hr^{-1}$, roughly an order of magnitude greater than the $SO_2$ oxidation rate.[17] In contrast to $H_2SO_4$, $HNO_3$ can remain present in the atmosphere as a gas-phase species. However, in the presence of substantial amounts of liquid water (as in clouds) and *a fortiori* in the presence of basic material (e.g., $NH_3$) nitric acid vapor is efficiently taken up by the condensed phase.[22,23]

The foregoing process involving the photochemically generated free radical species OH, is, of course, essentially entirely a daytime process. An alternative process [24-28] that does not require photochemically generated free radicals consists of reaction of $NO_2$ with $O_3$,

$$NO_2 + O_3 \rightarrow NO_3 + O_2 \tag{5}$$

followed by addition of $NO_3$ to $NO_2$ to form $N_2O_5$

$$NO_3 + NO_2 \rightarrow N_2O_5 \tag{6}$$

$N_2O_5$ is thought to react rapidly with liquid water to form nitric acid

$$N_2O_5 + H_2O(l) \rightarrow 2HNO_3(aq) \tag{7}$$

(The corresponding reaction of $N_2O_5$ with water vapor proceeds only quite slowly or perhaps not at all.[29]) The maximum rate of $NO_2$ oxidation by this route, evaluated under assumption that reactions [6] and [7] proceed with unity yield for each occurrence of [5], is $2k_5 p_{O_3}$. For ozone partial pressure equal to 30 ppb, this rate is about 15% $hr^{-1}$. Complete description of the reaction must take into account the reversibility of reaction [6],

$$N_2O_5 \rightarrow NO_3 + NO_2 \tag{-6}$$

as well as competing reactions of $NO_3$ that do not yield $N_2O_5$, e.g.,

$$NO_3 \xrightarrow{h\nu} NO + O_2 \text{ or } NO_2 + O \tag{8}$$

and

$$NO_3 + NO \rightarrow 2NO_2 \qquad [9]$$

Both of these reactions will be more important under daytime conditions than nighttime ones, and thus the sequence [5]–[7] would appear to be a more important process for acid formation at nighttime rather than daytime, thus complementing the OH addition reaction [4]. However, comparison of chemical reaction rates with mass-transport rates governing uptake of gases by cloud droplets[30] suggests that the sequence [5]–[7] may be significant even under daytime conditions.

In contrast to the fairly well-developed understanding of the gas-phase paths, understanding of the aqueous-phase path has been rather less well developed. Reasons for thinking that aqueous-phase oxidation of sulfur and nitrogen oxides in clouds may be important include the strong thermodynamic driving force for such reactions, the high liquid water content of clouds (providing a medium for reaction), high surface-to-volume ratio (promoting interphase mixing), long residence times (permitting time for reactions to proceed significantly), and the prevalence of clouds in the atmosphere. The lack of ability to describe oxidation reactions of sulfur and nitrogen oxides in clouds, which has been described as the "missing link" in our understanding of the chemistry of acid rain formation,[31] has been the impetus behind the work in our laboratory over the past several years directed at elucidation of these processes.

### *Cloud Properties*

Before describing research about chemical reactions in clouds, it is useful to review some of the physical properties pertinent to the uptake and reaction of acidic species. Properties of concern include the liquid-water content, which represents the volume available for aqueous-phase reaction, and the size of drops making up the cloud, which governs the mass-transport rate. For a more complete description of the microphysical properties of clouds reference is made to monographs by Pruppacher and Klett,[32] Mason,[33] and Rogers.[34] A very readable account of the role of clouds in the earth's hydrologic cycle is given by McDonald.[35] In brief: a cloud is a supersaturated suspension of liquid or solid water particles in air. Here "supersaturated" means that the partial pressure of water vapor exceeds the equilibrium partial pressure at the temperature of the cloud; the amount of this supersaturation is quite small, generally a fraction of a percent. Clouds may consist of solid water (i.e., ice) or liquid water. At temperatures above 0°C, of course, clouds consist entirely of liquid water (so-called "warm" clouds). However, liquid water droplets may persist in clouds to temperatures well below 0°C (supercooled clouds). Mixed ice and water clouds can maintain a temperature of 0°C over quite a large volume.[36] From the perspective of cloud chemistry our interest is focused on liquid water clouds because of the ability of liquid water droplets to serve as a medium for chemical reaction.

The most notable characteristic of a cloud is, of course, the presence of condensed-phase water, for it is this property of a cloud that makes it visible.

In an absolute sense, however, the amount of condensed-phase water that is present in a cloud is invariably quite small, typically 0.1 to 1 gram per cubic meter; since the density of liquid water is approximately 1 gram per cm$^3$, we see that the fraction of volume of the cloud that is occupied by liquid water is in the range 0.1 to 1 parts per million (0.1 to 1 cm$^3$/m$^3$, or liquid water volume fraction L = 0.1 to 1 $\times$ 10$^{-6}$). That is, clouds are mostly air! The air between the cloud droplets is denoted the "interstitial" air (i.e., in the interstices between the cloud droplets), although, in view of the fact that the cloud droplets are few and far between, this is a bit of a misnomer. The interstitial air is important in cloud chemistry as the reservoir of gaseous species that may react in the aqueous phase of the cloud. Despite the low fraction of total cloud volume occupied by liquid water, it is important to note in the context of clouds serving as the medium for aqueous-phase reaction that this liquid water content is nonetheless some three to five orders of magnitude greater than that associated with clear-air aerosol.[37]

The amount of water vapor in a cloud is closely approximated by the saturation water vapor content, which is shown as a function of temperature in FIGURE 9. It is seen that over the temperature range 0 to 25°C the saturation water vapor content substantially exceeds the amount of liquid water present in the cloud.

An important feature of clouds is the fact that the liquid water contained in them is in a high state of dispersion. This is illustrated schematically in FIGURE 10, which depicts the sizes of drops characteristic of clouds and of rain. Also indicated for comparison is a typical clear-air aerosol particle that may serve as a nucleus for cloud droplet formation or, alternatively, may remain present in the interstitial cloud air as an unactivated aerosol particle. As may be inferred from FIGURE 10, the drops within a cloud are not all of a single size, but exhibit a range of sizes. FIGURE 11 shows a spectrum of number

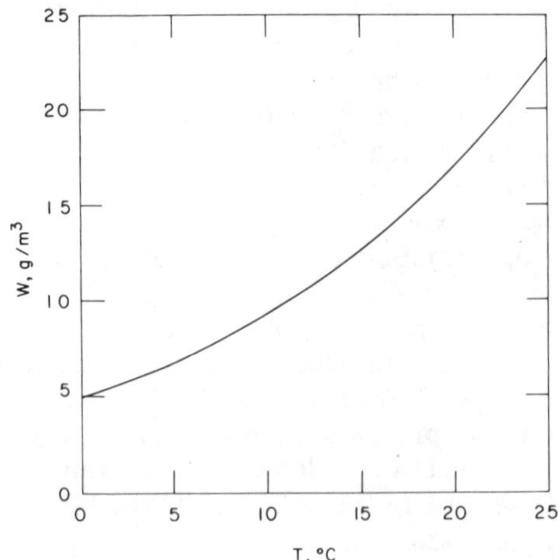


FIGURE 9. Saturation water vapor content of air as a function of temperature.

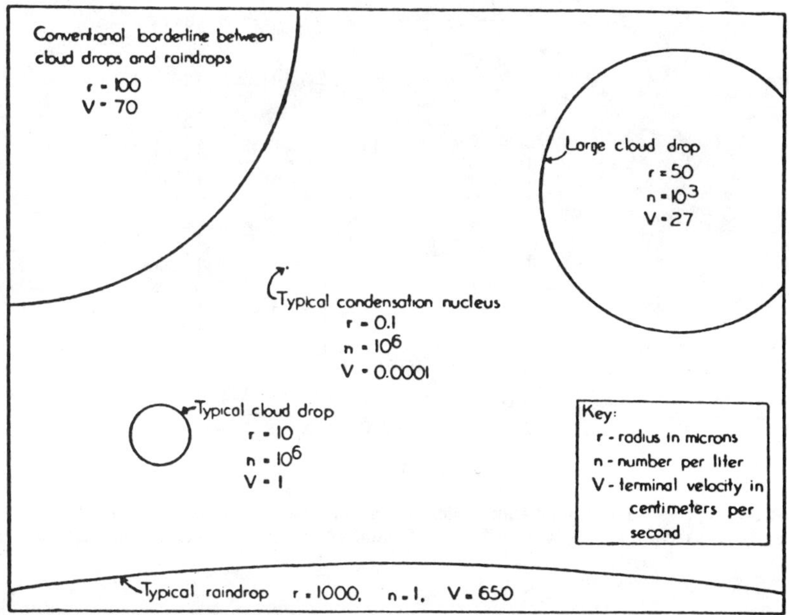


**FIGURE 10.** Comparative sizes, concentrations, and terminal fall velocities of some particles involved in condensation and precipitation processes. Note for comparison an aerosol particle that serves as nucleus for cloud droplet formation. (From McDonald.[35] Reprinted by permission of Academic Press, Inc.)

of drops as a function of drop radius for a stratus cloud; the sum of drop number density taken over all drop sizes represents the total number density of drops in the cloud, in this instance about 300 cm$^{-3}$. Also shown is the distribution of liquid water content within the cloud as a function of drop size; the sum of partial water content taken over all drop sizes represents the total liquid water content. In this example the fraction of the total volume occupied by liquid water, L, is about $0.3 \times 10^{-6}$.

The question arises: How does a cloud come to exist? It might be observed that a cloud is unstable thermodynamically with respect to a single large drop, in view of the excess free energy associated with the large amount of surface area. In brief answer to a complicated question, cloud formation is a consequence of cooling that results in the actual water vapor content exceeding the equilibrium value. Such a situation most commonly arises from the lifting of an air parcel and the resulting cooling as the rising parcel expands and thereby adiabatically performs work on its surroundings. Alternatively, as in fogs, a vapor-laden air parcel may come into contact with a colder surface. In response to the supersaturation of water vapor, condensation of water takes place on available existing aerosol particles; these particles consist largely of solutions of salts or partially neutralized acids that had been present in clear air prior to cloud formation. The fraction and size distribution of these aerosol parti-

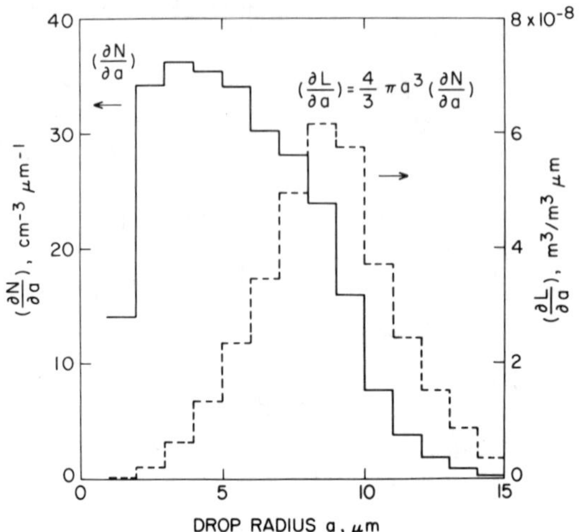


**FIGURE 11.** Spectra of droplet number density and liquid water volume fraction for aged stratus cloud. N = 300 cm$^{-3}$; L = 0.34 × 10$^{-6}$. (From Schwartz[30] on the basis of data from Knollenberg.[38])

cles that become activated into cloud droplets depend rather sensitively on the rate of cooling and on other properties of the condensing cloud. In any event, despite the thermodynamic instability of a cloud there is little mechanism for interaction of the cloud droplets, and the dispersion is quite stable. It might be observed that the characteristic ratio of interdrop distance to drop dimension is of order $L^{-1/3}$, where L is the liquid-water volume fraction; for $L = 1 \times 10^{-6}$ the characteristic interdrop spacing is 100 times the drop dimension. Cloud droplets can thus to good approximation be treated as noninteracting entities. Coalescence of cloud droplets into rain drops occurs for the most part only by the mechanism of scavenging of smaller drops by larger drops that have acquired an appreciable gravitationally induced fall velocity.

A trivial but important consequence of the low liquid water volume fraction of clouds upon cloud chemistry is the high degree of concentration that results upon dissolving a gaseous species in the aqueous phase. Consider a gaseous species X initially (prior to uptake by cloudwater) having a partial pressure $p_X$. If this species were to dissolve entirely in cloudwater having volume fraction L, the resulting concentration would be given by

$$[X] = p_X/LRT, \tag{1}$$

where R is the universal gas constant and T is the absolute temperature. (For $p_X$ in atmospheres (atm) and [X] in moles per liter (M), $RT \approx 25$ atm M$^{-1}$). It is seen that even a rather low initial partial pressure $p_X$ may lead to a rather high aqueous-phase concentration. Thus for $p_X = 1 \times 10^{-9}$ atm (1 ppb) and $L = 1 \times 10^{-6}$ (1 cm$^3$/m$^3$), [X] = 40 μM.

# LABORATORY STUDIES OF CLOUD CHEMISTRY

In this section I outline information that can be derived from laboratory studies that is pertinent to cloud chemistry and review some of this work pertinent to aqueous-phase reactions of $SO_2$ and $NO_2$ in clouds. Specifically the objective of such laboratory studies is to determine the fundamental physical and chemical properties necessary to evaluate the rate and extent of gas-aqueous reactions in clouds, based upon measured or modeled reagent concentrations and cloud properties. These fundamental physical and chemical properties include:

- overall chemical equilibrium;
- solubilities of reagent gases;
- mass-transport coefficients; and
- aqueous-phase reaction mechanisms and kinetic rate laws.

These properties are defined and described here with particular reference to the cloud chemistry of $SO_2$ and $NO_2$.


## *Chemical Equilibria*

At the outset we might mention, as was noted above, that in an atmosphere containing $O_2$ and liquid water, there is a strong thermochemical driving force for reaction of $SO_2$ and $NO_2$ to sulfuric and nitric acids, respectively, and this driving force is all the stronger in the presence of strong oxidants such as $O_3$ or $H_2O_2$. However, it must be stressed that while thermochemical driving force is a necessary condition for reaction to proceed, it is not sufficient. This was emphasized by Lewis and Randall in their 1923 monograph *Thermodynamics*[39] in a passage that remains very appropriate to those of us who are engaged in the study of gas-liquid reactions in the ambient environment:

> We see from the large negative free energy of formation of nitric acid that it should be producible directly from its elements. Even starting with water and air, we see by our equations that nitric acid should form until it reaches a concentration of about 0.1 M where the calculated equilibrium exists. It is to be hoped that nature will not discover a catalyst for this reaction, which would permit all of the oxygen and part of the nitrogen of the air to turn the oceans into dilute nitric acid.

While one shares the wish expressed here that nature not discover such a catalyst, one cannot help noting that should one discover such a catalyst it would be extraordinarily valuable in view of the large industrial usage of nitric acid. (For a delightful account of the history of nitric acid manufacture and of its present industrial manufacture and usage one is referred to the monograph of Chilton.[40]) In any event the foregoing passage from Lewis and Randall serves as a warning that one should not rely solely on thermochemical driving force as a basis for inferring reaction rates, a warning to which we shall return later. (On a longer — geological — time scale one must still face the question why $N_2$ persists in the atmosphere. As noted by Stedman and Shetter,[21] the presence and maintenance of $N_2$ in the earth's atmosphere must be attributed to geologic processes.)

### Gas Solubilities

The reversible dissolution of a nonreacting gas in a liquid generally obeys a limiting law, valid in the limit of low partial pressure of the gas, that the concentration of the dissolved gas in equilibrium with the gas phase is proportional to the partial pressure of the gas. We denote this proportionality by

$$[X] = H_x\, p_x \tag{2}$$

where $[X]$ represents the liquid-phase concentration of the dissolved gas, $p_x$ represents the partial pressure, and $H_x$ is the proportionality constant. Equation (2) is denoted Henry's law, and $H_x$ the Henry's law coefficient. It should be noted that a variety of units are employed for Henry's law coefficient and, as well, that the sense of the proportionality is often reversed. Consequently care must be exercised whenever Henry's law coefficients are encountered. We have adopted the convention that partial pressures are expressed in units of atmospheres (atm) and solution concentrations in units of moles per liter (M), whence the Henry's law coefficient has units $M\ atm^{-1}$. Specializing to aqueous solutions, we observe that $H_x$ represents the equilibrium constant of the reaction

$$X(g) = X(aq)$$

for standard states taken as 1 atm and 1 M. We observe further that the Henry's law coefficient is related to the free energy of dissolution in the usual way for any equilibrium constant,

**TABLE 2.** Henry's Law Coefficients of Some Atmospheric Gases Dissolving in Liquid Water[a]

| Gas | H (M atm$^{-1}$) |
|---|---|
| $O_2$ | $1.3(-3)$[b] |
| NO | $1.9(-3)$ |
| $C_2H_4$ | $4.9(-3)$ |
| $NO_2$[c] | $1\ \ (-2)$ |
| $O_3$ | $1.3(-2)$ |
| $N_2O$ | $2.5(-2)$ |
| $CO_2$[d] | $3.4(-2)$ |
| $SO_2$[d] | $1.3$ |
| $CH_3ONO_2$[e] | $2.6$ |
| PAN[c,e] | $3.6$ |
| $HNO_2$ | $4.9(1)$ |
| $NH_3$[d] | $6.2(1)$ |
| $H_2CO$ | $6.3(3)$ |
| $H_2O_2$ | $1\ \ (5)$ |
| $HNO_3$[d] | $2.1(5)$ |

NOTE: Adapted from Reference 41; PAN and $CH_3ONO_2$ are from Ref. 42.
[a] T = 25°C except as noted.
[b] The notation $1.3(-3)$ represents $1.3 \times 10^{-3}$.
[c] Physical solubility; reacts with liquid water.
[d] Physical solubility, i.e., exclusive of acid-base equilibria.
[e] T = 22°C.

$$H_x = \exp(-\Delta G_{sol}/\underset{\sim}{R}T) \tag{3}$$

For gases which dissolve nonreactively Henry's law coefficients may be determined by saturating the solution and measuring the dissolved gas concentration and the gas-phase partial pressure. For Henry's law coefficients of some atmospheric gases see TABLE 2.

In the case of a gas such as $SO_2$, which undergoes partial and reversible aqueous-phase reaction,

$$SO_2(aq) \overset{H_2O}{=} H^+ + HSO_3^- \tag{[10]}$$

both the Henry's law coefficient and the first dissociation constant of sulfurous acid $K_{al}$ may be determined by suitably varying the gas-phase partial pressure and/or altering the $H^+$ concentration of the solution with a nonvolatile acid.[43] The pH dependence of the total solubility (i.e., undissociated and ionic forms) of several atmospheric acids and $NH_3$ is shown in FIGURE 12. This solubility may be represented by an effective Henry's law coefficient, H*, which is useful if the aqueous solution is sufficiently well buffered that the dissolution of the gas in question does not greatly alter the pH. These effective Henry's law coefficients are conveniently denoted by the oxidation number of the species in question. Thus $H^*_{S(IV)}$ represents the equilibrium ratio of total

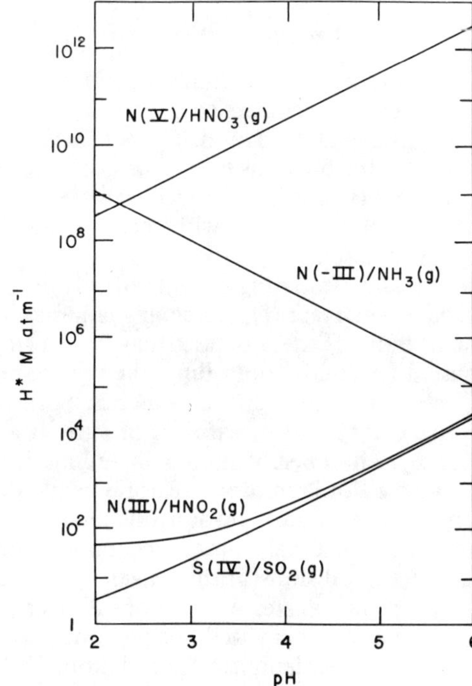


FIGURE 12. Effective Henry's law coefficients for gases that undergo rapid acid-base dissociation reactions in aqueous solution, as a function of solution pH. Buffer capacity of solution is assumed to greatly exceed incremental concentration from uptake of indicated gas. (From Schwartz.[41].)

sulfur(IV) concentration, $[S(IV)] \equiv [SO_2(aq)] + [HSO_3^-] + [SO_3^=]$, to $SO_2$ partial pressure.

In the case of gases such as $NO_2$, which react rapidly and irreversibly in liquid water, the Henry's law coefficient cannot be determined by a direct approach. In such cases Henry's law coefficients may be determined by means of thermochemical cycles and/or from kinetic studies, as discussed below.

It is useful to consider the distribution of material in a cloud between gas and aqueous phases under Henry's law equilibrium. The ratio of material in solution to that in the gas phase within a given volume of a cloud is

$$\frac{C(aq)}{C(g)} = \frac{L[X]}{P_X/\underset{\sim}{R}T} = H_X L \underset{\sim}{R} T \tag{4}$$

For $H_X \ll (L\underset{\sim}{R}T)^{-1}$ most of the material is present in the gas phase, whereas for $H_X \gg (L\underset{\sim}{R}T)^{-1}$ the opposite is the case. For $L \approx 1 \times 10^{-6}$, $(L\underset{\sim}{R}T)^{-1} \approx 4 \times 10^4$ M atm$^{-1}$, and correspondingly greater for lower values of L. Examination of the values of Henry's law coefficients in TABLE 2 shows that only for very soluble gases (e.g., $H_2O_2$) do Henry's law coefficients approach such a value, and thus that most nonreactive gases will be partitioned largely in the interstitial air within the cloud. On the other hand, as indicated in FIGURE 12, dissociative gases such as $HNO_3$ and $NH_3$ would be partitioned largely into cloudwater at representative values of pH and liquid water content.

### Mass-Transport Kinetics

The overall process of uptake and reaction of gases in cloud droplets may be considered to consist of a sequence of mass-transport processes followed by aqueous-phase reaction. The pertinent mass-transport processes are diffusion from the bulk gas to the surface of the drop, transport across the interface, and diffusion in the aqueous phase. It is useful to think of these processes as taking place in series with each other and as well with chemical reaction, as indicated schematically in FIGURE 13. By identifying the rate-determining process and evaluating the rate of this process it is possible to ascertain the overall uptake rate. The foregoing is somewhat of an oversimplification since one or more of the processes may have a similar rate. Moreover, if aqueous-phase diffusion is controlling, the uptake rate nonetheless does not become independent of the reaction kinetics.[44] Still, this picture is conceptually very useful. Quantitative treatment of various aspects of this problem has been given by a number of groups including our own.[30,44-47]

In terms of laboratory measurements, the important quantities to be determined are the gas- and aqueous-phase diffusion coefficients of the reagent gases, and the so-called mass accommodation coefficient ($\alpha$), the fraction of gas kinetic collisions of the reagent gas molecules at the air-water interface that result in transfer of the reagent gas across the interface. The diffusion coefficients are fairly well known either from direct measurement or by estimation from semiempirical correlations.[30] However, little is known about the

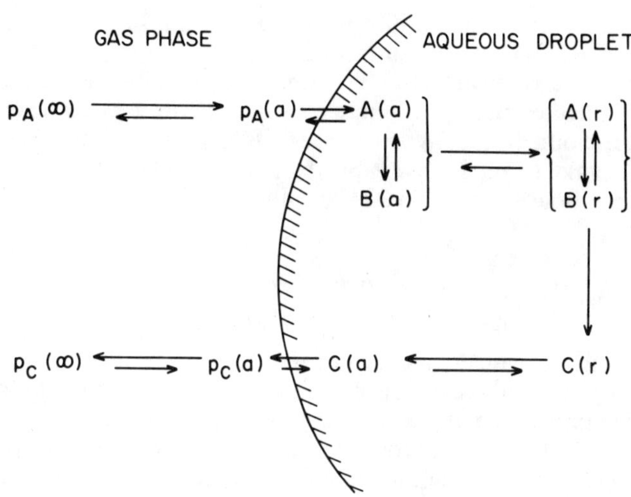



**FIGURE 13.** Schematic illustration of the subprocesses making up the overall process of reaction of gases in cloud droplets. A represents aqueous-phase reagent species transferred from the gas phase; B, species in rapid equilibrium with A; and C, product species, at the surface of the drop (a) or in the interior (r); $P_A$ and $P_C$ represent gas-phase partial pressures of A and C at the surface of the drop (a) and at large distances from the drop ($\infty$). (From Schwartz.[30])

values of mass-accommodation coefficients on theoretical grounds, and the few experimental measurements appear to be widely inconsistent.[30] My own view, based in part on some recent studies of water vapor on aqueous solution (e.g., Ref. 48), is that values of $\alpha$ are close to unity. However, others (e.g., Chameides[49]) consider $\alpha$ to be substantially lower, perhaps of order $10^{-4}$. Ultimately this question must be resolved by further experimental measurement with a variety of gases. It might be noted, though, that for droplets in the size range characteristic of cloud droplets, unless $\alpha$ is of order $10^{-2}$ or less, interfacial mass-transport is faster than gas-phase diffusion, and thus the overall rate of uptake and reaction is insensitive to the value of $\alpha$.

An important limiting case in the description of combined mass-transport and aqueous-phase reaction kinetics is attained when mass-transport processes are all sufficiently rapid that the aqueous-phase concentration of the dissolved reagent gas is essentially equal to the value given by Henry's law referred to the bulk gas-phase partial pressure. Under these circumstances description of the mixed-phase reaction kinetics is greatly simplified, since reaction rates can be evaluated using the equilibrium dissolved reagent gas concentration, and the overall rate is insensitive to mass-transport kinetics. Criteria to examine whether this condition is met have been given[30,44] as inequalities involving the Henry's law coefficient of the reacting gas, the aqueous-phase rate coefficient, the gas- and aqueous-phase diffusion coefficients, and the radius of the cloud droplet. As it happens, these criteria appear to be met for reactions of $SO_2$ and $NO_2$ in cloud droplets under most situations of interest, although departure from the equilibrium condition may be expected under certain extremes of reagent concentrations.[41]

At the other extreme, for highly soluble and/or highly reactive gases, uptake is controlled entirely by the rate at which the reagent gas can be transported to the cloud droplet by gas-phase diffusion. An example of this situation is the uptake of nitric acid vapor by cloud droplets, which is evaluated to occur on a time scale of several seconds.[23]

### *Kinetic Studies*

We examine here laboratory kinetic studies directed to determination of the mechanism and kinetic rate laws of reactions pertinent to aqueous-phase reactions in clouds and the use of such data to evaluate the rates of these reactions in the atmosphere. An underlying assumption here is that the kinetic rate laws are transferable from the laboratory to the ambient atmosphere. Successful application of this approach requires description of the overall reactions in terms of elementary reactions. In the case of aqueous-phase reactions, such reactions are potentially susceptible to the influence of trace species — catalysts or inhibitors — and this possibility must be borne in mind when applying laboratory rate data to atmospheric systems.

We focus our attention here on aqueous-phase oxidation of $SO_2$ and $NO_2$, with oxidants $O_2$, $O_3$, and $H_2O_2$, as possibly important acid-forming reactions in clouds.

### *$SO_2$ Oxidation*

We first consider oxidation of dissolved sulfur(IV) by $O_2$. As is well known, the aqueous oxidation of sulfur(IV) by $O_2$ is an example of a catalytic reaction; indeed it is questionable whether the uncatalyzed reaction proceeds at all.[50] Untold numbers of laboratory investigations of the kinetics of this reaction have been conducted. For a recent review, see Hoffmann.[51] While the details of the kinetics have yet to be resolved, it is apparent that the reaction can be catalyzed by various transition metals either introduced intentionally or present as impurities and is sensitive as well to the presence of other trace species, e.g., organics that may serve as chain terminators, thereby decreasing the reaction rate. Susceptibility to the influence of trace species is often, as is the case for this reaction, manifested by the significant variability in kinetic results from laboratory to laboratory. The sensitivity of this reaction to trace constituents brings into serious question the utility of efforts to model this reaction for real clouds. In view of this it is perhaps fortunate that estimates[52] of the rate of this reaction for representative concentrations of transition metal ions and solution pH indicate that under most conditions this reaction is not of great importance in the atmosphere.

The possible atmospheric importance of $O_3$ as an aqueous-phase oxidant of S(IV) was suggested by Penkett[53] in 1972 and the kinetics of this reaction have been the subject of several subsequent investigations.[46,52,54–57] The results

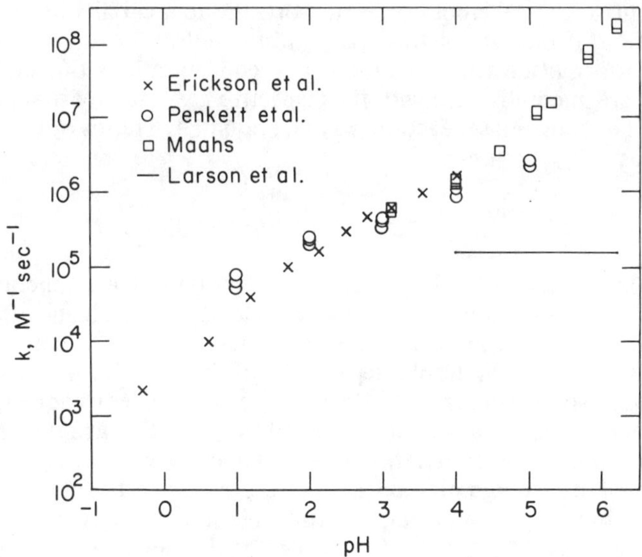


**FIGURE 14.** Second-order rate constant $k^{(2)}$ for oxidation of sulfur(IV) by ozone according to $d[S(VI)]/dt = k^{(2)}[O_3(aq)][S(IV)]$, as a function of solution pH at 25°C. Data are from References 46, 54, 55, and 57. (Modified from Schwartz.[41])

of these studies are summarized in FIGURE 14, which shows the pH dependence of the second-order rate constant defined as:

$$\frac{d[S(VI)]}{dt} = -\frac{d[S(IV)]}{dt} = k^{(2)} [S(IV)] [O_3(aq)] \tag{5}$$

It is seen that similar values and pH-dependence were found by Erickson *et al.*,[54] Penkett *et al.*,[46] and Maahs,[56,57] all of whom used a stopped-flow technique, whereas Larson *et al.*,[55] who used a bubbler technique, obtained a substantially lower value of $k^{(2)}$ with little indication of any pH-dependence. Examination of the mass-transport conditions of the latter study suggests that the observed reaction rate may have been restricted by mass transport.[41] As with the $O_2$ oxidation the question may be asked whether this reaction rate is susceptible to the influence of trace species serving as catalysts or inhibitors; it has been inferred on the basis of isotope exchange studies[58] that the reaction is not a simple bimolecular reaction, but proceeds by way of intermediates. Martin[52] and Maahs[57] report no influence of transition metal ions (up to $5 \times 10^{-5}$ M) on the rate of this reaction, whereas Harrison *et al.*[59] report a positive influence. Martin and Damschen[60] report no influence on this reaction of ethanol, acetone, formaldehyde, and acetic acid, whereas Maahs[57] finds ~25–50% apparent decrease in the $O^3$-S(IV) reaction rate by $(1-5) \times 10^{-5}$ M hydroquinone, an efficient free-radical scavenger. From these studies it is not unambiguous that catalytic or inhibition effects may be neglected for this reaction. On the other hand, the fairly close agreement of the kinetic results from the several laboratories supports the applicability of these results to evaluation of the rate of this reaction in clouds.

Under assumption that the cloudwater concentrations of both $O_3$ and sulfur(IV) are in equilibrium with the respective gas-phase partial pressures, the rate of aqueous-phase reaction may be evaluated in terms of these partial pressures as

$$\frac{d[S(VI)]}{dt} = k^{(2)}\, H^*_{S(IV)}\, H_{O_3}\, p_{SO_2}\, p_{O_3}. \tag{6}$$

The rate of S(IV) oxidation by $O_3$ is shown in FIGURE 15 as a function of pH for conditions representative of a liquid water cloud in the ambient atmosphere. In this evaluation[41] I have used the pH-dependence of the rate constant given in FIGURE 14 and, as well, the pH-dependence of the sulfur(IV) solubility given in FIGURE 12. The left-hand ordinate gives the instantaneous aqueous-phase reaction rate (molar per hour) for assumed $SO_2$ and $O_3$ partial pressures of 1 ppb and 30 ppb, respectively; this rate scales linearly with either partial pressure. It is seen that at high pH values the rate is quite significant in the context of acidification of precipitation, but that this rate strongly decreases with decreasing pH. The strong pH-dependence of the rate is a consequence of the pH-dependence of both the rate constant and the sulfur(IV) solubility. As a consequence of this strong pH-dependence this reaction tends to become self-limiting; i.e., the generation of strong acid as the reaction proceeds decreases the rate of further reaction. On the right-hand ordinate the reaction rate is expressed in ratio to the $SO_2$ partial pressure in units of percent $hr^{-1}$ as is commonly given for gas-phase rates. Here a liquid water volume fraction $L = 1$ $cm^3$ $m^{-3}$ has been assumed; the oxidation rate referred to the gas-phase reagent concentration scales linearly with L. (This rate also scales linearly with $p_{O_3}$, but is independent of $p_{SO_2}$). By reference to this ordinate scale it is seen that the oxidation rate can be quite large in comparison to gas-phase oxidation rates ($\sim 1\%$ $hr^{-1}$) at high pH, but of course continues to exhibit a strong pH-dependence. Thus from the perspective of both the rate of acidification of cloudwater and the rate of oxidation referred to gas-phase $SO_2$, this evaluation makes a strong case for the importance of aqueous-phase in-cloud oxidation of $SO_2$ by $O_3$.

A second strong oxidant of concern in the in-cloud oxidation of $SO_2$ is hydrogen peroxide, as suggested initially by Penkett et al.[46] The kinetics and mechanism of this reaction have been studied in a number of laboratories.[46,61-67] The reaction is catalyzed by $H^+$ and, as well, by a number of weak acids and is apparently an example of specific acid catalysis. The pH-dependence of the second-order reaction rate constant, defined according to

$$\frac{d[S(VI)]}{dt} = -\frac{d\,[S(IV)]}{dt} = k^{(2)}\, [S(IV)]\, [H_2O_2] \tag{7}$$

is shown in FIGURE 16. The kinetics of reaction are dependent on buffer concentration and on ionic strength; in preparing FIGURE 16 an attempt has been made to back out this dependence by extrapolation to zero buffer and ionic

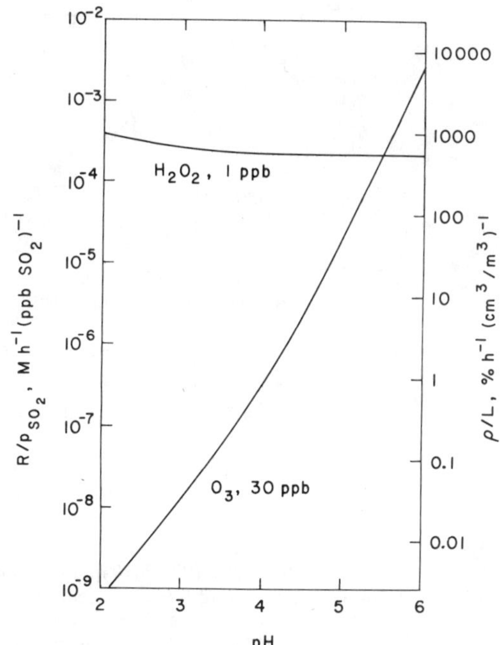



**FIGURE 15.**  Rate of aqueous-phase oxidation of S(IV) by $O_3$ (30 ppb) and $H_2O_2$ (1 ppb), as a function of solution pH. Gas-aqueous equilibria are assumed for all reagents. $R/P_{SO_2}$ represents rate of reaction referred to gas-phase $SO_2$ partial pressure per $cm^3 \cdot m^{-3}$ liquid water volume fraction. Temperature = 25°C. (Modified from Schwartz.[41])

concentration. The residual scatter of the data probably reflects inaccuracies in this procedure, as well as a slight dependence on temperature over the range indicated. The dashed line in FIGURE 16, which represents a slope of $-1$ arbitrarily placed through the data, shows the first-order dependence on $H^+$ concentration in the pH range in which $HSO_3^-$ is the dominant S(IV) species. The reproducibility of the data from the several laboratories and in particular the reproducibility of the $[H^+]$-dependence suggest the lack of susceptibility of this reaction to influence of trace constituents, and this inference is supported by isotope tracer studies[61] that are consistent with a molecular (i.e., non-free-radical) mechanism. We are thus encouraged to apply these data to evaluation of the S(IV)-$H_2O$ reaction rate in the ambient atmosphere. Moreover we would note that recent work in our laboratory[68] measuring the rate of this reaction in precipitation samples gives rates that are generally consistent with the data in FIGURE 16, albeit with considerable scatter.

Evaluation of the rate of in-cloud oxidation of $SO_2$ by $H_2O_2$ with the reaction rate constant represented by the dashed line in FIGURE 16 is given in FIGURE 15; here a gas-phase $H_2O_2$ concentration of 1 ppb is assumed, and the Henry's law coefficient of $H_2O_2$ is taken as $1 \times 10^5$ M atm$^{-1}$.[64] It is seen that a rather high reaction rate is indicated and that, in contrast to the situation for the $O_3$ reaction, there is only a slight pH-dependence. This near pH-independence is due to a cancellation of the pH-dependences of the S(IV) solubility and the reaction rate constant. Because of this pH-independence, the reaction can

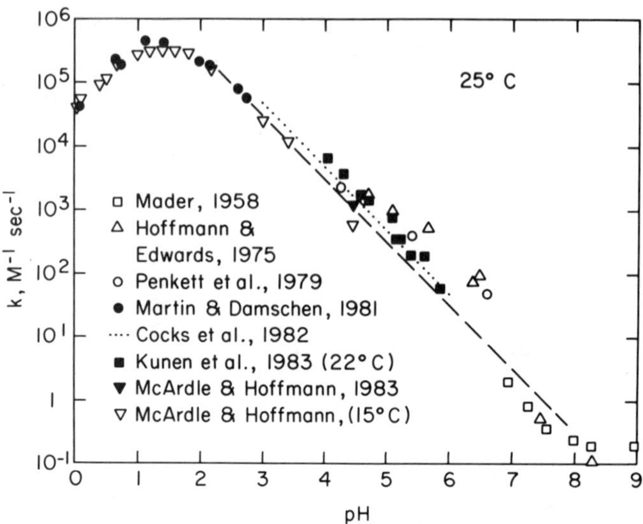


**FIGURE 16.** Second-order rate constant $k^{(2)}$ for oxidation of sulfur(IV) by hydrogen peroxide according to $d[S(VI)]/dt = k^{(2)}[H_2O_2][S(IV)]$, a function of solution pH. Data are from Refs. 46 and 62–64 (as corrected to zero buffer concentration in Ref. 64) and Refs. 65–67. Temperature 25°C except as indicated. *Dashed line* represents slope of $-1$ (i.e., $k^{(2)} \propto [H^+]$) arbitrarily drawn through the data. (Modified from Schwartz.[41])

continue to proceed at high rate even at low pH, and thus this reaction assumes major potential importance as a pathway for in-cloud formation of sulfuric acid.

### *NO₂ Oxidation*

I now turn to consideration of aqueous-phase oxidation of $NO_2$, which has been the subject of much research in our laboratory over the past several years. When we began this work little information was available to permit quantitative evaluation of the rate of this reaction in the ambient atmosphere or even to permit qualitative assessment of its importance. $NO_2$, of course, is well known[69,70] to react rapidly with liquid water to give a mixture of nitrous and nitric acids,

$$2NO_2 + H_2O(l) \rightarrow HNO_2 + HNO_3 \tag{11}$$

The nitrous acid produced in this reaction may undergo decomposition to yield additional nitric acid as well as nitric oxide,

$$3HNO_2 \rightarrow HNO_3 + 2NO + H_2O \tag{12}$$

corresponding to the overall reaction

$$3NO_2 + H_2O \rightarrow 2HNO_3 + NO \tag{13}$$

These reactions in fact serve as the basis for the industrial manufacture of nitric acid.[40] Yet another reaction whose importance under atmospheric conditions must be examined is the reaction of $NO_2$ with NO to form nitrous acid,

$$NO_2 + NO + H_2O \rightarrow 2HNO_2 \qquad [14]$$

The fact that these reactions occur rapidly at high concentrations and also the strong thermodynamic driving force for the production of $HNO_3$ are probably the basis for the assumption by quite a few investigators that these reactions are rapid also under ambient conditions, i.e., with cloud- or rainwater,[71-77] with liquid-water-containing clear-air aerosols,[78-82] with surface water,[83-86] or with liquid water contained in vegetation.[87-89] The acidity resulting from assumption of these equilibria has been shown as well in model calculations to exert a major influence upon aqueous-phase concentrations of sulfur(IV) and resultant sulfate formation rates as a consequence of the highly pH-dependent solubility of sulfur(IV).[79-81,90] It might be noted as well that reaction of $NO_2$ to form nitrous and nitric acids has been assumed to be the initial step responsible for the pulmonary toxicity of this substance.[91] In contrast to the studies cited, the work of a few investigators[92-94] has minimized the importance of reaction of $NO_2$ with liquid water at concentrations representative of the ambient atmosphere.

On the basis of the discussion presented above, it should be evident that the information necessary to evaluate the rate of one or another of these reactions, e.g. [11], in cloudwater is the Henry's law solubility of $NO_2$ and the aqueous-phase kinetic rate law. In order to determine these quantities we have followed several different approaches, namely, literature review, thermochemical measurements, and direct, mixed-phase chemical kinetic measurements.

As might be inferred from the industrial importance of reaction of $NO_2$ with liquid water, there has been quite a number of laboratory investigations of the reactive uptake of $NO_2$ by liquid water. However, such studies have been conducted not only because of practical application but also because of the intrinsic interest of this chemistry. In fact, some of these reactions served as an early proving ground for theories relating reaction kinetics and chemical thermodynamics.[95] In general, studies of reaction kinetics of this system are complicated not only by the interplay between mass transport and chemical reactions, but also by the dimerization of $NO_2$,

$$2NO_2 = N_2O_4 \qquad [15]$$

which takes place in both the gas and aqueous phases and affects the kinetics of mass transport as well as reaction. As noted above, recognition of the fact that the Henry's law coefficient is an equilibrium constant that must be related to other equilibrium constants by means of thermochemical cycles led us[96] to attempt to evaluate the Henry's law coefficient of $NO_2$ by such an approach, as summarized in FIGURE 17. In this figure the solid arrows represent reactions whose equilibrium constants we considered to be well established, the dashed arrows represent reactions for which equilibrium constants had been reported but which we did not consider well established, and the dotted arrows

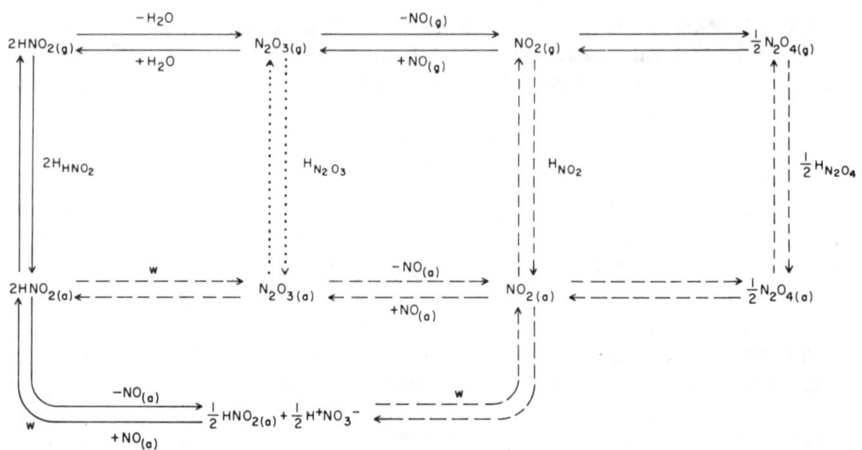

**FIGURE 17.** Thermodynamic cycles leading to evaluation of Henry's law coefficient of $NO_2$, $N_2O_3$ and $N_2O_4$. *Solid arrows* represent reactions whose equilibrium constants are considered well established. Equilibrium constants represented by *dashed arrows* have been measured directly, but are not considered firm. There are no direct measurements for the equilibrium represented by *dotted arrows*. (From Schwartz and White.[36])

represent a reaction for which no equilibrium constant data were available. Direct measurements of $NO_2$ solubility from kinetic studies[97,98] yielded values of $H_{NO_2}$ of $1.9 \times 10^{-2}$ and $4 \times 10^{-2}$ M atm$^{-1}$, whereas the several thermochemical paths yielded values of $1.0 \times 10^{-2}$, $1.9 \times 10^{-2}$ and $0.85$ M atm$^{-1}$. A value for $H_{NO_2}$ of order $10^{-2}$ M atm$^{-1}$ would be consistent with that expected from physical properties of the molecule (cf. $H_{O_3} = 1.3 \times 10^{-2}$ M atm$^{-1}$), whereas a value of order 1 M atm$^{-1}$ would be anomalously high. The large discrepancy of the latter value suggested to us that the then-accepted[99,100] value for the equilibrium constant for the hydrolysis of $N_2O_3$,

$$N_2O_3(aq) + H_2O = 2\ HNO_2(aq), \qquad [16]$$

might be incorrect. This prompted us to re-examine this equilibrium constant and in particular to extend the range of examination to lower acid concentrations than had been investigated previously. That study revealed an unexpected dependence of the equilibrium constant $K_{16}$ on $[H^+]$ in several molar perchloric acid, resolving the discrepancy in $H_{NO_2}$. On this basis we were then able to recommend[96] a value for the Henry's law coefficient of $NO_2$ of $(1.2 \pm 0.4) \times 10^{-2}$ M atm$^{-1}$. It is interesting to note that the new measurement of $K_{16}$ also resolved a discrepancy in the interpretation of the kinetics and mechanism of diazotization of amines, which reaction proceeds via $N_2O_3$ as an intermediate.[96,102]

In a second review [103] we examined studies pertinent to the kinetics of the dissolution reactions. Depending on the nature of the contact between the gas and aqueous phases, which governs the competition between mass transport and chemical kinetics, such studies typically yield products of various

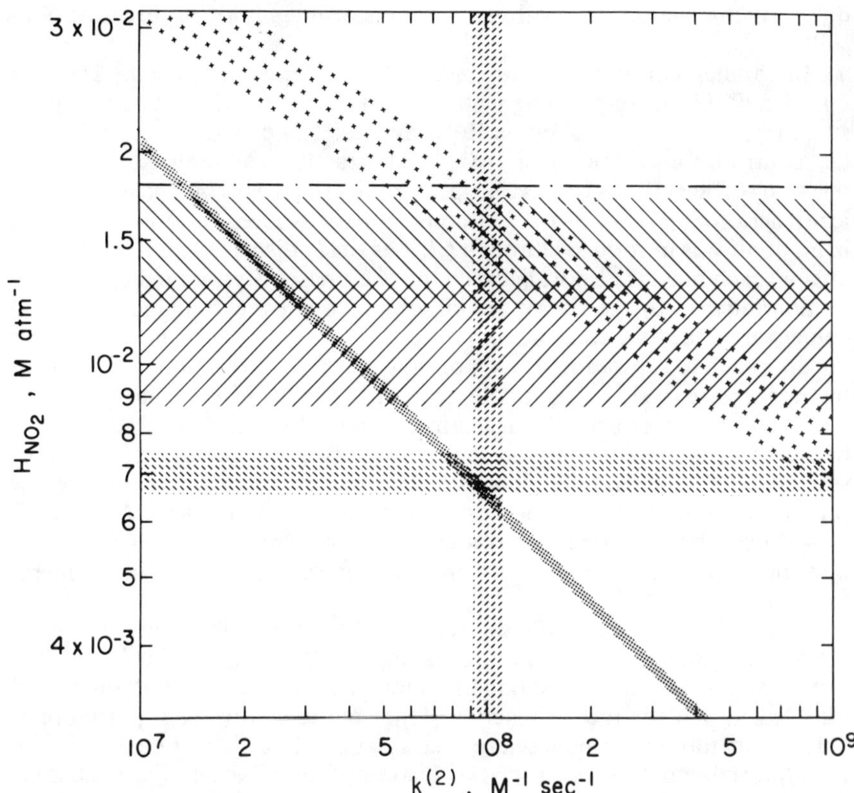



**FIGURE 18.** Plot of data pertinent to determination of $H_{NO_2}$ and second-order rate constant for reaction [11]. (From Schwartz and White.[103])

powers of the Henry's law coefficient and the reaction rate constant k. Such data, when plotted in a plane whose coordinates are log k and log H, yield straight lines, or, when experimental uncertainties are taken into account, bands whose width corresponds to the error bar on the measured quantity. An example of such a plot is given in FIGURE 18. Here $k^{(2)}$ represents the rate constant for reaction [11] defined as

$$\frac{d[HNO_2]}{dt} = \frac{d[HNO_3]}{dt} = k^{(2)} [NO_2(aq)]^2 \tag{8}$$

In the absence of error the several lines should intersect at a point whose coordinates represent the values of $k^{(2)}$ and $H_{NO_2}$; the breadth of the region of intersection shown in the figure is a measure of the uncertainty in values of these quantities to be derived from the several studies. On the basis of this analysis we recommended a value of the rate constant for reaction [11], as $(0.7 \pm 0.35) \times 10^8$ M$^{-1}$ sec$^{-1}$. A rate constant of this magnitude represents quite a high reaction rate, within two orders of magnitude of diffusion controlled.

In parallel with this literature review we embarked on a kinetic study of our own.[104] The distinguishing feature of this study was the low partial pressure of $NO_2$ employed, as low as $10^{-7}$ atm. Such low pressures both ruled out any contribution of the dimer $N_2O_4$ to the reaction and reduced the rate of reaction sufficiently that mass transport could effectively compete with chemical reaction. A sketch of the gas-liquid reactor is given in FIGURE 19. $NO_2$ in $N_2$ or air was introduced through a frit as finely dispersed bubbles. In such an approach mass transport between the two phases is promoted by the convection induced by the rise of the bubbles as well as by the high interfacial contact area; the time constant for convective mixing could be measured from the approach to saturation of $CO_2$ or $SO_2$. The rate of reaction was followed by the increase in electrical conductivity of the solution associated with formation of ionic products. By measuring the rate of reaction [11] as a function of $NO_2$ partial pressure and the convective mixing time constant it was possible to determine both the Henry's law coefficient of $NO_2$ and the rate law for the reaction. The rate law was established to be second order in $NO_2$ (equation 8) over the pressure range studied, and the Henry's law coefficient and rate constant were in good agreement with the values obtained in the literature review.

On the basis of this study we were able to place a high degree of confidence in the applicability of the rate law and the constants obtained to evaluation of rates of reaction of $NO_2$ in the ambient atmosphere. This evaluation[105] established that the rate of reaction [11] at representative $NO_2$ partial pressures is far too low to represent either an appreciable loss route of $NO_2$ or an appreciable process for the acidification of cloud water. The reason for this is that despite the high reaction rate constant, the second-order rate law and the low Henry's law coefficient conspire to make the overall rate very slow at low $NO_2$ partial pressures. Furthermore, examination of other possible aqueous-phase atmospheric reactions of $NO_2$ (e.g., with NO, with transition metal ions, or with $O_3$ or $H_2O_2$) has failed to reveal any such reactions that proceed with appreciable rate.[105,106] Thus, on the basis of laboratory studies it appears that the reactive uptake of $NO_2$ to form cloudwater nitric acid must be dominated by gas-phase rather than aqueous-phase oxidation.

Before concluding the discussion on the kinetics of reaction of $NO_2$ with water it might be useful to reflect on the shift of reaction order, from first order at high concentrations to second order at low concentrations. The reason for this is the shift in predominant species from $N_2O_4$ to $NO_2$. However, we must view this shift as apparent rather than real, since $[N_2O_4]$ is quadratic in $[NO_2]$ throughout the entire concentration range. Thus, the second-order rate law is entirely predictable from consideration of the sequence of elementary reactions that constitute the overall reaction. In fact, a quite accurate rate expression applicable to low $NO_2$ concentrations under conditions of Henry's law saturation may be derived from the work of Abel and Schmid[95] in the 1920s on the kinetics of decomposition of nitrous acid as interpreted correctly by those authors in terms of elementary reactions.

A spin-off of our laboratory investigations of $NO_2$ chemistry that might

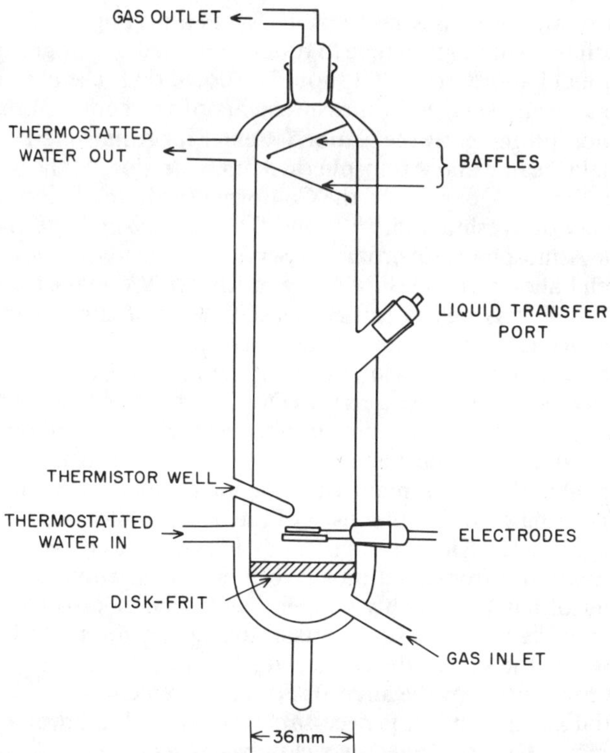



**FIGURE 19.** Gas-liquid contactor employed in study of reaction of $NO_2$ with water. (From Lee and Schwartz.[104])

be mentioned here is a revision of thinking of the mechanism of $NO_2$ toxicity in pulmonary systems. Again, in view of the low solubility and second-order rate law, reaction [11] seems improbable at low $NO_2$ partial pressures. As an alternative it has been suggested[107] that $NO_2$ may serve as a one-electron oxidizing agent leading to the formation of organic free radicals by abstraction of an allylic hydrogen from unsaturated molecules. The relevance of research conducted for a particular application to other areas of application is a major strength underlying the approach of carrying out research directed to determination of fundamental chemical and physical properties. Just as our research into cloud chemistry has benefited from a long history of research carried out for other purposes, so will our research serve as the basis for future work in still further areas of application.

## FIELD MEASUREMENTS

I turn now to an account of the second "side" of our examination into cloud chemistry, namely measurement of the chemical composition of clouds and evaluation of the processes responsible for establishing that composition. Again, emphasis will be given here to results obtained by our group at Brookhaven National Laboratory, but it should be noted that several other research groups are actively engaged in measurements of the composition of clouds and fogs, most prominently the National Center for Atmospheric Research,[108] the National Oceanic and Atmospheric Administration,[109] the State University of New York at Albany,[110,111] the California Institute of Technology,[112,113] the University of Washington,[114,115] and Sonoma Technology, Inc.[116] in this country, the Atmospheric Environment Service of Canada,[117] the Central Electric Research Laboratories in Britain,[118] and KEMA N.V. in the Netherlands.[119] Attention is called as well to earlier measurements of cloud composition to which reference is made in the articles cited here.

The platform for our measurements of cloud chemistry has been a twin-engine high-wing STOL-type aircraft (FIG. 20) that has been outfitted with sampling and measurement equipment suitable for characterization of the composition of cloudwater and interstitial air.[120] The aircraft is fully equipped for instrumented flight as is required for flying in clouds. Personnel onboard the aircraft during such flights consist of the pilot, copilot, and two members of our scientific staff. Cloud-sampling missions require considerable cooperation and assistance from aviation authorities. We have found that FAA Air Traffic Control has been highly cooperative whenever possible.

Air for the chemical measurements is ducted into the aircraft through intakes located on the side of the cabin (FIG. 21) under "ram" pressure induced by the forward motion of the aircraft. For measurements of the composition of interstitial air in clouds it is necessary to remove cloud droplets from the inlet air stream. This is effected by a cloudwater separator[121] mounted in the sampling line. This separator, which was developed at the Central Electric Research Laboratories in Britain, contains a set of static blades (FIG. 22) set at an angle to induce a swirl downstream that in effect centrifuges the cloud droplets to the walls while transmitting gases and aerosol particles that have

**FIGURE 20.** The Britten-Norman Islander research aircraft that has been employed in our cloud-composition measurements.

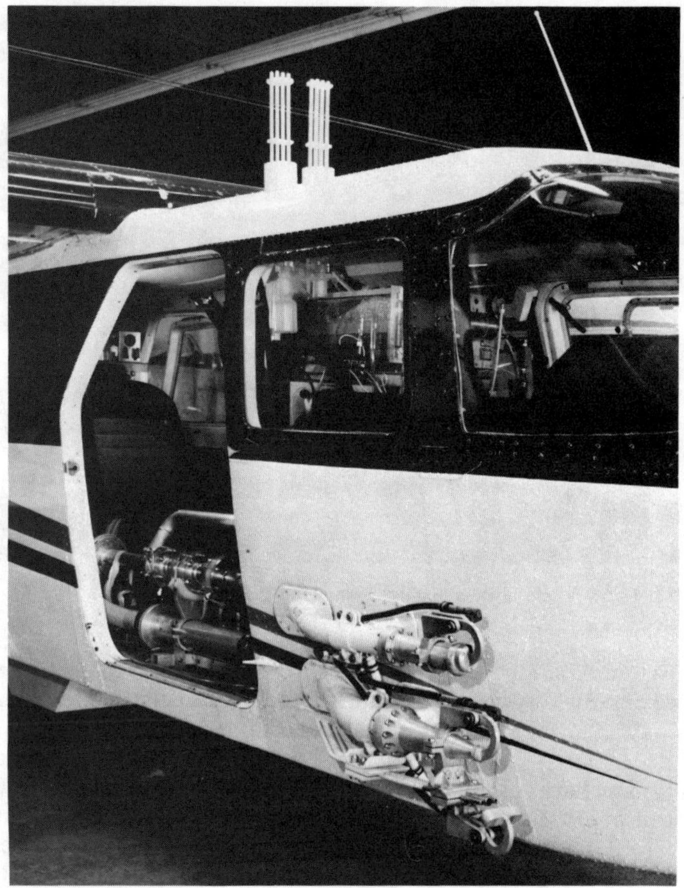

**FIGURE 21.** Detail of research aircraft showing inlets for air samples (*lower right*, with remote-operated covers), cloudwater separator (in cabin), cloudwater collectors (protruding through cabin roof) and collection bottles (in cabin).

not been incorporated into cloud droplets. A portion of this air stream is passed through a series of filters that collect aerosol particles and gaseous $HNO_3$ and $SO_2$ for subsequent analysis.[122] The time required for collection of sufficient sample for analysis is about half an hour. Additional portions of the interstitial air stream are sampled by various "real-time" instruments on board the aircraft. These instruments, which for the most part are modified from commercially available instruments[123,124] to enhance sensitivity or chemical selectivity, are mounted in two instrument racks located fore and aft of the scientific observers (FIGS. 23 and 24). Interstitial species measured continuously include $SO_2$, aerosol sulfate, NO, $NO_x$, $HNO_3$, $O_3$, and $NH_3$.

Cloud liquid water is collected by means of collectors projecting through the cabin roof (FIG. 21). These collectors, developed at the State University

**FIGURE 22.** Detail of cloudwater separator showing static blade assembly.

of New York at Albany,[125] consist of arrays of rods, each rod having a vertical slot milled into the forward surface and extending the length of the rod. Cloud droplets hit the face of the slot and are ducted under slight suction through tubing into the cabin. In order to obtain a real-time assessment of cloudwater composition we have designed flow-through cells that permit continuous measurement of pH (with a glass electrode) or conductivity (with platinum electrodes) before the sample is ducted into the collection bottle. Collection of sufficient sample for analysis (about 5 cm³) requires 5 to 15 minutes or more, depending on cloud liquid water content.

Physical and meteorologic measurements made onboard the aircraft include liquid water content (by a hot-wire indicator[126]), relative humidity, temperature, light-scattering, and total and ultraviolet insolation. Signals from the several real-time instruments, from the flowmeter used with the filter sampler, and from navigational instruments are recorded digitally for subsequent processing. Additionally, strip-chart records are available for display of the several chemical concentrations and liquid water content during the conduct of the flight. The continuous measurements, having time-resolution typically of a few seconds, provide much more detailed information on the spatial variability of cloud composition than can be obtained from the collected samples.

### Cloud Composition Measurements

I present here some of the highlights of our measurements of cloud composition; for details of these measurements see References 127–129. To date

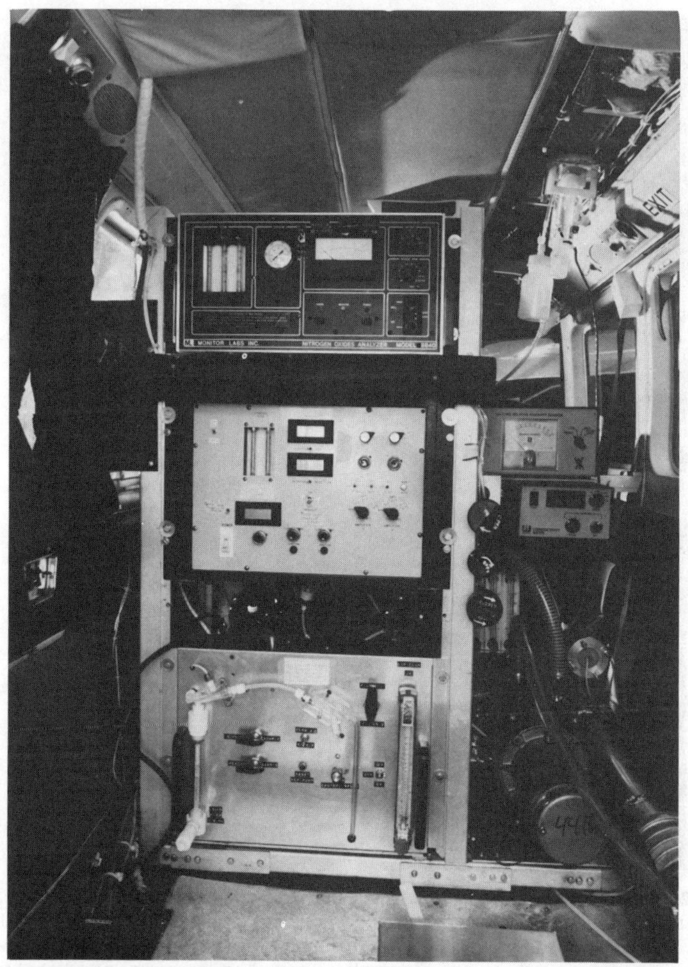

**FIGURE 23.** View of front instrument rack in aircraft, showing chemiluminescence $NO_x$ analyzer, dual-channel flame photometric sulfur analyzer, and fluorescence derivatization $NH_3$ analyzer. Mounted to *right* of rack are read-out units for relative humidity and conductivity. Collected bottles and pH/conductivity cell for use with cloudwater collector are visible at *upper right*.

we have concentrated our efforts on warm (i.e., having temperatures greater than 0°C) stratiform clouds, and the results presented here are restricted to such clouds. This type of cloud has been selected for our initial study because these clouds are prevalent, responsible for delivery of much precipitation to the surface, amenable to forecasting, widespread and spatially uniform, and subject to dynamic interpretation.[130] Measurements have been made at several locations in the eastern United States, including Long Island, NY; Salisbury, MD; Charleston, SC; and Birmingham, AL. Altitudes were typically

**FIGURE 24.** View of rear instrument rack in the aircraft, showing on the *left*, the data acquisition system, navigational instrument read-outs, Doppler radar, and the integrating-nephelometer. On the *right* are the chemiluminescence ozone monitor, the high-volume flowmeter and integrator, radiation thermometer and the turbulence monitor. Strip-chart recorders are on top of the rack.

in the range 500–1500 m above mean sea level, i.e., within the lower portion of the planetary boundary layer.

### *Cloudwater Composition*

As an initial means of displaying concentration measurements of cloudwater species, it is convenient to present these data as frequency distributions, the percent of the samples measured exhibiting concentrations of the several ionic species within specified ranges. Such frequency distributions are given in FIGURE 25 for $H^+$, $SO_4^=$, $NH_4^+$, and $NO_3^-$. Concentrations of other cations ($Na^+$, $K^+$, $Ca^{++}$, $Mg^{++}$) and anions ($Cl^-$) have been almost invariably well below

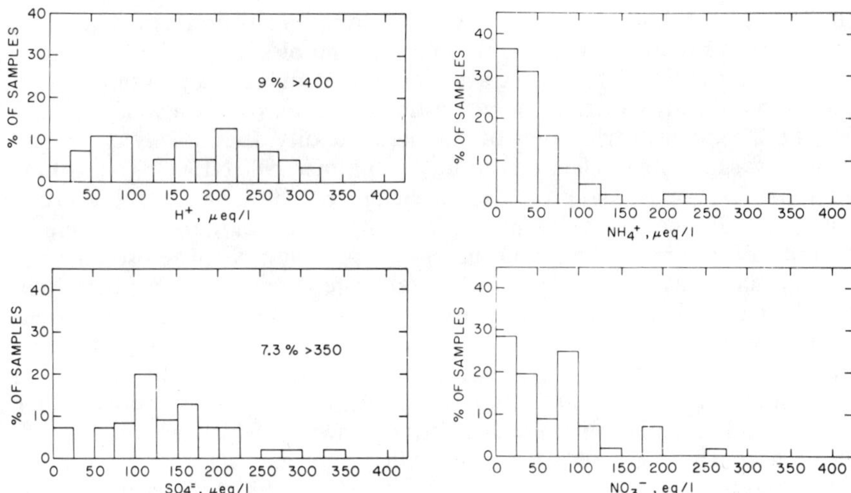

**FIGURE 25.** Frequency distributions of concentrations of principal ions in cloudwater samples. (From Daum *et al.*[127] Reprinted by permission.)

those of these principal ionic species. Examination of cloudwater composition and comparison with that of rainwater samples indicates that the composition of cloudwater is, not surprisingly, similar to that of rainwater, with a possible indication that for the clouds sampled concentrations were somewhat greater than are typical in precipitation samples.

Further insight into cloudwater composition may be gained by examination of fractional acidity, $[H^+]/([NO_3^-] + 2[SO_4^=])$. The frequency distribution of this quantity for cloudwater samples given in FIGURE 26 indicates that for most samples $H^+$ is the principal cation associated with sulfate and nitrate anions. Comparison of the distribution of fractional acidity of cloudwater samples with those for rainwater samples (FIGS. 6 and 7) shows a general

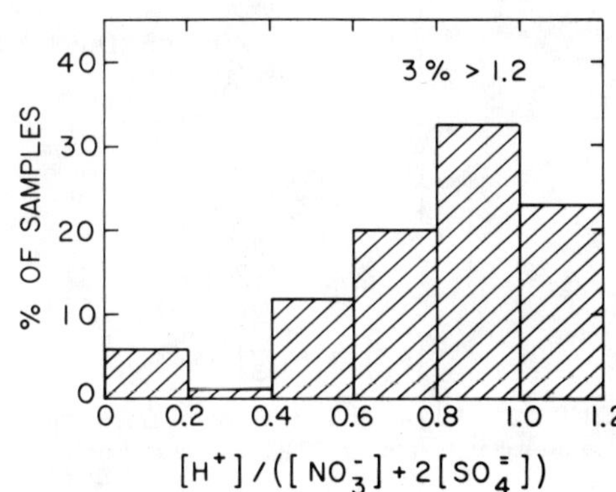

**FIGURE 26.** Frequency distribution of the fractional acidity of cloudwater samples. (From Daum *et al.*[129] Reprinted by permission.)

similarity, but with cloudwater samples, if anything, exhibiting a somewhat greater fractional acidity than the rainwater samples.

The relatively high fractional acidity of cloud and rainwater samples contrasts dramatically with that characteristic of clear-air (i.e., noncloud) aerosol. FIGURE 27 shows distributions of fractional acidity obtained by our group in aircraft sampling of clear air during the summer 1980 NEROS (NorthEast Regional Oxidant Study) project in the vicinity of Baltimore, MD.[131] The distribution in FIGURE 27a represents the fractional acidity of the collected aerosol sample. It is clear that the sulfate and nitrate present in these aerosol samples are not mainly associated with $H^+$, but rather represent acids that have been mostly neutralized by $NH_3$. It is clear further that the composition of rainwater (cf. FIG. 6 and 7; the latter representing summer precipitation at nearby Lewes, DE, is particularly pertinent) cannot be derived simply by dissolution into rainwater of aerosol of composition exhibited by these samples. This conclusion holds all the more strongly for cloudwater, which, as noted above, appears to exhibit greater fractional acidity than rainwater. These observations therefore give strong indication of the participation of gaseous species

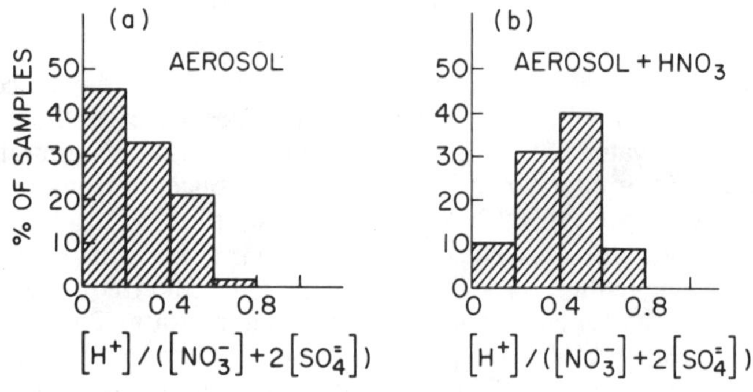


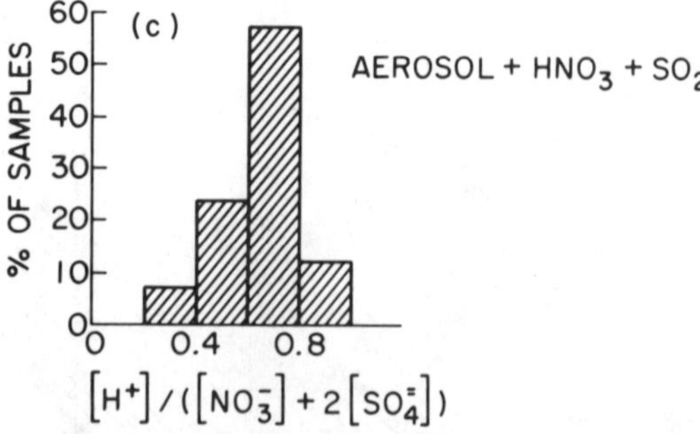


**FIGURE 27.** Frequency distribution of fractional acidity of clear-air samples taken in Baltimore, Maryland, summer 1980. **(a)** Aerosol only; **(b)** aerosol + $HNO_3$; **(c)** aerosol + $HNO_3$ + $SO_2$, with $SO_2$ assumed to yield $2H^+$ + $SO_4^=$. (From Daum.[131] Reprinted by permission.)

in establishing the composition of the cloudwater samples. To explore this further, FIGURE 27b shows the frequency distribution of fractional acidity for the clear-air NEROS samples evaluated under the assumption that nitric acid (which was sampled simultaneously with the aerosol) would contribute both to the $H^+$ and $NO_3^-$ concentrations upon dissolution of the material present in the clear air in cloud- or rainwater. This assumption seems most reasonable in view of the fact (as discussed above and also in Ref. 23) that $HNO_3$ would completely and rapidly partition into cloudwater. It is seen that the distribution of fractional acidity is shifted to somewhat greater values, but still does not match that of the cloud- or rainwater distributions. However, if the $SO_2$ that was present is also included in evaluating the fractional acidity under assumption that $SO_2$ reacts to form $H_2SO_4$ (FIG. 27c), then a distribution of fractional acidity is obtained that much more closely matches those of cloud- and rainwater. Considerations such as these give strong support to the hypothesis (e.g., Newman[132]) that rainwater in the eastern United States must derive its sulfate and acidity largely from aqueous-phase oxidation of $SO_2$.

I now turn to consideration of reasons for variability in concentrations of cloudwater-dissolved substances. A major reason for this variability is the variability of liquid water content within clouds. As may be seen by reference to FIGURE 9, slight changes in temperature may lead to changes in liquid water content that are quite large relative to typical liquid water content of 0.1 to 1 cm³/m³. An instance of the effect of variable liquid water content on dissolved species concentrations can be seen in FIGURE 28, which presents ionic concentration data obtained in a series of measurements in Charleston, SC, in February 1982. FIGURE 28a presents aqueous-phase concentrations, whereas FIGURE 28b presents these concentrations multiplied by liquid water content. Attention is called in particular to the measurements on missions 1 and 2, which were made in morning and afternoon of the same day, the afternoon measurements being in clouds of lower liquid water content. From compar-

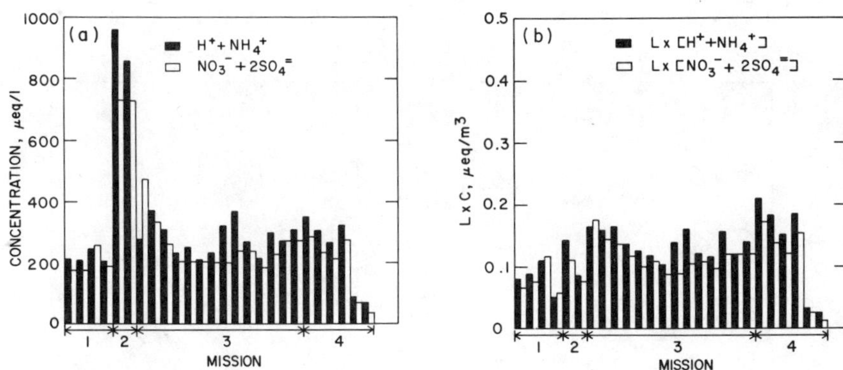

**FIGURE 28.** Cation ($[H^+] + [NH_4^+]$) and anion ($[NO_3^-] + 2[SO_4^=]$) concentrations for cloudwater samples collected in Charleston, South Carolina, February 1982. **(a)** Aqueous concentrations; **(b)** concentrations per cubic meter of air. (From Daum *et al.*[128] Reprinted by permission.)

isons of the two figures it is seen that much of the apparent increase in concentrations between the two missions is due simply to a decrease in liquid water content.

In order to remove the effect of variable liquid water content on reported concentrations and as well to permit comparison of concentrations of species in the several phases (gas, aerosol, aqueous solution), we have taken to expressing the concentrations of cloudwater-dissolved species as potential partial pressure, the partial pressure the species would exhibit as an ideal gas-phase species. From Equation (1) it is seen that this potential partial pressure is evaluated as

$$\hat{P}_X = LRT[X] \tag{9}$$

This quantity is equivalent to the partial pressure of a gaseous constituent or the mole fraction of an aerosol constituent and thus constitutes a common currency for comparison of concentrations in the several phases.

Having defined the potential partial pressure, we proceed to examine frequency distributions of cloudwater concentrations expressed in this way. Such frequency distributions are given in FIGURE 29. The potential partial pressures of the principal ions are found to exhibit values up to several parts per billion, comparable to nonurban gas-phase concentrations of $SO_2$ and $NO_x$ in these regions. This comparability offers no support, at least for the stratiform clouds

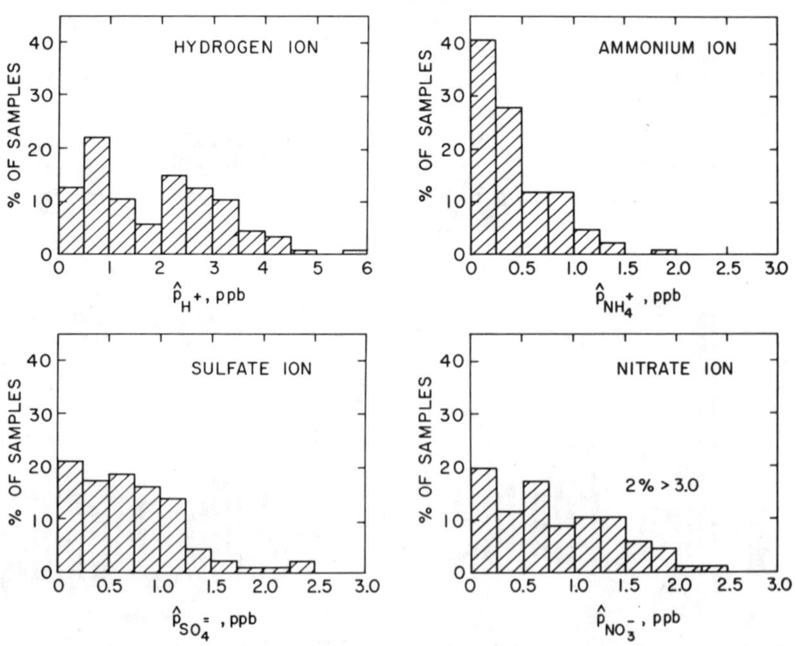


**FIGURE 29.** Frequency distributions of the potential partial pressures of the major ions in cloudwater samples. Number of samples (N) = 88. (From Daum *et al.*[129] Reprinted by permission.)

that have been studied to date, to the suggestion[133,134] that clouds may process and accumulate within their cloudwater materials from a volume of air substantially greater than that contained within their physical boundaries. Rather, these observations are consistent with the more customary picture of cloud droplets remaining with and interacting with only the air in which the droplets are initially formed.

Of course, variability in species concentrations may be due simply to variability in the amount of solute present. An instance of this is found in the final sets of samples collected on mission 4 (FIG. 28) in comparison to the earlier samples obtained on that mission. For the earlier samples the aircraft was at an altitude-of 0.6 km above mean sea level, whereas for the later samples the aircraft had climbed to 0.9 km. The difference in concentrations between the two altitudes is reflected in both the aqueous solution concentrations (FIG. 28a) and the absolute concentrations (FIG. 28b). The altitude-dependence is reflected as well in the pH trace registered on this mission (FIG. 30), which

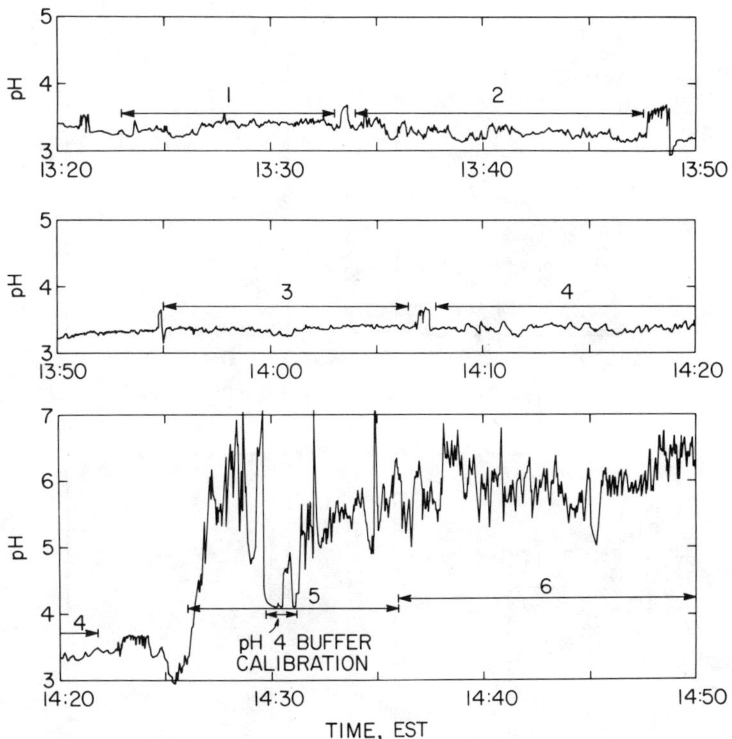


FIGURE 30. Continuous cloudwater pH obtained during aircraft measurements in the vicinity of Charleston, SC, February 28, 1982 (cf. FIGURE 28, mission 4). Values from 1320 to 1425 were obtained at an altitude of 600 m above mean sea level. Values from 1425 to 1450 were obtained at an altitude of 900 m above mean sea level. *Arrows* indicate collection times and pH values of collected samples. Note pH calibration at 1430. (From Daum *et al.*[128] Reprinted by permission.)

shows a sharp increase in pH at about 1425, when the aircraft made the ascent. In fact, the dramatic change in pH led us initially to question the performance of the pH meter; however, in-flight calibration (at 1430) confirmed that the pH meter was behaving properly, as was later substantiated by the chemical analysis of the collected samples.

### Interstitial Composition

I have emphasized that the great majority of the volume of a cloud consists of air. Our interest in cloud composition is thus directed as much to the composition of this air as to the composition of the cloudwater. As with clear air, our interest is focused both on gaseous and particulate species.

Concentrations of principal ionic species in aerosol particles present within the interstitial air in clouds are shown in FIGURE 31. (Here, a negative concen-

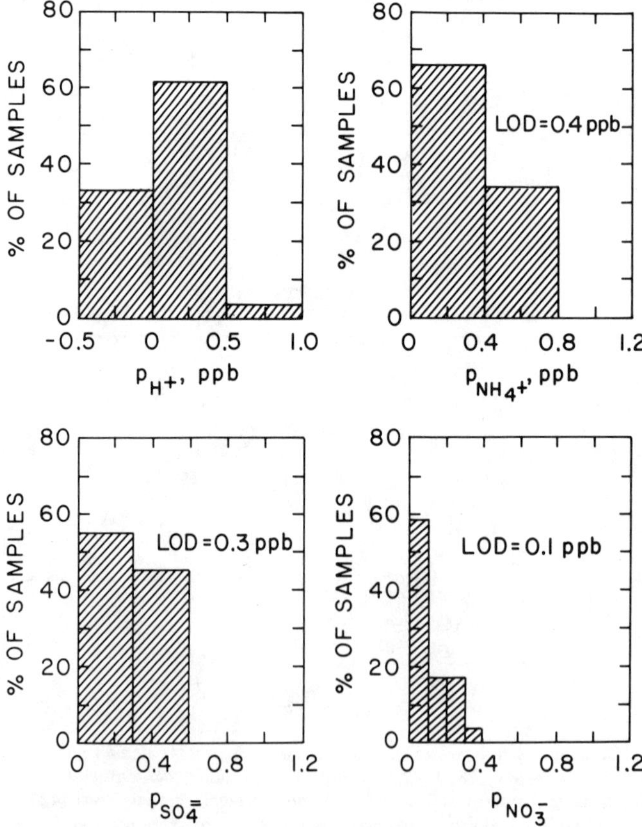



**FIGURE 31.** Frequency distributions of the concentrations of interstitial aerosol constituents. N = 29. (From Daum *et al.*[129])

tration represents net basicity in comparison to blanks for $H^+$ determined by Gran [pH] titration[135] following leach of the filter by $10^{-4}$ $N$ acid solution.) Interstitial aerosol composition is quite distinct from that of cloudwater, e.g., in the ratios of concentrations of the several ionic constituents, being similar in that respect rather to composition of clear-air aerosol. Concentrations of interstitial aerosol species are seen also to be typically quite low in comparison to concentrations of the same species in cloudwater, FIGURE 29. This point is illustrated by FIGURE 32, which presents the fractional distribution $\eta_x$ of each of the several ionic species, defined as

$$\eta_x = \frac{\hat{p}_x(c.w.)}{\hat{p}_x(c.w.) + p_x(a.p.)} \tag{10}$$

where c.w. and a.p. denote cloudwater and aerosol particle, respectively. (In these diagrams, the hatched areas represent $\eta$ values that could be determined unambiguously, whereas the open values represent situations where only a lower limit value could be determined for $\eta$ because the interstitial concentration was below the limit of detection. Thus these diagrams must be interpreted such that an open area might appear anywhere to the right of the indi-

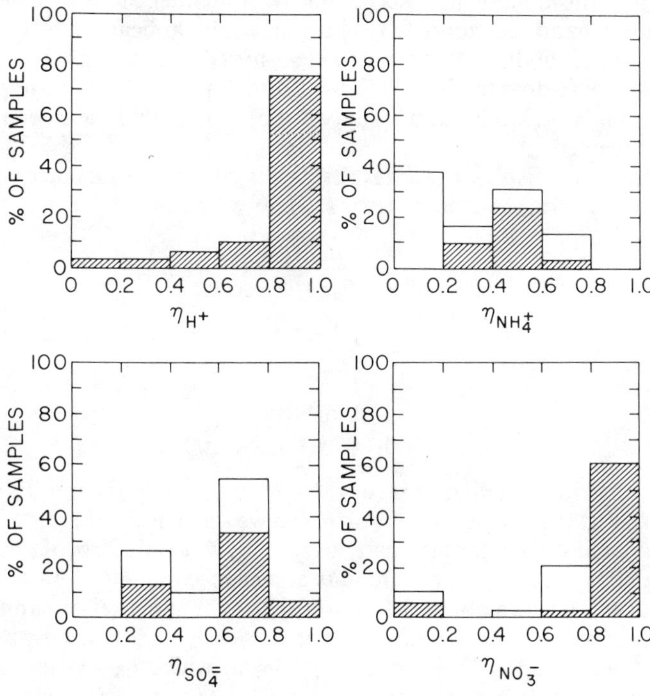

FIGURE 32. Frequency distribution of the scavenging efficiencies $\eta_x$, for $H^+$ $NH_4^+$, $SO_4^=$, and $NO_3^-$. $\eta_x = \hat{p}_x/(\hat{p}_x + p_x)$ where $\hat{p}_x$ is the cloudwater potential partial pressure of species X and $p_x$ is the concentration of the interstitial aerosol expressed as a partial pressure. N = 29. (From Daum *et al.*[129] Reprinted by permission.)

cated location.) Despite the ambiguity in these figures introduced by limits of detection, it is nonetheless quite clear with the possible exception of $NH_4^+$, that the several ionic species are preferentially distributed into the cloudwater. This preferential distribution may be a consequence of uptake of ionic species into the cloudwater from pre-existing aerosol particles and/or of uptake and reaction of gases, which we consider presently.

Interest in concentrations of interstitial gases is focused on the role of these gases as precursors of cloudwater acidity, as neutralizing agents, and as oxidants. $HNO_3$ and $NH_3$ have invariably been below detection limits (0.4 ppb), consistent with rapid and complete uptake of these gases by cloudwater as expected from solubility and mass-transport considerations as discussed above. At the other extreme, in-cloud $O_3$ concentrations have been found to be comparable to those in clear air, consistent with the low solubility of this gas. It must be conceded, however, that since concentrations of $O_3$ are typically much greater ($\sim 40$ ppb) than concentrations of potential reducing species with which it might react (e.g., $SO_2$, typically in the range up to several ppb), it would be difficult to discern any depletion of $O_3$ due to in-cloud reaction. In-cloud concentrations of nitrogen oxides (mainly $NO_2$) have been found to be comparable to clear-air concentrations, establishing that NO and $NO_2$ can persist in clouds for considerable periods of time (of the order of an hour or more). On the other hand, in-cloud $SO_2$ concentrations appear quite variable— sometimes rather high, comparable to clear-air concentrations, and at other times quite low (below the limit of detection 0.2 ppb), often in the presence of several ppb $NO_x$, which would suggest that $SO_2$ should have been present had it not reacted.

It is useful to display these results as distribution efficiencies, as were presented above for ionic constituents. Here we define

$$\eta_S \equiv \frac{\hat{p}_{SO_4^=}(\text{c.w.})}{\hat{p}_{SO_4^=}(\text{c.w.}) + p_{SO_2}(g)} \tag{11}$$

the fraction of sulfur present at cloudwater sulfate versus gaseous $SO_2$, and similarly for nitrogen,

$$\eta_N \equiv \frac{\hat{p}_{NO_3^-}(\text{c.w.})}{\hat{p}_{NO_3^-}(\text{c.w.}) + p_{NO_x}(g)} \tag{12}$$

Frequency distribution diagrams for these quantities are given in FIGURE 33. Looking first at the distribution of nitrogen we see that the weight of the distribution is largely in favor of the gaseous species, with only 20% of the samples exceeding 40% cloudwater nitrate and none exceeding 60%. In contrast it is seen that the fractional uptake of sulfur for the same set of samples was broadly distributed over the entire range of possible values; i.e., there is a significant percentage of the samples (14%) in which $\eta_S$ was less than 0.2 and, as well, a significant percentage in which the aqueous fraction exceeded 0.6 (at least 10% and perhaps as great as 50%). The broad distribution for sulfur must reflect differing conditions characterizing the various samples, e.g., the extent of oxidation that had occurred in pre-cloud air, the amount

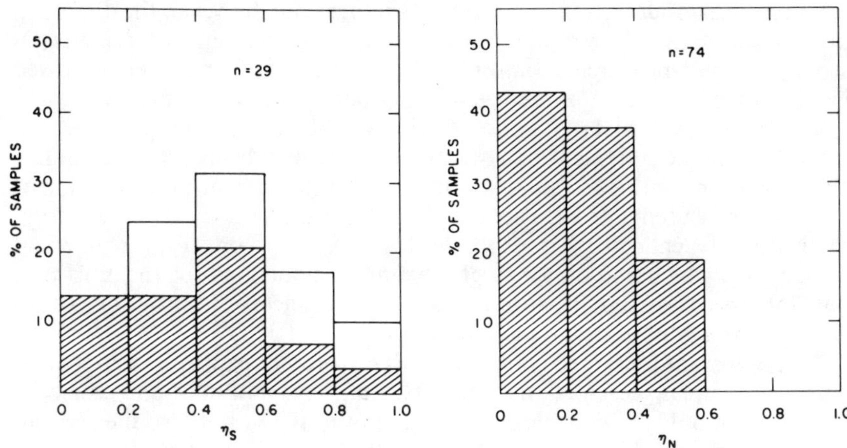

**FIGURE 33.** Frequency distributions of the scavenging efficiencies $\eta_S$ for sulfur compounds (N = 29) and $\eta_N$ for nitrogen compounds (N = 74). $\eta_S = \hat{p}_{SO_4^=}/(\hat{p}_{SO_4^=} + p_{SO_2})$ where $\hat{p}_{SO_4^=}$ is the cloudwater potential partial pressure of $SO_4^=$ and $p_{SO_2}$ is the interstitial $SO_2$ partial pressure; $\eta_N = \hat{p}_{NO_3^-}/(\hat{p}_{NO_3^-} + p_{NO_x})$. (From Daum et al.[129] Reprinted by permission.)

of time in cloud prior to sampling, or the amount of oxidant available for reaction with $SO_2$. Comparison of the frequency distributions for sulfur and nitrogen establishes that $NO_x$ is much less efficiently converted to cloudwater nitrate than $SO_2$ is to cloudwater sulfate in the clouds sampled here and/or in the pre-cloud air. Since available information (e.g., Refs. 17, 136, 137) suggests that oxidation of $NO_x$ to $HNO_3$ in clear air is more rapid than oxidation of $SO_2$ to $SO_4^=$, these measurements give strong evidence, at least under some circumstances, for the rapid and complete in-cloud oxidation of $SO_2$ to $SO_4^=$.

As discussed above, laboratory experiments suggest that strong oxidants, $O_3$ and $H_2O_2$, may be responsible for this oxidation, so it is of interest to examine concentrations of these species in clouds. As noted already, measurement of $O_3$ in interstitial cloud air presents no problem. On the other hand, measurement of $H_2O_2$ in air or in cloud- or rainwater has proved to be a vexatious undertaking, plagued with artifacts in sampling and interferences in determination. In view of the magnitude of the Henry's law coefficient of $H_2O_2$, which leads to the presence of comparable amounts of this material in the gas and aqueous phases, one might consider determination of this species in either of the two phases. Because of unresolved problems in sampling gaseous $H_2O_2$, we have chosen to measure this species in collected cloudwater samples.

A fairly convincing approach to the determination of $H_2O_2$ in cloud- or rainwater is that taken by Guilbault et al.[138] as modified for flow systems by Lazrus et al.,[139] which makes use of the specificity of enzyme reactions to eliminate interferences. The approach is to assay $H_2O_2$ by means of the peroxidase-catalyzed dimerization of p-hydroxyphenylacetic acid (PHOPAA) and to determine the amount of dimer by spectrofluorimetry. This technique is quite sensitive (limit of detection $\sim 0.2\ \mu M$). As it turns out, however, the dimerization is induced as well by organic peroxides; in order to obtain a specific $H_2O_2$ signal, a second measurement must be made in which $H_2O_2$ has been destroyed by addition of a second enzyme, catalase. From the difference of the two signals, a specific measure of $H_2O_2$ concentration is obtained.

We began measurement of $H_2O_2$ in collected cloudwater samples in late 1982 and have continued such measurements to the present time. In some of the early measurements we were assisted by personnel and equipment from the National Center for Atmospheric Research. Later on, using equipment of our own, which had only a single channel, we obtained for the most part only total peroxide signals, although in a few instances concentrations of organic peroxides have been comparable to those of $H_2O_2$.[140]

A synopsis of measured concentrations of peroxide expressed as potential partial pressure is presented in FIGURE 34. In the evaluation of $\hat{p}_{H_2O_2}$ we have included not only the concentration of dissolved $H_2O_2$, but also the concentration of gaseous $H_2O_2$ in equilibrium with the measured aqueous concentration, evaluated as

$$\hat{p}_{H_2O_2} = [H_2O_2]\ (L\underset{\sim}{R}T + H\bar{H}_{H_2O_2})\tag{13}$$

It is seen that $H_2O_2$ concentrations that we have observed are of the order of 0.1 up to occasionally as great as $\sim 1$ ppb. Such values are within the range expected to be present as a consequence of gas-phase photochemical reactions,[141,142] although such a coincidence of values should not be taken as proof of the predominance of the gas-phase pathway for formation of $H_2O_2$.

It is possible on the basis of measured $O_3$ and $H_2O_2$ concentrations to evaluate rates of $SO_2$ oxidation by either of these species. Such evaluations yield instantaneous oxidation rates for $SO_2$ of the order of 100% $hr^{-1}$ by $H_2O_2$ for the higher $H_2O_2$ concentrations in this range, and some two to three orders

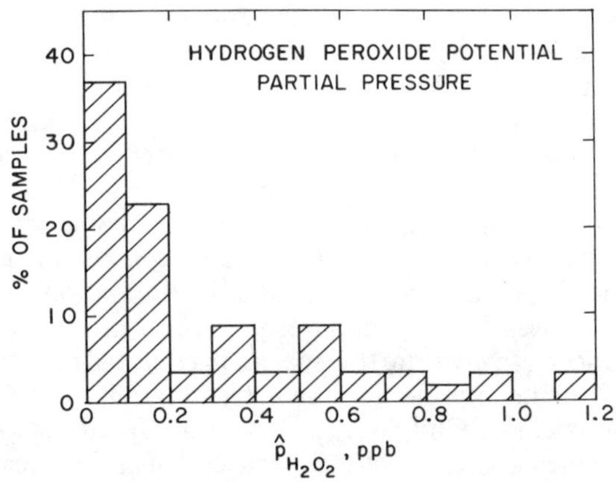


FIGURE 34. Frequency distribution of the potential partial pressure of peroxide in cloudwater samples, $\hat{p}_{H_2O_2}$, evaluated as discussed in the text. N = 57. (From Daum *et al.*[129] Reprinted by permission.)

of magnitude lower for $O_3$.[124,140] On this basis it would appear that aqueous-phase oxidation of $SO_2$ by $H_2O_2$ is a rapid process and perhaps the major pathway for in-cloud oxidation of $SO_2$.

In order to examine this process further we have plotted a "scatter diagram" of gas-phase $SO_2$ partial pressures against aqueous-phase $H_2O_2$ concentrations (expressed as potential partial pressure) (FIG. 35). Most striking in this figure is the division of the data points into two classes, those with appreciable $H_2O_2$ but with very low $SO_2$, near or below the limit of detection, and those with appreciable $SO_2$ but with very low $H_2O_2$, near or below the limit of detection for that species. These data suggest a near mutual exclusivity of the two species in clouds, which we interpret as indicating that reaction had proceeded until the reagent initially present in lesser concentration (the "limiting reagent") has been exhausted. We consider these findings to be an exciting and important confirmation from field measurements of the occurrence of in-cloud reaction of $SO_2$ and $H_2O_2$, as has been anticipated from evaluations based on laboratory studies. It must be stressed that our measurements thus far have been restricted to warm stratiform clouds, for the most part not in the immediate vicinity of sources, and it will thus be interesting to see how generally the mutual exclusivity of $SO_2(g)$ and $H_2O_2(aq)$ holds. We would note that Richards et al.[116] have found these species to be simultaneously present in clouds in the Los Angeles urban area.

An additional consideration that is suggested in these data concerns the capacity of ambient concentrations of $H_2O_2$ for oxidation of $SO_2$. Since reaction of $SO_2$ with $H_2O_2$ depletes $H_2O_2$ on a one-to-one (molar) basis for every $SO_2$ oxidized, it is seen from FIGURE 34 that $H_2O_2$ concentrations such as we have observed are capable of oxidizing only a limited amount of $SO_2$. It thus seems plausible that in regions of high $SO_2$ concentrations (i.e., near sources) $H_2O_2$ would be exhausted by reaction with $SO_2$, whereas in regions well re-

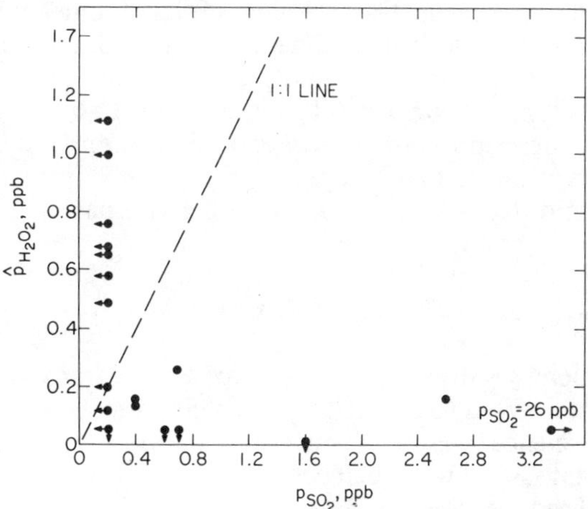


FIGURE 35. Scatter diagram of cloudwater peroxide potential partial pressure versus interstitial $SO_2$ partial pressure. (From Daum et al.[129] Reprinted by permission.)

moved from sources $H_2O_2$ would be present in excess. Thus the rate or extent of $SO_2$ oxidation and incorporation into precipitation would be dependent on the $SO_2$ concentration in relation to the $H_2O_2$ concentration prior to reaction. Considerations such as these bear directly on the development of strategies for control of acid deposition and in particular on the question of linearity of acid deposition in response to changes in $SO_2$ emissions such as might be imposed by regulation.[2,143] Unraveling the details of these processes is the task of future research.

## CONCLUSION

This article has covered a lot of ground, ranging from fundamental laboratory studies of chemical thermodynamics and kinetics to highly empirical measurements of the concentrations of chemicals in the ambient atmosphere. However, the unifying theme of the article, and for that matter of our work on cloud chemistry, has been an underlying faith in the pertinence and transferability of laboratory-determined quantities to description of chemical reactions in the atmosphere. Of course, this approach must be taken with caution and wisdom. In the case of chemical thermodynamics, we have noted that a strong driving force for a given reaction does not necessarily imply that the reaction will proceed to completion. Similarly, in the case of chemical kinetics one must be careful to describe the system in terms of appropriate elementary reactions in order to permit confident application of laboratory-determined rate expressions to field evaluations. Additionally, one must always be alert to the possibility of catalysts or inhibitors that can alter the rate of reaction from that determined in laboratory studies. However, such processes are themselves amenable to laboratory study; they just represent a more complicated sequence of reactions.

With respect to the chemistry of clouds, a level of understanding of cloud acidification processes has been attained, from both laboratory studies and field measurements, that is substantially greater than that previously available. Ultimately, information derived from studies such as these will be incorporated into the description of source receptor relationships in acid deposition and, it is hoped, lead to development of more efficient strategies for the mitigation of acid deposition that might otherwise have been obtained.

## SUMMARY

The process whereby sulfur and nitrogen oxides are transformed in the atmosphere and deposited to the surface in precipitation is of current interest in atmospheric science because of concern over apparent widespread environmental effects associated with such deposition. A key component of this process is the oxidation of $SO_2$ and $NO_x$ ($\equiv NO + NO_2$) to form sulfuric and nitric acids that are deposited in precipitation. This article presents two complementary approaches to the study of this chemistry as it occurs in liquid water clouds. The first approach relies upon laboratory determination of fundamental physical and chemical properties (such as overall reaction equilibria, gas solubilities, chemical kinetic rate expressions, and diffusion coefficients) and evaluation of rates of dissolution and reaction for representative reagent concentrations and physical situations. The second approach consists of measuring concentrations of relevant reagent and product species and other pertinent quantities (most importantly, cloud liquid water content) in and about clouds and of drawing inferences from these measurements about the rate and extent of processes responsible for establishing cloudwater composition.

On the basis of laboratory studies, the following inferences may be made: (1) Aqueous-phase oxidation of $SO_2$ by $H_2O_2$ or $O_3$ may be sufficiently rapid to contribute significantly to cloudwater acidity for representative concentrations of these oxidants; however, the rate of the $O_3$ reaction decreases strongly with decreasing pH and is therefore of relatively little importance below about pH 4.5. (2) Despite strong thermochemical driving force for oxidation of $NO_2$ to nitric acid in cloudwater, this reaction appears to be negligibly slow for representative concentrations of $NO_2$, largely because of the low Henry's law solubility of this species in water. These inferences gain support from field measurements of the composition of liquid water stratiform clouds at various locations in the eastern United States, which indicate that the fractional uptake of $SO_2$ (as $SO_4^=$) by cloudwater is frequently quite high, whereas a corresponding high fractional uptake of $NO_x$ as $NO_3^-$ is never observed. A near mutual exclusivity of gaseous $SO_2$ and dissolved $H_2O_2$ in clouds supports the inference that reaction of these species in clouds is rapid and can represent a major process for cloudwater acidification.

[**Note added in proof:** The period since June 1984, when this article was submitted for publication, has been a time of extraordinary activity and productivity in research directed to both "sides" of cloud chemistry. Attention is called here to pertinent new papers dealing with laboratory studies (References 144–184) and field studies (References 185–224) and as well to interpretive and modeling studies that deal with cloud chemistry (References 225–270). Attention is called as well to two recent textbooks that extensively discuss cloud chemistry (References 271 and 272).]

# ACKNOWLEDGMENTS

It is a pleasure to acknowledge the participation in various aspects of this work of numerous valued colleagues: Leonard Newman, Peter Daum, Thomas Kelly and Yin-Nan Lee, all at Brookhaven; John Freiberg, Greenwich, Connecticut; Mark Hjelmfelt, National Center for Atmospheric Research; Boulder, Colorado; George Markovits, Technical Engineering College, Beersheva, Israel, and Warren White, Washington University, St. Louis, Mo. Invaluable assistance in field measurements has been received from Robert Brown, Seymour Fink, and Daniel Leahy. Allan Lazrus, John Lind, and Scott McLaren of NCAR provided assistance with some of the $H_2O_2$ measurements. Finally, I would like to thank Frances Sterrett for her encouragement.

## REFERENCES


1. MITCHELL, J. 1967. "Both Sides Now." Siquomb Publishing Corp. New York, NY.
2. NATIONAL RESEARCH COUNCIL (U.S.), COMMITTEE ON ATMOSPHERIC TRANSPORT AND CHEMICAL TRANSFORMATION IN ACID PRECIPITATION (J. Calvert, Chairman). 1983. Acid Deposition—Atmospheric Processes in Eastern North America. A Review of Current Scientific Understanding. National Academy Press. Washington DC.
3. ENVIRONMENTAL PROTECTION AGENCY (U.S.). 1983. The Acidic Deposition Phenomenon and Its Effects: Critical Assessment Review Papers. Vol. 1, Atmospheric Sciences. A. P. Altshuller, Ed. Report EPA 600/8-83-016A.
4. UNITED STATES—CANADA MEMORANDUM OF INTENT ON TRANSBOUNDARY AIR POLLUTION. 1982. Final Report of Atmospheric Sciences and Analysis Work Group 2, Report 2F. Available from U.S. EPA, Office of Environmental Processes and Effects Research. Washington DC.
5. COWLING, E. B. 1982. Acid precipitation in historical perspective. Environ. Sci Technol. **16:** 110A–123A.
6. GORDON G. 1984. Acid rain . . . What is it? Resources (Resources for the Future, Washington D.C.) No. 75: 6–8.
7. ROBINSON, E. & J. B. HOMOLYA. Natural and anthropogenic emissions sources. *In* Ref. 3:2-1—2-119.
8. LOVETT, G. M., W. A. REINERS & R. K. OLSON. 1982. Cloud droplet deposition in subalpine balsam fir forests. Science **218:** 1303–1304.
9. ENVIRONMENTAL PROTECTION AGENCY (U.S.). 1983. The Acidic Deposition Phenomenon and its Effects: Critical Assessment Review Papers. Vol. 2, Effects Sciences. R. A. Linthurst, Ed. Report EPA-600/8-83-016B.
10. UNITED STATES—CANADA MEMORANDUM OF INTENT ON TRANSBOUNDARY AIR POLLUTION. 1982. Final Report of Impact and Assessment Work Group 1. Available from U.S. EPA, Office of Environmental Processes and Effects Research. Washington, DC.
11. NATIONAL RESEARCH COUNCIL (U.S.), COMMITTEE ON THE ATMOSPHERE AND THE BIOSPHERE (D. W. Schindler, Chairman). 1981. Atmosphere-Biosphere Interactions: Toward a Better Understanding of the Consequences of Fossil Fuel Combustion. National Academy Press. Washington, DC.
12. HOFFMANN, M. R. 1984. Reply to "Comment on 'Acid Fog'". Environ. Sci. Technol. **18:** 61–64.
13. INTERAGENCY TASK FORCE ON ACID PRECIPITATION (U.S.). 1982. Generic Project Descriptors for Acid Deposition Research. Washington DC.
14. KINSMAN, J. D. & J. WISNIEWSKI. 1983. Inventory of Acid Deposition Research Projections Funded by the Private Sector. Electric Power Research Institute Report EA-2889. Palo Alto, CA.
15. RAYNOR, G. S. & J. HAYES. 1984. Applications of within-event precipitation chemistry measurements. *In* Sampling and Analysis of Rain. S. A. Campbell, Ed.: 50–60. Special Technical Publication 823, American Society for Testing and Materials, Philadelphia, PA.
16. DAUM, P. M. 1983. Brookhaven National Laboratory, private communication. Evaluations were made using data from the MAP3S/RAINE Data Base; see MAP3S/RAINE Research Community, 1982. The MAP3S/RAINE precipitation chemistry network. Statistical overview for the period 1976–1980. Atmos. Environ. **16:** 1603–1631.
17. CALVERT, J. G. & W. R. STOCKWELL. 1984. Mechanism and rates of the gas-phase oxidations of sulfur dioxide and nitrogen oxides in the atmosphere. *In* SO₂,

NO and $NO_2$ Oxidation Mechanisms: Atmospheric Considerations, J.G. Calvert, Ed.: 1–62. Butterworth. Boston MA.

18. STOCKWELL, W. R. & J. G. CALVERT. 1983. The mechanism of the $HO-SO_2$ reaction. Atmos. Environ. **18:** 2231–2235.

19. NEWMAN, L. 1981. Atmospheric oxidation of sulfur dioxide: A review as viewed from power plant studies and smelter plume studies. Atmos. Environ. **15:** 2231–2239.

20. DIMITRIADES, B., Ed. 1977. Proceedings of an International Conference on Photochemical Oxidant Pollution and Its Control. U.S. EPA Report 600/3-77-001a and b (two volumes).

21. STEDMAN, D. H. & R. E. SHETTER. 1983. The global budget of atmospheric nitrogen species. Advan. Environ. Sci. Technol. **12:** 411–454.

22. STELSON, A. W. & J. H. SEINFELD. 1982. Relative humidity and pH dependence of the vapor pressure of ammonium nitrate-nitric acid solutions at 25°C. Atmos. Environ. **16:** 93–100.

23. LEVINE, S. Z. & S. E. SCHWARTZ. 1982. In-cloud and below-cloud scavenging of nitric acid vapor. Atmos. Environ. **16:** 1725–1734.

24. ANTELL, M. R. 1978. Correspondence. Environ. Sci. Technol. **12:** 1438.

25. RICHARDS, L. W. 1983. Comments on the oxidation of $NO_2$ to nitrate—day and night. Atmos. Environ. **17:** 397–402.

26. JONES, C. L. & J. H. SEINFELD. 1983. The oxidation of $NO_2$ to nitrate—day and night. Atmos. Environ. **17:** 2370–2373.

27. COX, R. A. & G. B. COKER. 1983. Kinetics of the reaction of nitrogen dioxide with ozone. J. Atmos. Chem. **1:** 53–63.

28. HEIKES, B. G. & A. M. THOMPSON. 1983. Effects of heterogeneous processes on $NO_3$, HONO and $HNO_3$ chemistry in the troposphere. J. Geophys. Res. **88:** 10883–10895.

29. TUAZON, E. C., R. ATKINSON, C. N. PLUM, A. M. WINER, & J. N. PITTS, JR. 1983. The reaction of gas phase $N_2O_5$ with water vapor. Geophys. Res. Lett. **10:** 953–956.

30. SCHWARTZ, S. E. 1986. Mass-transport considerations pertinent to aqueous-phase reactions of gases in liquid-water clouds. *In* Chemistry of Multiphase Atmospheric Systems, W. Jaeschke, Ed.: 416–471. Springer. Berlin and Heidelberg.

31. BUDIANSKI, S. 1980. Acid rain and the missing link. Environ. Sci. Technol. **14:** 1172–1173.

32. PRUPPACHER, H. R. & J. D. KLETT. 1978. Microphysics of Clouds and Precipitation. D. Reidel. Hingham, MA.

33. MASON, B. J. 1971. The Physics of Clouds. Clarendon Press. Oxford.

34. ROGERS, R. R. 1976. A Short Course in Cloud Physics. Pergamon Press. Oxford.

35. McDONALD, J. E. 1958. The physics of cloud modification. Advan. Geophys. **5:** 223–303.

36. STEWART, R. E. 1984. Deep 0°C isothermal layers within precipitation bands over southern Ontario. J. Geophys. Res. **89:** 2567–2572.

37. HO, W. W., G. M. HIDY & R. M. GOVAN. 1980. Microwave measurements of the liquid water content of atmospheric aerosols. Advan. Environ. Sci. Technol. **9:** 215–236.

38. KNOLLENBERG, R. G. 1981. Techniques for probing cloud microstructure. *In* Clouds—Their Formation, Optical Properties, and Effects. P. V. Hobbs & A. Deepak, Eds.: 15–91. Academic Press. New York, NY.

39. LEWIS, G. N. & M. RANDALL. 1923. Thermodynamics.:568 McGraw-Hill. New York, NY.

40. CHILTON, T. H. 1968. Strong Water. MIT Press. Cambridge, MA.

41. SCHWARTZ, S. E. 1984. Gas-aqueous reactions of sulfur and nitrogen oxides in liquid water clouds. *In* $SO_2$, NO, and $NO_2$ Oxidation Mechanisms: Atmospheric Considerations. J. G. Calvert, Ed.:173–208. Butterworth. Boston, MA.

42. LEE, Y.-N., G. I. SENUM & J. S. GAFFNEY. 1983. Peroxyacetyl nitrate (PAN) stability, solubility, and reactivity — Implications for tropospheric nitrogen cycles and precipitation chemistry. Fifth International Conference of the Commission on Atmospheric Chemistry and Global Pollution. Symposium on Tropospheric Chemistry, Oxford, England, August 28–September 2, 1983.

43. JOHNSTONE, H. F. & P. W. LEPPLA. 1934. The solubility of sulfur dioxide. J. Am. Chem. Soc. **56**: 2233–2238.

44. SCHWARTZ, S. E. & J. E. FREIBERG. 1981. Mass-transport limitation to the rate of reaction in liquid droplets: Application to oxidation of $SO_2$ in aqueous solutions. Atmos. Environ. **15**: 1129–1144.

45. BEILKE, S. & G. GRAVENHORST. 1978. Heterogeneous $SO_2$ oxidation in the droplet phase. Atmos. Environ. **12**: 231–239.

46. PENKETT, S. A., B. M. R. JONES, K. A. BRICE & A. E. J. EGGLETON. 1979. The importance of atmospheric ozone and hydrogen peroxide in oxidising sulphur dioxide in cloud and rainwater. Atmos. Environ. **13**: 123–137.

47. SEINFELD, J. H. 1980. Lectures in Atmospheric Chemistry, Monograph Series, Vol. 76. American Institute of Chemical Engineers. New York, NY.

48. WAGNER, P. E. 1982. Aerosol growth by condensation. *In* Aerosol Microphysics II: Chemical Physics of Microparticles. W. H. Marlow, Ed.: 129–178. Springer-Verlag. Berlin.

49. CHAMEIDES, W. L. & D. D. DAVIS. 1982. The free radical chemistry of cloud droplets and its impact upon the composition of rain. J. Geophys. Res. **87**: 4863–4877.

50. DASGUPTA, P. K. 1980. Discussions. Atmos. Environ. **14**: 272–274.

51. HOFFMANN, M. R. & S. C. BOYCE. Catalytic autoxidation of aqueous sulfur dioxide in relationship to atmospheric systems. Advan. Environ. Sci. Technol. **12**: 147–189.

52. MARTIN, L. R. 1984. Kinetic studies of sulfite oxidation in aqueous solution. *In* $SO_2$, NO, and $NO_2$ Oxidation Mechanisms: Atmospheric Conditions. J. G. Calvert, Ed.: 63–100. Butterworth. Boston, MA.

53. PENKETT, S. A. 1972. Oxidation of $SO_2$ and other atmospheric gases by ozone in aqueous solution. Nature Phys. Sci. **240**: 105–106.

54. ERICKSON, R. E., L. M. YATES, R. L. CLARK & D. McEWEN. 1977. The reaction of sulfur dioxide with ozone in water and its possible atmospheric significance. Atmos. Environ. **11**: 813–817.

55. LARSON, T. V., H. R. HORIKE & H. HARRISON. 1978. Oxidation of $SO_2$ by $O_2$ and $O_3$ in aqueous solution: A kinetic study with significance to atmospheric rate processes. Atmos. Environ. **12**: 1597–1611.

56. MAAHS, H. G. 1983. Measurements of the oxidation rate of sulfur(IV) by ozone in aqueous solution and their relevance to $SO_2$ conversion in nonurban tropospheric clouds. Atmos. Environ. **17**: 341–345.

57. MAAHS, H. G. 1983. Kinetics and mechanism of the oxidation of S(IV) by ozone in aqueous solution with particular reference to $SO_2$ conversion in nonurban tropospheric clouds. J. Geophys. Res. **88**: 10721–10723.

58. ESPENSON, J. H. & H. TAUBE. 1965. Tracer experiments with ozone as oxidizing agent in aqueous solution. Inorg. Chem. **4**: 704–709.

59. HARRISON, H., T. V. LARSON & C. S. MONKMAN. 1982. Aqueous phase oxidation of sulfites by ozone in the presence of iron and manganese. Atmos. Environ. **16**: 1039–1041.

60.  MARTIN, L. R. & D. E. DAMSCHEN. Kinetic studies of sulfite oxidation in aqueous solution. Presented at the 185th National Meeting of the American Chemical Society, Las Vegas, Nevada, March 18–April 2, 1982.

61.  HALPERIN, J., & H. TAUBE. 1952. The transfer of oxygen atoms in oxidation-reduction reactions. IV. The reaction of hydrogen peroxide with sulfite and thiosulfate, and oxygen, manganese dioxide and of permanganate with sulfite. J. Am. Chem. Soc. **74:** 380–382.

62.  MADER, P. M. 1958. Kinetics of the hydrogen peroxide-sulfite reaction in alkaline solution. J. Am. Chem. Soc. **80:** 2634–2639.

63.  HOFFMANN, M. R. & J. O. EDWARDS. 1975. Kinetics of oxidation of sulfite by hydrogen peroxide in acidic solution. J. Phys. Chem. **79:** 2096–2098.

64.  MARTIN, L. R. & D. E. DAMSCHEN. 1981. Aqueous oxidation of sulfur dioxide by hydrogen peroxide at low pH. Atmos. Environ. **15:** 1615–1621.

65.  COCKS, A. T., W. J. MCELROY & P. G. WALLIS. 1982. The Oxidation of Sodium Sulphite Solutions by Hydrogen Peroxide. Central Electricity Research Laboratories Report RD/L/2215N81.

66.  KUNEN, S. M., A. L. LAZRUS, G. L. KOK & B. G. HEIKES. 1983. Aqueous oxidation of $SO_2$ by hydrogen peroxide. J. Geophys. Res. **88:** 3671–3674.

67.  MCARDLE, J. V. & M. R. HOFFMANN. 1983. Kinetics and mechanism of the oxidation of aquated sulfur dioxide by hydrogen peroxide at low pH. J. Phys. Chem. **87:** 5425–5429.

68.  LEE, Y.-N., J. SHEN, P. J. KLOTZ, S. E. SCHWARTZ & L. NEWMAN. 1986. Kinetics of the hydrogen peroxide-sulfur(IV) reaction in rainwater collected at a northeastern U.S. site. J. Geophys. Res. **91:** 13264–13274.

69.  LATIMER, W. M. & J. H. HILDEBRAND. 1940. Reference Book of Inorganic Chemistry (revised ed.) :199, 475. Macmillan. New York, NY.

70.  COTTON, F. A. & G. WILKINSON. 1980. Advanced Inorganic Chemistry, 4th ed.: 426, 1205–1216. John Wiley. New York, NY.

71.  GEORGII, H.-W. 1963. Oxides of nitrogen and ammonia in the atmosphere. J. Geophys. Res. **68:** 3963–3970.

72.  COOPER, H. G. H., J. A. LOPEZ & J. M. DEMO. 1976. Chemical composition of acid precipitation in central Texas. Water Air Soil Pollut. **6:** 351–359.

73.  SÖDERLUND, R. & B. H. SVENNSON. 1976. The global nitrogen cycle, *In* Nitrogen, Phosphorus and Sulfur — Global Cycles. B. H. Svennson and R. Söderlund, Eds. SCOPE Rep. **7:** 23–73.

74.  MATTESON, M. J. 1979. Capture of atmospheric gases by water vapor condensation on carbonaceous particles. Carbonaceous Particles in the Atmosphere. Conference Proceedings, Berkeley, California, March 20, 1978, Report LBL-9037. Lawrence Berkeley Labs. Berkeley, CA.

75.  HERRMANN, J. P. & M. J. MATTESON. 1980. Nitrogen-dioxide absorption in evaporating and condensing water droplets. Advan. Environ. Sci. Eng. **3:** 92–99.

76.  RODHE, H., P. CRUTZEN & A. VANDERPOL. 1980. Formation of sulfuric and nitric acid in the atmosphere during long-range transport. Symposium on the Long-Range Transport of Air Pollutants and its Relation to General Circulation.:165–172. Report 538, World Meteorological Organization, Geneva.

77.  MARTIN, L. R., D. E. DAMSCHEN & H. S. JUDEIKIS. 1981. The reactions of nitrogen oxides with $SO_2$ in aqueous aerosols. Atmos. Environ. **15:** 191–195.

78.  HIDY, G. M. & C. S. BURTON. 1975. Atmospheric aerosol formation by chemical reactions. Int. J. Chem. Kinet. Symp. **1:** 509–541.

79.  OREL, A. E. & J. H. SEINFELD. 1977. Nitrate formation in atmospheric aerosols, Env. Sci. Technol. **11:** 1000–1007.

80.  PETERSON, T. W. & J. H. SEINFELD. 1979. Sulfate and nitrate levels in aqueous,

atmospheric aerosols. *In* Nitrogenous Air Pollutants: Chemical and Biological Implications. D. Grosjean, Ed.: 259–268. Ann Arbor Science. Ann Arbor, MI.

81. MIDDLETON, P. & C. S. KIANG. 1979. Relative importance of nitrate and sulfate aerosol production mechanisms in urban atmospheres. *In* Nitrogenous Air Pollutants: Chemical and Biological Implications. D. Grosjean, Ed.: 269–288. Ann Arbor Science. Ann Arbor, MI.

82. PARUNGO, F. P. & R. F. PUESCHEL. 1980. Conversion of nitrogen oxide gases to nitrate particles in oil refinery plumes. J. Geophys. Res. **85:** 4507–4511.

83. LISS, P. S. 1976. The exchange of gases across lake surfaces. J. Great Lakes Res. **2**(Suppl. 1): 88–100, 126.

84. HICKS, B. B. & P. S. LISS. 1976. Transfer of $SO_2$ and other reactive gases across the air-sea interface. Tellus **28:** 348–354.

85. KABEL, R. L. 1976. Atmospheric impact on nutrient budgets. J. Great Lakes Res. **2**(Suppl. 1): 114–126.

86. SLINN, W. G. N., L. HASSE, B. B. HICKS, A. W. HOGAN, D. LAL, P. S. LISS, K. O. MUNNICH, G. A. SEHMEL & O. VITTORI. 1978. Some aspects of the transfer of atmospheric trace constituents past the air-sea interface. Atmos. Environ. **12:** 2055–2087.

87. HILL, A. C. 1971. Vegetation: A sink for atmospheric pollutants. J Air. Pollut. Control Assoc. **21:** 2055–2087.

88. HILL, A. C. & E. M. CHAMBERLAIN, JR. 1976. The removal of water-soluble gases from the atmosphere by vegetation. Atmosphere-Surface Exchange of Particulate and Gaseous Pollutants (1974). ERDA Symp Ser. **38:** 153–170.

89. WESELY, M. L., J. A. EASTMAN, D. H. STEDMAN & E. D. YALVAC. 1982. An eddy-correlation meansurement of $NO_2$ flux to vegetation and comparison to $O_3$ flux. Atmos. Environ. **16:** 815–820.

90. BEYAK, R. A. & T. W. PETERSON. 1980. Modeling of aerosol dynamics: Aerosol size and composition. Ann. N.Y. Acad. Sci. **338:** 174–189.

91. GOLDSTEIN, E., N. F. PEEK, N. J. PARKS, H. H. HINES, E. P. STEFFEY & B. TARKINGTON. 1977. Fate and distribution of inhaled nitrogen dioxide in rhesus monkeys. Am. Rev. Respir. Dis **115:** 403–412.

92. NASH, T. 1970. Absorption of nitrogen dioxide by aqueous solution. J. Chem. Soc. A: 3023–3024.

93. GRAEDEL, T. E., L. A. FARROW & T. A. WEBER. 1975. The influence of aerosols on the chemistry of the troposphere. Int. J. Chem. Kinet. Symp. **1:** 581–594.

94. BARNES, R. A. 1979. The long range transport of air pollution: A review of European experience. J. Air Pollut. Control Assoc. **29:** 1219–1235.

95. ABEL, E. & H. SCHMID. 1928. Kinetik der salpetrigen Säure. VI. Gleichgewicht der Salpetrigsäure-Salpetersäure-Stickoxyd-Reaktion im Zusammenhang mit deren Kinetik. Z. Phys. Chem **136:** 430–436.

96. SCHWARTZ, S. E. & W. H. WHITE. 1981. Solubility equilibria of the nitrogen oxides and oxyacids in dilute aqueous solution. Advan. Environ. Sci. Eng. **4:** 1–45.

97. BUNTON, C. A. & G. STEDMAN. 1958. The absorption spectra of nitrous acid in aqueous perchloric acid. J. Chem. Soc.:2440–2444.

98. TURNEY, T. A. 1960. The nitrous acid-dinitrogen trioxide equilibrium in aqueous perchloric acid. J. Chem. Soc.: 4263–4265.

99. SCHMID, H. & P. KRENMAYR. 1967. Der Aktivitätskoeffizient der salpetrigen Säure and die Gleichgewichtskonstante der Distickstofftrioxidbildung. Monatshefte Chem. **98:** 417–422.

100. SHAW, A. W. & D. J. VOSPER. 1972. Dinitrogen trioxide. Part XI. The electronic spectrum of dinitrogen trioxide. J. Chem. Soc. (Dalton): 961–964.

101. MARKOVITS, G. Y., S. E. SCHWARTZ & L. NEWMAN. 1981. Hydrolysis equilibrium of dinitrogen trioxide in dilute acid solution. Inorg. Chem. **20:** 445–450.

102. CASADO, J., A. CASTRO, J. R. LEIS, M. A. LOPEZ QUANTELA & M. MOSQUERA. 1983. Kinetic studies on the formation of $N$-nitroso compounds VI. The reactivity of $N_2O_3$ as a nitrosating agent. Monatshefte für Chemie **114:** 639–646.

103. SCHWARTZ, S. E. & W. H. WHITE. 1983. Kinetics of reactive dissolution of nitrogen oxides into aqueous solution. Advan. Environ. Sci. Technol. **12:** 1–116.

104. LEE, Y.-N. & S. E. SCHWARTZ. 1981. Reaction kinetics of nitrogen dioxide with liquid water at low partial pressure. J. Phys. Chem. **85:** 840–848.

105. LEE, Y.-N. & S. E. SCHWARTZ. 1981. Evaluation of the rate of uptake of nitrogen dioxide by atmospheric and surface liquid water. J. Geophy. Res. **86:** 11971–11983.

106. LEE, Y.-N. 1984. Atmospheric aqueous-phase reactions of nitrogen species. *In* Gas-Liquid Chemistry of Natural Waters. L. Newman, Ed. Brookhaven National Laboratory Report, BNL 51757,: 20-1 – 20-10.

107. PRYOR, W. A. & J. W. LIGHTSEY. 1981. Mechanisms of nitrogen dioxide reactions: Initiation of lipid peroxidation and the production of nitrous acid. Science **214:** 435–437.

108. LAZRUS, A. L., P. L. HAAGENSON, G. L. KOK, B. J. HUEBERT, C. W. KREITZBERG, G. E. LIKENS, V. A. MOHNEN, W. E. WILSON & J. W. WINCHESTER. 1983. Acidity in air and water in a case of warm frontal precipitation. Atmos. Environ. **17:** 581–591.

109. PARUNGO, F., C. NAGAMOTO, I. NOLT, M. DIAS & E. NICKERSON. 1982. Chemical analysis of cloud water collected over Hawaii. J. Geophys. Res. **87:** 8805–8810.

110. FALCONER, R. F. & P. D. FALCONER. 1980. Determination of cloud water acidity at a mountain observatory in the Adirondack Mountains of New York State. J. Geophys. Res. **85:** 7465–7470.

111. CASTILLO, R. A., J. E. JIUSTO & E. McLAREN. 1983. The pH and ionic composition of stratiform cloud water. Atmos. Environ. **17:** 1497–1505.

112. WALDMAN, J. M., J. W. MUNGER, D. J. JACOB, R. C. FLAGAN & M. R. HOFFMANN. 1982. Chemical composition of acid fog. Science **218:** 677–680.

113. MUNGER, J. W., D. J. JACOB, J. M. WALDMAN & M. R. HOFFMANN. 1983. Fogwater chemistry in an urban atmosphere. J. Geophys. Res **88:** 5109–5121.

114. HEGG, D. A. & P. V. HOBBS. 1981. Cloud water chemistry and production of sulfates in clouds. Atmos. Environ. **15:** 1957–1604.

115. HEGG, D. A. & P. V. HOBBS. 1982. Measurements of sulfate production in natural clouds. Atmos. Environ. **16:** 2663–2668.

116. RICHARDS, L. W., J. A. ANDERSON, D. L. BLUMENTHAL, J. A. McDONALD, G. L. KOK, & A. L. LAZRUS. 1983. Hydrogen peroxide and sulfur(IV) in Los Angeles cloud water. Atmos. Environ. **17:** 911–914.

117. LEAITCH, W. R., J. W. STRAPP, H. A. WIEBE & G. A. ISAAC. 1983. Measurements of scavenging and transformation of aerosol inside cumulus. *In* Precipitation Scavenging, Dry Deposition and Resuspension. Vol. 1. H. R. Pruppacher, R. G. Semonin & W. G. N. Slinn, Eds.:53–69. Elsevier. New York, NY.

118. KALLEND, A. S., A. R. W. MARSH, G. M. GLOVER, A. H. WEBB, D. J. MOORE, P. A. CLARK, B. E. A. FISHER, D. J. A. DEAR, P. LIGHTMAN & C. K. LAIRD. 1982. Studies of the fate of atmospheric emissions in power plant plumes over the North Sea. *In* Physico-Chemical Behaviour of Atmospheric Pollutants. B. Versino & H. Ott, Eds.: 482–491. D. Reidel. Hingham, MA.

119. RÖMER, F. G., J. W. VILJEER, L. VAN DEN BELD, H. J. SLANGEWAL, A. A. VELDKAMP & H. F. R. REIJNDERS. 1983. Preliminary measurements from an air-

craft into the chemical composition of clouds. *In* Acid Deposition. S. Beilke & A. J. Elshout, Eds.: 195–203. D. Reidel. Boston, MA.

120. Leahy, D. 1981. The Brookhaven National Laboratory Atmospheric Sciences Aircraft, BNL 28102R, Brookhaven National Laboratory. Upton, New York, NY.

121. Walters, P. T., M. J. Moore & A. H. Webb. 1983. A separator for obtaining samples of cloud water in aircraft. Atmos. Environ. **17:** 1083–1091.

122. Daum, P. H. & D. F. Leahy. 1982. The Brookhaven National Laboratory filter pack system for collection and determination of air pollutants, Rep. BNL 31381, Brookhaven National Laboratory. Upton, New York, NY.

123. Tanner, R. L., P. H. Daum & T. J. Kelly. 1983. New instrumentation for airborne acid rain research. Int. J. Environ. Anal. Chem. **13:** 323–335.

124. Garber, R. W., P. H. Daum, R. F. Doering, T. D'Ottavio & R. L. Tanner. 1983. Determination of ambient aerosol and gaseous sulfur using a continuous FPD. III: Design and characterization of a monitor for airborne applications. Atmos. Environ. **17:** 1381–1385.

125. Winters, W., A. Hogan, V. Mohnen & S. Barnard. 1979. ASRC Airborne Cloud Water Collection System. ASRC-SUNY. Publication 728, Atmospheric Science Research Center, State University of New York at Albany. Albany, NY.

126. Neel, C. B. Measurement of cloud liquid-water content with a heated wire. 19th International Instrumentation Symposium, Instrument Society of America, Las Vegas, Nevada, May 21–23, 1973.

127. Daum, P. H., S. E. Schwartz & L. Newman. 1983. Studies of the gas- and aqueous-phase composition of stratiform clouds. *In* Precipitation Scavenging, Dry Deposition, and Resuspension, Vol. 1. H. R. Pruppacher, R. G. Semonin & W. G. N. Slinn, Eds.:31–52. Elsevier. New York, NY.

128. Daum, P. H., S. E. Schwartz & L. Newman. 1984. Acidic and related constituents in liquid water stratiform clouds. J. Geophys. Res. **89:** 1447–1458.

129. Daum, P. H., T. J. Kelly, S. E. Schwartz & L. Newman. 1984. Measurements of the chemical composition of stratiform clouds. Atmos. Environ. **18:** 2671–2684.

130. Schwartz, S. E., P. H. Daum, M. R. Hjelmfelt & L. Newman. 1983. Cloud-water acidity measurements and formation mechanisms — experimental design. *In* Precipitation Scavenging, Dry Deposition, and Resuspension, Vol. 1. H. R. Pruppacher, R. G. Semonin & W. G. N. Slinn, Eds.:15–30. Elsevier, New York, NY.

131. Daum, P. H. 1984. Brookhaven National Laboratory. Private communication. Evaluations were made using data from J. Ticher, R. Brown, P. Daum and D. Leahy. 1982. Data Collected Aboard the BNL Aircraft and Van during NEROS 1980. Brookhaven National Laboratory Report BNL 31044.

132. Newman, L. General considerations of how rainwater must obtain sulfate, nitrate, and acid. Chemical Congress, American Chemical Society Chemical Society of Japan, Honolulu, April, 1979.

133. Engelmann, R. J. 1970. Scavenging prediction using ratios of concentrations in air and precipitation. *In* Precipitation Scavenging. R. J. Engelmann & W. G. N. Slinn, Eds.:475–484. USAEC Symposium Series 22, CONF 700601. U.S. Atomic Energy Commission. Oak Ridge, TN.

134. Carmichael, G. R., T. Kitada & L. K. Peters. 1983. The effects of in-cloud and below-cloud scavenging on the transport and gas phase reactions of $SO_x$, $NO_x$, $HC_x$, $H_xO_y$ and $O_3$ compounds. *In* Precipitation Scavenging, Dry Deposition and Resuspension, Vol. 1. H. R. Pruppacher, R. G. Semonin & W. G. N. Slinn, Eds.: 675–686. Elsevier. New York, NY.

135. ASKNE, C. 1973. Method of Determining Strong Acid in Precipitation Based on Coulometric Titration with Application of Gran's Plot. Institutet för Vatten-och Luftvårdsforskning, Report B152. Gothenburg, Sweden.

136. RICHARDS, L. W., J. A. ANDERSON, D. L. BLUMENTHAL, A. A. BRANDT, J. A. McDONALD, N. WATERS, E. S. MACIAS & P. S. BHARDWAJA. 1981. The chemistry, aerosol physics, and optical properties of a western coal-fired power plant plume. Atmos. Environ. **15:** 2111–2134.

137. FORREST, J., R. W. GARBER & L. NEWMAN. 1981. Conversion rates in power plant plumes based on filter pack data: The coal-fired Cumberland plume. Atmos. Environ. **15:** 2273–2282.

138. GUILBAULT, G. G., P. J. BRIGNAC & M. JUNEAU. 1968. New Substrate for fluorometric determination of oxidative enzymes. Anal. Chem. **40:** 1256–1263.

139. LAZRUS, A. L., G. L. KOK, S. N. GITLIN, J. A. LIND & S. E. McLAREN. 1985. Automated fluorimetric method for hydrogen peroxide in atmospheric precipitation. Anal. Chem. **57:** 917–922.

140. KELLY, T. J., P. H. DAUM & S. E. SCHWARTZ. 1985. Measurements of peroxides in cloudwater and rain. J. Geophys. Res. **90:** 7861–7871.

141. CALVERT, J. G. & W. R. STOCKWELL. 1983. Acid generation in the troposphere by gas-phase chemistry. Environ. Sci. Technol. **17:** 428A–442A.

142. HOV, Ø. 1983. One-dimensional vertical model for ozone and other gases in the atmospheric boundary layer. Atmos. Environ. **17:** 535–549.

143. HIDY, G. M. 1984. Source-receptor relationships for acid deposition: Pure and simple? Air Pollut. Control Assoc. J. **34:** 518–531.

144. PENKETT, S. A. 1986. Laboratory studies of the multiphase $S^{IV} \rightarrow S^{VI}$ conversion rate. *In* Chemistry of Multiphase Atmospheric Systems. W. Jaeschke, Ed. :507–539. Springer-Verlag. Berlin and Heidelberg.

145. RADOJEVIC, M. 1984. On the discrepancy between reported studies of the uncatalysed aqueous oxidation of $SO_2$ by $O_2$. Environ. Technol. Ltr. **5:** 549–566.

146. CLARKE, A. G. & M. RADOJEVIC. 1984. Oxidation rates of $SO_2$ in sea-water and sea-salt aerosols. Atmos. Environ. **18:** 2761–2767.

147. BERRESHEIM, H. & W. JAESCHKE. 1986. Study of metal aerosol systems as a sink for atmospheric $SO_2$. J. Atmos. Chem. **4:** 311–334.

148. KLEINMAN, M. T., R. F. PHALEN, R. MANNIX, M. AZIZIAN & R. WALTERS. 1985. Influence of Fe and Mn ions on the incorporation of radioactive $^{35}SO_2$ by sulfate aerosols. Atmos. Environ. **19:** 607–610.

149. HUIE, R. E. & P. NETA. 1984. Chemical behavior of $SO_3^-$ and $SO_5^-$ radicals in aqueous solutions. J. Phys. Chem. **88:** 5665–5669.

150. NETA, P. & R. E. HUIE. 1985. Free-radical chemistry of sulfite. Environ. Health Persp. **64:** 209–217.

151. OVERTON, J. H., JR. 1985. Validation of the Hoffmann and Edwards' S(IV)-$H_2O_2$ mechanism. Atmos. Environ. **19:** 687–690.

152. JAESCHKE, W. A. & G. J. HERRMANN. 1987. $SO_2$ oxidation by hydrogen peroxide in suspended droplets. *In* Acid Rain: Sources and Atmospheric Chemistry. R. W. Johnson, G. E. Gordon, W. H. Calkins & A. W. Elzerman, Eds. American Chemical Society Symposium Series. In press.

153. BENNER, W. H., P. M. McKINNEY & T. NOVAKOV. 1985. Oxidation of $SO_2$ in fog droplets by primary oxidants. Atmos. Environ. **19:** 1377–1383.

154. TANAKA, S., K. YAMANAKA & Y. HASHIMOTO. 1987. Measurement of concentration and oxidation rate of S(IV) in rain water in Yokahama, Japan. *In* Acid Rain: Sources and Atmospheric Chemistry. R. W. Johnson, G. E. Gordon, W. H. Calkins & A. W. Elzerman, Eds. American Chemical Society Symposium Series. In press.

155. HOIGNÉ, J., H. BADER, W. R. HAAG & J. STAEHELIN. 1985. Rate constants of reactions of ozone with organic and inorganic compounds in water-III. Water Res. **19:** 993–1004.

156. HOFFMANN, M. R. 1986. On the kinetics and mechanism of oxidation of aquated sulfur dioxide by ozone. Atmos. Environ. **20:** 1145–1154.

157. MILLER, D. F., A. W. GERTLER, M. R. WHITBECK & D. LAMB. 1986. Ozone-induced oxidation of $SO_2$ in simulated clouds. J. Geophys. Res. **91:** 14439–14444.

158. HOLDREN, M. W., C. W. SPICER & J. M. HALES. 1984. Peroxyacetyl nitrate solubility and decomposition rate in acidic water. Atmos. Environ. **18:** 1171–1173.

159. LEE, Y.-N. & J. A. LIND. 1986. Kinetics of aqueous-phase oxidation of nitrogen-(III) by hydrogen peroxide. J. Geophys. Res. **91:** 2793–2800.

160. GOLDBERG, R. N. & V. B. PARKER. 1985. Thermodynamics of solution of $SO_2(g)$ in water and of aqueous sulfur dioxide solutions. J. Res. Natl. Bureau Stand. **90:** 341–358.

161. AYERS, G. P., R. W. GILLETT & E. R. CAESER. 1985. Solubility of ammonia in water in the presence of atmospheric $CO_2$. Tellus **37B:** 35–40.

162. AYERS, G. P., J. L. GRAS, A. ADRIAANSEN & R. W. GILLETT. 1984. Solubility of ammonia in rainwater. Tellus **36B:** 85–91.

163. CLEGG, S. L. & P. BRIMBLECOMBE. 1986. The dissociation constant and Henry's law constant of HCl in aqueous solution. Atmos. Environ. **20:** 2483–2485.

164. HWANG, H. & P. K. DASGUPTA. 1985. Thermodynamics of the hydrogen peroxide-water system. Environ. Sci. Technol. **19:** 255–258.

165. LIND, J. A. & G. L. KOK. 1986. Henry's law determinations for aqueous solutions of hydrogen peroxide, methylhydroperoxide, and peroxyacetic acid. J. Geophys. Res. **91:** 7889–7895.

166. DONG, S. & P. K. DASGUPTA. 1986. Solubility of gaseous formaldehyde in liquid water and generation of trace standard gaseous formaldehyde. Environ. Sci. Technol. **20:** 637–40.

167. DEISTER, U., R. NEEB, G. HELAS & P. WARNECK. 1986. Temperature dependence of the equilibrium $CH_2(OH)_2 + HSO_3^- = CH_2(OH)SO_3^- + H_2O$ in aqueous solution. J. Phys. Chem. **90:** 3213–3217.

168. DONG, S. & P. K. DASGUPTA. 1986. On the formaldehyde-bisulfite-hydroxymethane-sulfonate equilibrium. Atmos. Environ. **20:** 1635–1637.

169. KOK, G. L., S. N. GITLIN & A. L. LAZRUS. 1986. Kinetics of the formation and decomposition of hydroxymethanesulfonate. J. Geophys. Res. **91:** 2801–2804.

170. OLSON, T. M. & M. R. HOFFMANN. 1986. On the kinetics of formaldehyde-S(IV) adduct formation in slightly acidic solution. Atmos. Environ. **20:** 2277–2278.

171. HOFFMANN, M. R. & J. G. CALVERT. 1985. Chemical transformation modules for Eulerian acid deposition models. Vol. II. The aqueous-phase chemistry. U.S. EPA Report 600-3-85/036.

172. STAEHELIN, J. & J. HOIGNÉ, 1985. Decomposition of ozone in water in the presence of organic solutes acting as promoters and inhibitors of radical chain reactions. Environ. Sci. Technol. **19:** 1206–1213.

173. GERTLER, A. W., M. R. WHITBECK & D. F. MILLER. $NO_2$ transformation in cloud droplets. Presented at the 78th Annual Meeting, Air Pollution Control Association, Detroit, Michigan, June 16–21, 1985.

174. HUIE, R. E. & P. NETA. 1986. Kinetics of one-electron transfer reactions involving $ClO_2$ and $NO_2$. J. Phys. Chem. **90:** 1193–1198.

175. NETA, P. & R. E. HUIE. 1986. Rate constant for reactions of $NO_3$ radicals in aqueous solutions. J. Phys. Chem. **90:** 4644–4648.

176. BIELSKI, B. H., D. E. CABELLI, R. L. ARUDI & A. B. ROSS. 1985. Reactivity of $HO_2/O_2^-$ radicals in aqueous solution. J. Phys. Chem. Ref. Data. **14:** 1041–1100.

177. SCHWARZ, H. A. & R. W. DODSON. 1984. Equilibrium between hydroxyl radicals and thallium(II) and the oxidation potential of OH(aq). J. Phys. Chem. **88:** 3643–3647.

178. KLÄNING, U. K., K. SEHESTED & J. HOLCMAN. 1985. Standard Gibbs energy of formation of the hydroxyl radical in aqueous solution. Rate constants for the reaction $ClO_2^- + O_3 \rightleftharpoons O_3^- + ClO_2$. J. Phys. Chem. **89:** 760–763.

179. TOPALIAN, J. H., S. MITRA, D. C. MONTAGUE, A. QUINTANAR & H. R. PRUPPACHER. 1984. On the scavenging of organic gases by cloud and raindrops: I. A theoretical and experimental study of acetaldehyde absorption and desorption for water drops in air. J. Atmos. Chem. **1:** 325–334.

180. WALCEK, C. J., H. R. PRUPPACHER, J. H. TOPALIAN & S. K. MITRA. 1984. On the scavenging of $SO_2$ by cloud and raindrops: II. An experimental study of $SO_2$ absorption and desorption for water drops in air. J. Atmos. Chem. **1:** 291–306.

181. MOZURKEWICH, M. 1986. Aerosol growth and the condensation coefficient of water: A review. Aerosol Sci. Technol. **5:** 223–236.

182. MOZURKEWICH, M., P. H. MCMURRY, A. GUPTA & J. G. CALVERT. 1987. Mass accommodation coefficient for $HO_2$ radicals on aqueous particles. J. Geophys. Res. **92:** 4163–4170.

183. RICHARDSON, C. B., H.-B. LIN, R. MCGRAW & I. N. TANG. 1986. Growth rate measurements for single suspended droplets using the optical resonance method. Aerosol Sci. Technol. **5:** 103–112.

184. TANG, I. N. & J. H. LEE. 1987. Accommodation coefficients of ozone and sulfur dioxide: Their implications on $SO_2$ oxidation in cloud water. *In* Acid Rain: Sources and Atmospheric Chemistry. R. W. Johnson, G. E. Gordon, W. H. Calkins & A. W. Elzerman, Eds. American Chemical Society Symposium Series. In press.

185. BROWN, R. M., P. H. DAUM, S. E. SCHWARTZ & M. R. HJELMFELT. 1984. Variations in the chemical composition of clouds during frontal passage. *In* Meteorology of Acid Deposition, Transactions of Specialty Conference, Hartford, Connecticut, October 1983. P. J. Samson, Ed. :202–212. Air Pollution Control Association. Pittsburgh, PA.

186. ISAAC, G. A. & P. H. DAUM. 1987. A winter study of air, cloud and precipitation chemistry in Ontario, Canada. Atmos. Environ. In press.

187. DAUM, P. H., T. J. KELLY, J. W. STRAPP, W. R. LEAITCH, P. JOE, R. S. SCHEMENAUER, G. A. ISAAC, K. G. ANLAUF & H. A. WIEBE. 1987. Chemistry and physics of a winter stratus cloud layer: A case study. J. Geophys. Res. In press.

188. TEN BRINK, H. M., S. E. SCHWARTZ & P. H. DAUM. 1987. Efficient scavenging of aerosol sulfate by liquid-water clouds. Atmos. Environ. In press.

189. KELLY, T. J., S. E. SCHWARTZ & P. H. DAUM. On the detectability of acid formation in clouds. Presented at the 80th Annual Meeting, Air Pollution Control Association, New York, New York, June 1987.

190. DAUM, P. H. 1987. Processes determining cloudwater composition: Inferences from field measurements. *In* Processes of Acidic Deposition in Mountainous Terrain. M. Unsworth, Ed. In press.

191. HOBBS, P. V. 1986. Field studies of cloud chemistry and the relative importance of various mechanisms of the incorporation of sulfate and nitrate into cloud water. *In* Chemistry of Multiphase Atmospheric Systems. W. Jaeschke, Ed. :191–211. Springer-Verlag. Berlin and Heidelberg.

192. HEGG, D. A., P. V. HOBBS & L. F. RADKE. 1984. Measurements of the scavenging of sulfate and nitrate in clouds. Atmos. Environ. **18:** 1939–1946.

193. HEGG, D. A. & P. V. HOBBS. 1986. Sulfate and nitrate chemistry in cumuliform clouds. Atmos. Environ. **20:** 901–909.

194. SCHWARTZ, S. E. 1987. Discussion: Sulfate and nitrate chemistry in cumuliform clouds. Atmos. Environ. In press.

195. MUNGER, J. W., D. J. JACOB & M. R. HOFFMANN. 1984. The occurrence of bisulfite-aldehyde addition products in fog- and cloudwater. J. Atmos. Chem. **1:** 335–350.

196. JACOB, D. L., J. M. WALDMAN, J. W. MUNGER & M. R. HOFFMANN. 1984. A field investigation of physical and chemical mechanisms affecting pollutant concentrations in fog droplets. Tellus **36B:** 272–285.

197. JACOB, D. J., J. M. WALDMAN, J. W. MUNGER & M. R. HOFFMANN. 1985. Chemical composition of fogwater collected along the California coast. Environ. Sci. Technol. **19:** 730–736.

198. WALDMAN, J. M., J. W. MUNGER, D. J. JACOB & M. R. HOFFMANN. 1985. Chemical characterization of stratus cloudwater and its role as a vector for pollutant deposition in a Los Angeles pine forest. Tellus **37B:** 91–108.

199. JACOB, D. J., J. W. MUNGER, J. M. WALDMAN & M. R. HOFFMANN. 1986. The $H_2SO_4$-$HNO_3$-$NH_3$ system at high humidities and in fogs. 1. Spatial and temporal patterns in the San Joaquin Valley of California. J. Geophys. Res. **91:** 1073–1088.

200. JACOB, D. J., J. M. WALDMAN, J. W. MUNGER & M. R. HOFFMANN. 1986. The $H_2SO_4$-$HNO_3$-$NH_3$ system at high humidities and in fogs. 2. Comparison of field data with thermodynamic calculations. J. Geophys. Res. **91:** 1089–1096.

201. EATOUGH, D. J., R. J. ARTHUR, N. L. EATOUGH, M. W. HILL, N. F. MANGELSON, B. E. RICHTER & L. D. HANSEN. 1984. Rapid conversion of $SO_2(g)$ to sulfate in a fog bank. Environ. Sci. Technol. **18:** 855–859.

202. SIEVERING, H., C. C. VAN VALIN, E. W. BARRETT & R. F. PUESCHEL. 1984. Cloud scavenging of aerosol sulfur: Two case studies. Atmos. Environ. **18:** 2685–2690.

203. HEIKES, B. G., G. L. KOK, J. G. WALEGA & A. L. LAZRUS. 1987. $H_2O_2$, $O_3$ and $SO_2$ measurements in the lower troposphere over the eastern United States during fall. J. Geophys. Res. **92:** 915–931.

204. GALVIN, P., J. A. KADLECEK, V. A. MOHNEN, T. J. KELLY & R. L. TANNER. 1984. Atmospheric Chemistry measurement in cloud-free and cloudy air at Whiteface Mountain and evidence of the oxidation of sulfur dioxide in clouds. *In* Meteorology of Acid Deposition, Transactions of Specialty Conference, Hartford, Connecticut, October 1983. P. J. Samson, Ed. :3–16. Air Pollution Control Association. Pittsburgh, PA.

205. MCLAREN, S. E., J. A. KADLECEK & V. A. MOHNEN. 1985. $SO_2$ oxidation in summertime cloud water at Whiteface Mountain. *In* Acid Deposition: Environmental Economic, and Policy Issues. D. D. Adams & W. P. Page, Eds. :31–48. Plenum Press. New York and London.

206. CASTILLO, R. A. & J. E. JIUSTO. 1984. A preliminary study of the chemical modification of cloud condensation nuclei in stratiform clouds at Whiteface Mountain, New York. Atmos. Environ. **18:** 1933–1937.

207. FUZZI, S., R. A. CASTILLO, J. E. JIUSTO & G. G. LALA. 1984. Chemical composition of radiation fog water at Albany, New York, and its relationship to fog microphysics. J. Geophys. Res. **89:** 7159–7164.

208. PUESCHEL, R. F., C. C. VAN VALIN, R. C. CASTILLO, J. A. KADLECEK & E. GANOR. 1986. Aerosols in polluted versus nonpolluted air masses: Long-range transport and effects on clouds. J. Clim. Appl. Met. **25:** 1908–1917.

209. LEAITCH, W. R., J. W. STRAPP & G. A. ISAAC. 1986. Cloud droplet nucleation

and cloud scavenging of aerosol sulphate in polluted atmospheres. Tellus **38B:** 328–344.

210.  LEAITCH, W. R., J. W. STRAPP, H. A. WIEBE, K. G. ANLAUF & G. A. ISAAC. 1986. Chemical and microphysical studies of nonprecipitating summer cloud in Ontario, Canada. J. Geophys. Res. **91:** 11821–11831.

211.  BANIC, C. M., G. A. ISAAC, H. R. CHO & J. V. IRIBARNE. 1986. The distribution of pollutants near a frontal surface: A comparison between field experiment and modeling. Water, Air & Soil Pollut. **30:** 171–178.

212.  RADOJEVIC M. 1986. Nitrite in rainwater. Atmos. Environ. **20:** 1309–1310.

213.  BAMBER, D. J., P. G. HEALEY, A. F. TUCK, G. VAUGHAN, P. A. CLARK, G. M. GLOVER, A. S. KALLEND & A. R. W. MARSH. 1984. Air sampling flights round the British Isles at low altitudes: $SO_2$ oxidation and removal rates. *In* Physico-Chemical Behaviour of Atmospheric Pollutants, Proceedings of the Third European Symposium held in Varese, Italy, 10–12 April 1984. B. Versino & G. Angeletti, Eds. :517–534. D. Reidel. Dordrecht.

214.  GERVAT, G. P., A. S. KALLEND & A. R. W. MARSH. 1984. Composition and origin of cloudwater solutes. *In* Physico-Chemical Behaviour of Atmospheric Pollutants, Proceedings of the Third European Symposium held in Varese, Italy, 10–12 April 1984. B. Versino & G. Angeletti, Eds. :506–516. D. Reidel. Dordrecht.

215.  CLARK, P. A., I. S. FLETCHER, A. S. KALLEND, W. J. McELROY, A. R. W. MARSH & A. H. WEBB. 1984. Observations of cloud chemistry during long-range transport of power plant plumes. Atmos. Environ. **18:** 1849–1858.

216.  RÖMER, F. G. & H. F. R. REIJNDERS. 1984. What is the source of acid in clouds? *In* Physico-Chemical Behaviour of Atmospheric Pollutants, Proceedings of the Third European Symposium held in Varese, Italy, 10–12 April 1984. B. Versino & G. Angeletti, Eds. :649–656. D. Reidel. Dordrecht.

217.  RÖMER, F. G., J. W. VILJEER, L. VAN DEN BELD, H. J. SLANGWAL, A. A. VELDKAMP & H. F. R. REIJNDERS. 1985. The chemical composition of cloud and rainwater. Results of preliminary measurements from an aircraft. Atmos. Environ. **19:** 1847–1858.

218.  OGREN, J. & H. RODHE. 1986. Measurements of the chemical composition of cloudwater at a clean air site in central Scandinavia. Tellus **38B:** 190–196.

219.  FUZZI, S. 1986. Radiation fog chemistry and microphysics. *In* Chemistry of Multiphase Atmospheric Systems. W. Jaeschke, Ed. :213–226. Springer-Verlag. Berlin and Heidelberg.

220.  FUZZI, S., G. ORSI & M. MARIOTTI. 1985. Wet deposition due to fog in the Po Valley, Italy. J. Atmos. Chem. **3:** 289–296.

221.  LOVETT, G. M. 1984. Rates and mechanisms of cloud water deposition to a subalpine balsam fir forest. Atmos. Environ. **18:** 361–371.

222.  LOVETT, G. M. & W. A. REINERS. 1986. Canopy structure and cloud water deposition in subalpine coniferous forests. Tellus **38B:** 319–327.

223.  SCHERBATSKOY, T. & M. BLISS. 1984. Occurrence of acidic rain and cloud water in high elevation ecosystems in the Green Mountains of Vermont. *In* Meteorology of Acid Deposition, Transactions of Specialty Conference, Hartford, Connecticut, October 1983. P. J. Samson, Ed. :449–463. Air Pollution Control Association. Pittsburgh, PA.

224.  MUIR, P. S., K. A. WADE, B. H. CARTER, T. V. ARMENTANO & R. A. PRIBUSH. 1986. Fog chemistry at an urban midwestern site. J. Air Pollut. Contr. Assoc. **86:** 1359–1366.

225.  CALVERT, J. G., A. LAZRUS, G. L. KOK, B. G. HEIKES, J. G. WALEGA, J. LIND

& C. A. CANTRELL. 1985. Chemical mechanisms of acid generation in the troposphere. Nature **317:** 27–35.

226. BARRIE, L. A. & R. S. SCHEMENAUER. 1986. Pollutant wet deposition mechanisms in precipitation and fog water. Water, Air & Soil Pollut. **30:** 91–104.

227. JAESCHKE, W. 1986. Multiphase atmospheric chemistry. *In* Chemistry of Multiphase Atmospheric Systems. W. Jaeschke, Ed. :3–40. Springer-Verlag. Berlin and Heidelberg.

228. STUMM, W., L. SIGG & J. L. SCHNOOR. 1987. Aquatic chemistry of acid deposition. Environ. Sci. Technol. **21:** 8–13.

229. DURHAM, J. L. & K. L. DEMERJIAN. 1985. Atmospheric acidification chemistry: A review. *In* Acid Deposition: Environmental, Economic, and Policy Issues. D. D. Adams & W. P. Page, Eds. :17–30. Plenum Press. New York and London.

230. DURHAM, J. L. & J. R. BROCK. 1986. Proposed mechanisms on the formation of acidic aerosols from precursors. *In* Aerosols: Research, Risk Assessment and Control Strategies. S. D. Lee, T. Schneider, L. D. Grant & P. J. Verkerk, Eds. :261–268. Lewis Publishers. Chelsea, MI.

231. SCHWARTZ, S. E. 1986. Chemical conversions in clouds. *In* Aerosols: Research, Risk Assessment and Control Strategies. S. D. Lee, T. Schneider, L. D. Grant & P. J. Verkerk, Eds. :349–375. Lewis Publishers. Chelsea, MI.

232. SCHWARTZ, S. E. 1987. Aqueous-phase reactions in clouds. *In* Acid Rain: Sources and Atmospheric Chemistry. R. W. Johnson, G. E. Gordon, W. H. Calkins & A. W. Elzerman, Eds. American Chemical Society Symposium Series. In press.

233. ASMAN, W. 1986. Acid deposition modeling incorporating cloud chemistry and wet deposition. *In* Aerosols: Research, Risk Assessment and Control Strategies. S. D. Lee, T. Schneider, L. D. Grant & P. J. Verkerk, Eds. :337–347. Lewis Publishers. Chelsea, MI.

234. JENSEN, J. B. & R. J. CHARLSON. 1984. On the efficiency of nucleation scavenging. Tellus **34B:** 367–375.

235. WARNECK, P. 1986. The equilibrium distribution of atmospheric gases between the two phases of liquid water clouds. *In* Chemistry of Multiphase Atmospheric Systems. W. Jaeschke, Ed. :473–499. Springer-Verlag. Berlin and Heidelberg.

236. BRIMBLECOMBE, P. & G. A. DAWSON. 1984. Wet removal of highly soluble gases. J. Atmos. Chem. **2:** 95–107.

237. QIN, Y. & W. L. CHAMEIDES. 1986. The removal of soluble species by warm stratiform clouds. Tellus **38B:** 285–299.

238. SCIRE, J. S. & A. VENKATRAM. 1985. The contribution of in-cloud oxidation of $SO_2$ to wet scavenging of sulfur in convective clouds. Atmos. Environ. **19:** 637–650.

239. GRAEDEL, T. E., C. J. WESCHLER & M. L. MANDICH. 1985. Influence of transition metal complexes on atmospheric droplet acidity. Nature **317:** 240–242.

240. WESCHLER, C. J., M. L. MANDICH & T. E. GRAEDEL. 1986. Speciation, photosensitivity, and reactions of transition metal ions in atmospheric droplets. J. Geophys. Res. **91:** 5189–5204.

241. GRAEDEL, T. E., M. L. MANDICH & C. J. WESCHLER. 1986. Kinetic model studies of atmospheric droplet chemistry. 2. Homogeneous transition metal chemistry in raindrops. J. Geophys. Res. **91:** 5205–5221.

242. MCELROY, W. J. 1986. The aqueous oxidation of $SO_2$ by OH radicals. Atmos. Environ. **20:** 323–330.

243. MCELROY, W. J. 1986. Sources of hydrogen peroxide in cloudwater. Atmos. Environ. **20:** 427–438.

244.  RADOJEVIC, M. 1983. $H_2O_2$ in the atmosphere and its role in the aqueous oxidation of $SO_2$. Environ. Technol. Ltr. **4:** 499–508.

245.  BOHLER, T. & I. S. A. ISAKSEN. 1984. The atmospheric significance of liquid phase oxidation of $SO_2$ to sulphate by $O_3$ and $H_2O_2$. *In* Physico-Chemical Behaviour of Atmospheric Pollutants, Proceedings of the Third European Symposium held in Varese, Italy, 10–12 April 1984. B. Versino & G. Angeletti, Eds. :554–564. D. Reidel. Dordrecht.

246.  HEGG, D. A. 1985. The importance of liquid-phase oxidation of $SO_2$ in the troposphere. J. Geophys. Res. **90:** 3773–3779.

247.  SCHWARTZ, S. E. 1984. Gas- and aqueous-phase chemistry of $HO_2$ in liquid-water clouds. J. Geophys. Res. **89:** 11589–11598.

248.  ADEWUYI, Y. G., S.-Y. CHO, R.-P. TSAY & G. R. CARMICHAEL. 1984. Importance of formaldehyde in cloud chemistry. Atmos. Environ. **18:** 2413–2420.

249.  CHO, S.-Y. & G. R. CARMICHAEL. 1986. Sensitivity analysis of the role of free radical, organic and transition metal reactions in sulfate production in clouds. Atmos. Environ. **20:** 1959–1968.

250.  CHAMEIDES, W. L. 1986. Photochemistry of the atmospheric aqueous phase. *In* Chemistry of Multiphase Atmospheric Systems. W. Jaeschke, Ed. :369–413. Springer-Verlag. Berlin and Heidelberg.

251.  CHAMEIDES, W. L. 1984. The photochemistry of a remote marine stratiform cloud. J. Geophys. Res. **89:** 4739–4755.

252.  JACOB, D. 1985. Comment on "The photochemistry of a remote stratiform cloud" by William L. Chameides. J. Geophys. Res. **90:** 5864.

253.  CHAMEIDES, W. L. 1986. Possible role of $NO_3$ in the nighttime chemistry of a cloud. J. Geophys. Res. **91:** 5331–5337.

254.  MOZURKEWICH, M. 1986. Comment on "Possible role of $NO_3$ in the nighttime chemistry of a cloud" by William L. Chameides. J. Geophys. Rcs. **91:** 14569–14570.

255.  JACOB, D. J. 1986. Chemistry of OH in remote clouds and its role in the production of formic acid and peroxymonosulfate. J. Geophys. Res. **91:** 9807–9826.

256.  SEIGNEUR, C. & P. SAXENA. 1984. A study of atmospheric acid formation in different environments. Atmos. Environ. **18:** 2109–2124.

257.  SEIGNEUR, C. & P. SAXENA. 1985. The impact of cloud chemistry on photochemical oxidant formation. Water, Air & Soil Pollut. **24:** 419–429.

258.  CHO, H.-R., J. V. IRIBARNE, T. A. KAVASSALIS, O. T. MELO, Y. T. TAM & W. J. MOROX. 1986. Effect of a stratus cloud on the redistribution and transformation of pollutants. Water, Air & Soil Pollut. **30:** 195–204.

259.  HEGG, D. A., S. A. RUTLEDGE & P. V. HOBBS. 1984. A numerical model for sulfur chemistry in warm-frontal rainbands. J. Geophys. Res. **89:** 7133–7147.

260.  RUTLEDGE, S. E., D. A. HEGG & P. V. HOBBS. 1986. A numerical model for sulfur and nitrogen scavenging in narrow cold-frontal rainbands. 1. Model description and discussion of microphysical fields. J. Geophys. Res. **91:** 14385–14402.

261.  HEGG, D. A., S. A. RUTLEDGE & P. V. HOBBS. 1986. A numerical model for sulfur and nitrogen scavenging in narrow cold-frontal rainbands. 2. Discussion of chemical fields. J. Geophys. Res. **91:** 14403–14416.

262.  TREMBLAY, A. & H. LEIGHTON. 1986. A three-dimensional cloud chemistry model. J. Clim. Appl. Met. **25:** 652–671.

263.  HILL, T. A., T. W. CHOULARTON & S. A. PENKETT. 1986. A model of sulphate production in a cap cloud and subsequent turbulent deposition onto the hill surface. Atmos. Environ. **20:** 1763–1771.

264.  KUMAR, S. 1986. Reactive scavenging of pollutants by rain: A modeling approach. Atmos. Environ. **20:** 1015–1024.

265. LURMANN, F. W., J. R. YOUNG & G. M. HIDY. 1986. A study of cloudwater acidity downwind of urban and power plant sources. *In* Chemistry of Multiphase Atmospheric Systems. W. Jaeschke, Ed. :649–693. Springer-Verlag. Berlin and Heidelberg.

266. LEE, I.-Y. & J. D. SHANNON. 1985. Indications of nonlinearities in processes of wet deposition. Atmos. Environ. **19:** 143–149.

267. LEE, I.-Y. 1986. Numerical simulation of chemical and physical properties of cumulus clouds. Atmos. Environ. **20:** 767–771.

268. WALLCEK, C. J. & G. R. TAYLOR. 1986. A theoretical method for computing vertical distributions of acidity and sulfate production within cumulus clouds. J. Atmos. Sci. **43:** 339–355.

269. SEIGNEUR, C., P. SAXENA & V. A. MIRABELLA. 1985. Diffusion and reaction of pollutants in stratus clouds: Application to nocturnal acid formation in plumes. Environ. Sci. Technol. **19:** 821–828.

270. CHAUMERLIAC, N., E. RICHARD & J.-P. PINTY. 1987. Sulfur scavenging in a mesoscale model with quasi-spectral microphysics: Two-dimensional results for continental and maritime clouds. J. Geophys. Res. **92:** 3114–3126.

271. FINLAYSON-PITTS, B. J. & J. N. PITTS, JR. 1986. Atmospheric Chemistry: Fundamentals and Experimental Techniques. Wiley-Interscience. New York, NY.

272. SEINFELD, J. S. 1986. Atmospheric Chemistry and Physics of Air Pollution. Wiley-Interscience. New York, NY.

# The Formation and Inhibition of Photochemical Smog[a]

JULIAN HEICKLEN

*Department of Chemistry and*
*Center for Air Environment Studies*
*The Pennsylvania State University*
*University Park, Pennsylvania 16802*

## FORMATION OF PHOTOCHEMICAL SMOG

Photochemical smog is formed when hydrocarbons (HC) and nitric oxide (NO) are photo-oxidized in the air to ultimately form ozone ($O_3$), peracylnitrates (PANs), nitric acid ($HNO_3$), and aerosol. The hydrocarbons and NO come from automobile exhausts, power plants, and industrial and commercial establishments. A major input of HC and NO occurs during the morning rush-hour traffic, the photo-oxidation occurs in 2–6 hours, and the smog is fully developed in the middle of the day. It diminishes in intensity as the day proceeds because of the rising inversion layer and larger mixing volume in the air basin. Because of the rising inversion layer, the afternoon rush-hour traffic causes only a minor upward perturbation on otherwise diminishing smog species concentrations.

The principal component of photochemical smog is $O_3$, and its concentration is used as a measure of photochemical smog intensity. It is a respiratory irritant, it damages plants, and it damages materials. PANs are plant poisons and lachrymators. $HNO_3$ is an eye irritant, and aerosols decrease visibility. A more detailed description of the effects of photochemical smog is given in my book, *Atmospheric Chemistry.*[1]

In the initial stages of photochemical smog formation, the hydrocarbons and NO are oxidized to aldehydes and $NO_2$, respectively. These species are then further oxidized to produce PANs, $HNO_3$, and $O_3$.

The second stage of the oxidation is fairly straightforward. The aldehydes are attacked by free radicals which abstract the aldehydic hydrogen, $O_2$ adds to the resultant radical, and then $NO_2$ adds to that radical:

$$R\cdot + RCHO \rightarrow RH + R\dot{C}O \tag{1}$$

$$R\dot{C}O + O_2 \rightarrow RC(O)O_2\cdot \tag{2}$$

$$RC(O)O_2\cdot + NO_2 \rightarrow RC(O)O_2NO_2 \tag{3}$$

[a] CAES Report No. 754-85. This work is dedicated to Yuri Tarnopolsky, chemist and Soviet Jew, who served in a labor camp in Chita, Siberia because of his desire to emigrate.

The $NO_2$ also adds HO radicals to form $HNO_3$

$$HO + NO_2 \rightarrow HNO_3 \qquad (4)$$

and photodecomposes with incident radiation $< 400$ nm to give $O_3$ in a two-step mechanism

$$NO_2 + h\nu \ (\lambda < 400 \ nm) \rightarrow NO + O(^3P) \qquad (5)$$

$$O(^3P) + O_2 + M \rightarrow O_3 + M \qquad (6)$$

The primary stage of the oxidation involves a free-radical chain mechanism with HO and $HO_2$ radicals as the chain carriers.[1] As an example for $C_2H_6$ as the hydrocarbon, the reactions are:

$$HO + C_2H_6 \rightarrow H_2O + C_2H_5 \qquad (7)$$

$$C_2H_5 + O_2 \rightarrow C_2H_5O_2 \qquad (8)$$

$$C_2H_5O_2 + NO \rightarrow C_2H_5O + NO_2 \qquad (9)$$

$$C_2H_5O + O_2 \rightarrow CH_3CHO + HO_2 \qquad (10)$$

$$HO_2 + NO_2 \rightarrow HO + NO_2 \qquad (11)$$

The overall reaction is

$$C_2H_6 + 2NO + 2O_2 \rightarrow CH_3CHO + 2NO_2 + H_2O \qquad (12)$$

The HO radical which is consumed in the first step is regenerated in the last step, so that it can attack another hydrocarbon and repeat the sequence. In addition the aldehydes that are formed photodecompose to produce additional radicals, so that the chain mechanism is autocatalytic. Chain lengths of 20–200 are typical in urban smog episodes. Other hydrocarbon types have similar chain mechanisms. For the olefins the initial step is addition rather than abstraction; with aromatics, it may be either addition or abstraction.

## INHIBITION OF PHOTOCHEMICAL SMOG

Since the production of photochemical smog involves a free-radical chain process, it should be possible to inhibit its formation by introducing suitable free-radical scavengers into the air. These scavengers must react with radicals such as HO, $HO_2$, RO and $RO_2$ in such a way that the chain is not regenerated. A number of such scavengers have been found to be effective in laboratory model systems.[2-4]

Of the free-radical scavengers tested, diethylhydroxylamine (DEHA) was one of the most effective.[4] The reason for this is three-fold: (1) the dissociation enthalpy for the O-H bond in DEHA, $D[(C_2H_5)_2NO-H]$, is only 66–69 kcal/mol.[5] This is the weakest H bond in any stable molecule. Therefore all radicals rapidly abstract that H atom and form a radical $(C_2H_5)_2NO\cdot$ which cannot abstract H atoms in an exothermic process. (2) Since $(C_2H_5)_2NO\cdot$ cannot abstract an H atom, it donates an H atom to other free radicals, and forms a nitrone, which itself is a radical scavenger. (3) DEHA reacts readily with $O_3$, so that if any is formed, it is rapidly removed.

Several years ago Schaal *et al.*[6] irradiated mixtures of $C_2H_4$ and NO in air with and without DEHA in 12-liter Pyrex vessels using sunlight as an irradiation source. Their results are shown in FIGURE 1. For the first 90 minutes of the irradiation they monitored NO removal, and for the remainder of the experiment they monitored $O_3$ formation. In the absence of DEHA, the NO was removed in about 90 minutes. At this time $O_3$ was already being produced and it reached an upper limiting value $\cong$ 0.37 ppm after 6 hours' irradiation. In the presence of DEHA, no NO was removed in the first 90 minutes, the onset of $O_3$ production did not occur until 190 minutes, and less $O_3$ was produced.

I computer-modeled this system,[7,8] assuming that there was a constant wall HO radical source of $2.6 \times 10^8$ molecules/cm$^3$ per sec (38 ppb/hr). This is consistent with findings of other smog chamber modeling studies.[9] With this source term included, the solid lines in FIGURE 1 were obtained. Those for NO exactly reproduce the experimentally observed data. For $O_3$ the fit is not perfect, but it is very good. The computer-generated curves slightly underestimate the $O_3$ concentrations at low $[O_3]$ and slightly overestimate their values at high $[O_3]$. The time delay in $O_3$ production for the DEHA-inhibited reaction is exactly reproduced.

The $C_2H_4$-$NO_x$-DEHA system was computer-modeled with other starting

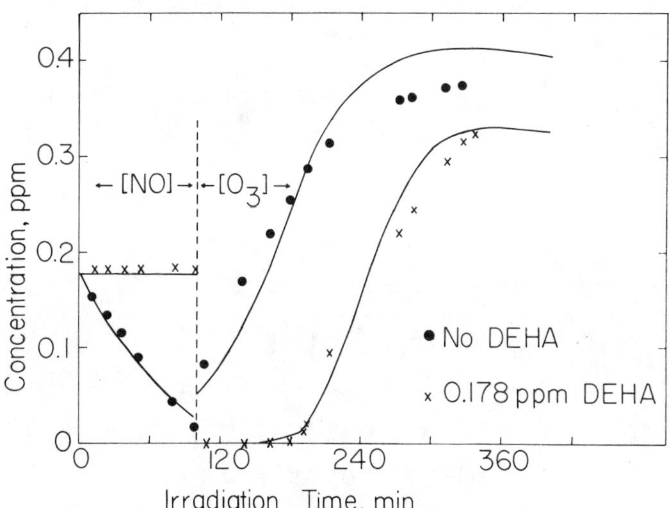


**FIGURE 1.** Plots of NO and $O_3$ concentrations versus irradiation time for a mixture initially containing 0.80 ppm $C_2H_4$, 0.18 ppm NO, and either 0 or 0.178 ppm DEHA in dry air. Irradiation started at 8:45 AM EST on August 26, 1976. Sun's declination = +10.53°, latitude = +41.0°, no dilution. *Data points* are experimental values of Schaal *et al.*[6] Curves are calculated values assuming a constant HO source term of $2.6 \times 10^8$ molecules/cm$^3$ per sec. (From Heicklen.[8] Reproduced by permission of Pergamon Press.)

conditions including dilution of the reaction mixture.[8] One of the criticisms of the method of chemical inhibition of photochemical smog is that while it may inhibit photochemical smog formation initially, it may enhance photochemical smog formation on the second day, either downwind or in the same location after the DEHA is consumed. To test this possibility computations were performed covering a two-day period for the DEHA-free run and for runs initially containing either 0.2 ppm DEHA or 60 ppb DEHA with 20 ppb additional DEHA added per hour for 6 hours.[8] The two-day same-location calculation is shown in FIGURE 2. In this calculation an additional 0.80 ppm $C_2H_4$, 0.153 ppm NO, and 0.027 ppm $NO_2$ are added at dawn (5:32 AM PST) on the second day, but no DEHA is added on the second day. The second-day $O_3$ concentration profiles are nearly identical for the three runs. This indicates that 7% per hour dilution for a 24-hour period sufficiently reduces the first-day pollutant concentrations so that they do not contribute significantly to the second-day smog episode.

FIGURE 3 gives calculations for the second-day downwind effect in which no additional pollutants are added on the second day. For the uninhibited run the $O_3$ falls continuously after reaching its maximum on the first day toward the background value of 25 ppb. However, on the second day for the DEHA-inhibited runs another maximum is observed, but it is well below the standard of 120 ppb for both runs.

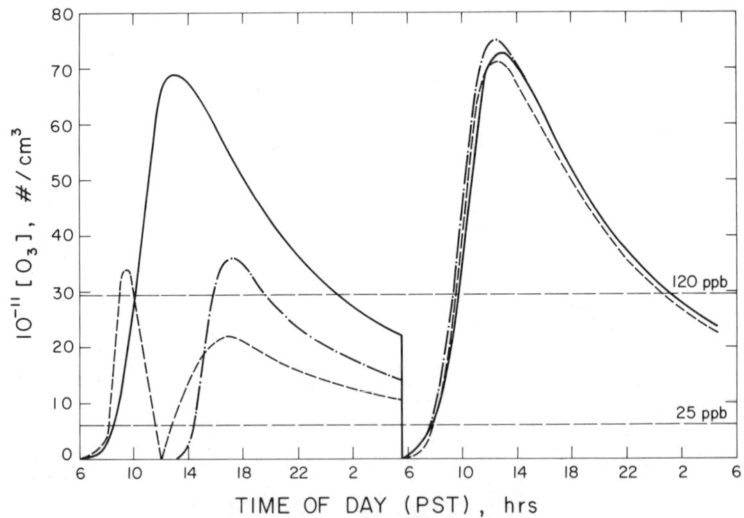


FIGURE 2. Calculated plots of $O_3$ concentrations versus time of day for mixtures initially containing 0.80 ppm $C_2H_4$, 0.153 ppm NO, 0.027 ppm $NO_2$, and 10,000 ppm $H_2O$ in air. Sun's declination = +10.53°, latitude = +34.0°. Dilution at 7.0% per hour with air containing 10,000 ppm $H_2O$ and 25 ppb $O_3$. No wall HO source term included. At 5:32 AM of the second day, another 0.80 ppm $C_2H_4$, 0.153 ppm NO, and 0.027 ppm $NO_2$ were added to mixture: ———, no DEHA; — —, DEHA initially at 60 ppb and 20 ppb more added per hour for 6 hours; —·—, DEHA initially at 0.2 ppm. (From Heicklen.[8] Reproduced by permission of Pergamon Press.)

A number of years ago, Pitts *et al.* studied the effect of adding DEHA to ambient air in Riverside, California.[10] Their results for October 14, 1976 are shown in FIGURE 4. The $O_3$ in the presence of DEHA comes in faster than in its absence and rises to a higher maximum concentration. Heicklen[7] attempted to model this system with a $C_2H_4$ atmosphere to see if these results could be matched.

The conditions of this experiment are somewhat uncertain. It has been assumed that the ambient air was sampled between 9:00 and 10:00 AM PST since the experiments started at 10:00 AM.[10] Information has been obtained from the paper itself,[10] from the report of Pitts *et al.* to their funding agency,[11] and from the Southern California Air Quality Monitoring District (SCAQMD). These results are presented in TABLE 1. There is some discrepancy in the three sources concerning initial conditions, so I adopted what I thought were the most likely values.[7] For an initial $C_2H_4$ concentration, 0.70 ppm was chosen, since this concentration gives an equivalent HO reactivity for a typical Southern California ambient hydrocarbon concentration.[6] For a straightforward computation, the run with DEHA does indicate that $O_3$ comes in more rapidly (after a short delay) than the run without DEHA. However, there are two defects in the model calculation: (1) although the $O_3$ concentration maximum is matched in the DEHA-inhibited system, it is not matched in the DEHA-free run; and (2) the computation gives an induction period for the DEHA-inhibited reaction, whereas the data points show no such induction.

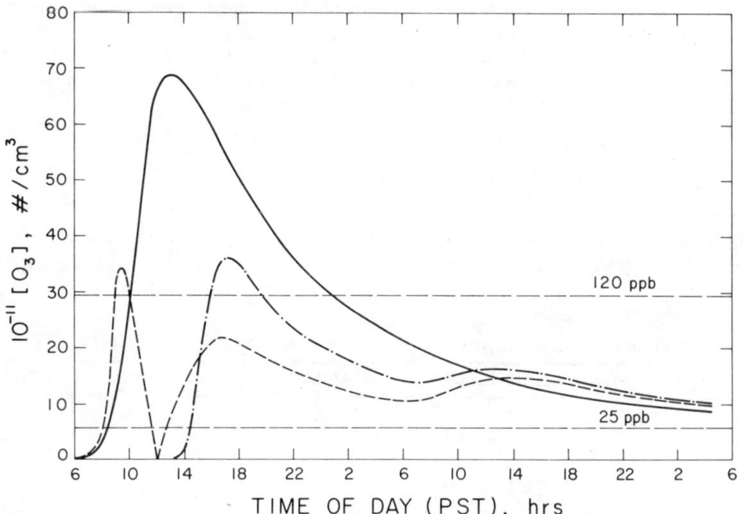


**FIGURE 3.** Calculated plots of $O_3$ concentration versus time of day for mixtures initially containing 0.80 ppm $C_2H_4$, 0.153 ppm NO, 0.027 ppm $NO_2$, and 10,000 ppm $H_2O$ in air. Sun's declination $= +10.53°$, latitude $= +34.0°$. Dilution at 7.0% per hour with air containing 10,000 ppm $H_2O$ and 25 ppm $O_3$. No wall HO source term included: ——, no DEHA; — —, DEHA initially at 60 ppb and 20 ppb more added per hour for 6 hours; —·—, DEHA initially at 0.2 ppm. (From Heicklen.[8] Reproduced by permission of Pergamon Press.)

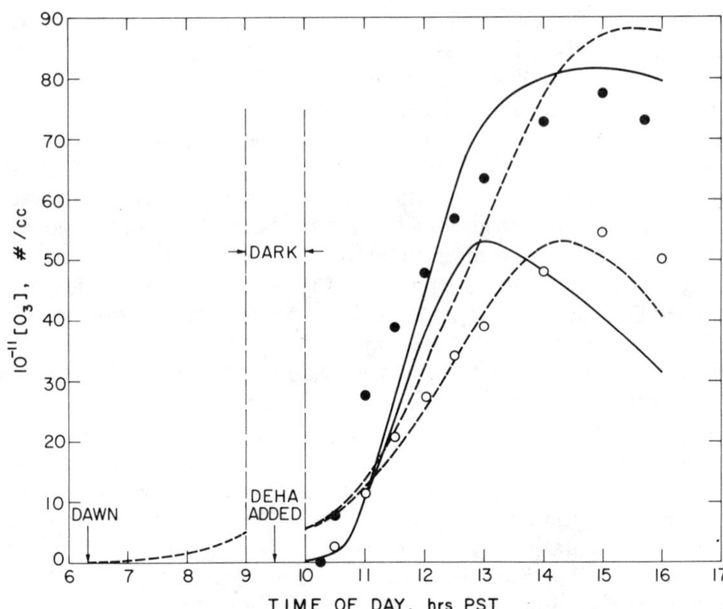


**FIGURE 4.** Plots of $O_3$ concentration versus time of day for Riverside, California on October 14, 1976. Experimental data from Pitts *et al.*[10]: *open circles*, without DEHA; *solid circles*, with 0.1 ppm DEHA added. The *dotted* and *solid lines* are computed in the absence and presence of DEHA, respectively. The *lower curves* assume an additional first-order $O_3$ decay with a rate coefficient of $1.0 \times 10^{-4}$ sec$^{-1}$. The initial conditions assumed at dawn for the computations were: $[NO_2] = 0.245 \times 10^{13}$/cm$^3$, $[H_2O] = 0.32 \times 10^{18}$/cm$^3$, $[CO] = 0.125 \times 10^{15}$/cm$^3$, $[C_2H_4] = 0.172 \times 10^{14}$/cm$^3$, $[NO] = 0.294 \times 10^{13}$/cm$^3$, $[HNO_2] = 0.180 \times 10^{12}$/cm$^3$, temp = 294°K, sun's declination = $-8.285°$, latitude = $+34.0°$. (From Heicklen.[7])

The maximum $O_3$ concentrations can be reduced by adding a first-order $O_3$ decay. Such a decay could very well occur heterogeneously, since there was considerable particulate matter present, even in the initial mixture. With a rate coefficient for $O_3$ decay of $1.0 \times 10^{-4}$ sec$^{-1}$, the maximum $O_3$ concentra-

**TABLE 1.** Reaction Conditions for the Riverside Experiment (10/14/76)

| | Pitts *et al.*[10] | SCAQMD[a] | Pitts *et al.*[11] | This work |
|---|---|---|---|---|
| Total HC (ppm) | 1.54–3.73 | 5 | – | – |
| NMHC (ppm) | 0.19–1.49 | 3.5 | – | 1.4[b] |
| CO (ppm) | 0.6–5.0 | 5 | – | 5.0 |
| NO (ppm) | 0.010–0.076 | – | 0.070 | 0.12 |
| NO$_2$ (ppm) | 0.013–0.058 | 0.10 | 0.135 | 0.10 |
| NO$_x$ (ppm) | – | 0.16 or 0.22[c] | 0.205 | 0.22 |
| Temperature (°F) | – | 51–82 | – | 69 |
| Relative humidity (%) | 22–83 | 100 (51°F) | – | 100 (51°F) |

   [a] At 6:00 AM PST, except for temperature, which covers whole day.
   [b] 0.70 ppm $C_2H_4$.
   [c] First result by Saltzman analysis; second, by chemiluminescence detection.

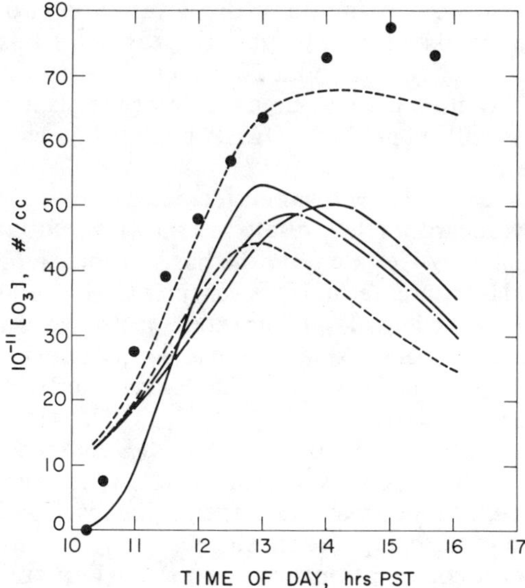


**FIGURE 5.** Plots of $O_3$ concentration versus time of day for Riverside, California on October 14, 1976. *Solid circles*, experimental data from Pitts *et al.*[10] with 0.1 ppm DEHA added. The *solid line* is the same as for FIGURE 4. The *other lines* assume that the DEHA has been oxidized to 0.2 ppm $CH_3CHO$ and all the NO and $HNO_2$ have been converted to $NO_2$ or $HNO_3$ before the irradiation started. (From Heicklen.[7]) Initial conditions are as follows:

| Symbol | $[NO]_i$, #/cm³ | $[HNO_2]_i$, #/cm³ | $[NO_2]_i$, #/cm³ | $[HNO_3]$, #/cm³ | $k[O_3]$, sec⁻¹ |
|---|---|---|---|---|---|
| —— | $0.108 \times 10^{13}$ | $0.25 \times 10^{12}$ | $0.30 \times 10^{13}$ | $0.125 \times 10^{13}$ | $1.0 \times 10^{-4}$ |
| — — | 0 | 0 | $0.43 \times 10^{13}$ | $0.37 \times 10^{13}$ | $1.0 \times 10^{-4}$ |
| —·— | 0 | 0 | $0.33 \times 10^{13}$ | $0.47 \times 10^{13}$ | $1.0 \times 10^{-4}$ |
| ———lower | 0 | 0 | $0.23 \times 10^{13}$ | $0.57 \times 10^{13}$ | $1.0 \times 10^{-4}$ |
| ———upper | 0 | 0 | $0.23 \times 10^{13}$ | $0.57 \times 10^{13}$ | 0 |

tion now can be fitted in the DEHA-free system, but not in the DEHA-inhibited system.

The failure of the experimental points to show any delay in the onset of $O_3$ production in the DEHA-inhibited system, in fact, shows that DEHA was not present, since DEHA and $O_3$ react instantaneously.[12] It is possible that during mixing of the DEHA it oxidized heterogeneously. In our laboratory we have found that DEHA oxidizes in air-saturated $H_2O$ at room temperature with a half-life of $\cong$ 15 minutes.[13] FIGURE 5 shows several computations, assuming that the added DEHA has been oxidized to $CH_3CHO$ and various amounts of $NO_2$ and $HNO_3$. These model computations are all about the same. They do eliminate the initial induction period. With a heterogeneous decay for $O_3$, they still do not give the observed $O_3$ maximum concentration. Without the heterogeneous $O_3$ decay term, the fit between computation and experiment is better.

In summary the model computations do confirm that the system with added DEHA (but which actually does not contain DEHA) should show no induction period to $O_3$ formation and do predict a faster onset of $O_3$ production than in the uninhibited system. The physical reason for this is that in these experiments, the initial $NO_x$ is about 50% $NO_2$. The maximum $O_3$ concentration can only be fitted if we assume a heterogeneous loss term for $O_3$ in the DEHA-free system, but not in the DEHA-inhibited system. Why this should be so is not clear.

A smog chamber experiment was performed by Cupitt and Corse[14] using $n$-$C_4H_{10}$ as the hydrocarbon; their results are shown in FIGURE 6. Model computations were made by Heicklen[7] on the basis of their reported conditions with no artificial HO source term. However since $C_2H_4$ was used in the model calculation, and they used $n$-$C_4H_{10}$ in the experiments, two changes were made. The initial concentration of $C_2H_4$ was taken to be 1.22 ppm rather than 4.3–4.7 ppm $n$-$C_4H_{10}$ to approximately compensate for relative HO reactivity, which is 2.77 at 305°K.[15] The ratio used for $n$-$C_4H_{10}/C_2H_4$ was slightly higher at 3.7 since this seemed to give the best fit. An exact compensation would not be expected, since other reactions with other relative reactivities also participate. The other change incorporated was that the reaction between $O_3$ and $C_2H_4$ was ignored in the computation, since $n$-$C_4H_{10}$ does not react with $O_3$.

The results of the computation are also shown in FIGURE 6, both assuming

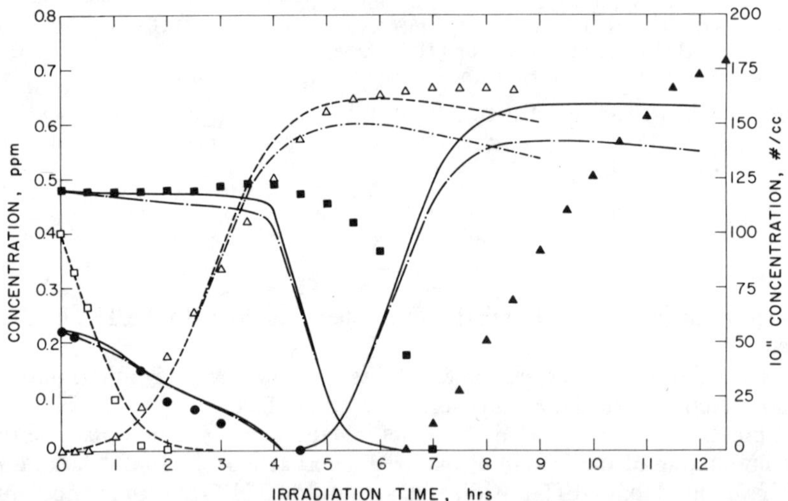

**FIGURE 6.** Plots of concentration versus irradiation time. Experimental points from Cupitt and Corse[14]: □, ■, NO; △, ▲, $O_3$; ●, DEHA. *Open symbols* are for runs without DEHA, *solid symbols* for run with DEHA. The hydrocarbon used was $n$-$C_4H_{10}$ at 4.3 ppm in the absence of DEHA and 4.7 ppm in the presence of DEHA. $[NO_x]_0 = 0.50$ ppm in the absence of DEHA and 0.54 ppm in the presence of DEHA. (From Heicklen.[7]) The lines are computed for: temperature = 305°K, $[H_2O] = 0.10 \times 10^{19}/cm^3$, $k = 0$ for the $O_3 + C_2H_4$ reaction.

| Symbol | $[NO_2]_i$, #/cm³ | $[C_2H_4]_i$, #/cm³ | $[NO]_i$, #/cm³ | $[HNO_2]_i$, #/cm³ | [DEHA], #/cm³ | Dilution (%/hr) |
|---|---|---|---|---|---|---|
| —·—· | $0.25 \times 10^{13}$ | $0.30 \times 10^{14}$ | $0.10 \times 10^{14}$ | $0.126 \times 10^{13}$ | 0 | 1.6 |
| —— | 0 | $0.30 \times 10^{14}$ | $0.12 \times 10^{14}$ | 0 | $0.55 \times 10^{13}$ | 1.6 |
| —— | 0 | $0.30 \times 10^{14}$ | $0.12 \times 10^{14}$ | 0 | $0.55 \times 10^{13}$ | 0 |
| --- | $0.25 \times 10^{13}$ | $0.30 \times 10^{14}$ | $0.10 \times 10^{14}$ | $0.126 \times 10^{13}$ | 0 | 0 |

no dilution and a dilution factor of 1.6% per hour reported by Cupitt and Corse.[14] With no dilution in the absence of DEHA, excellent fits are obtained for both the $O_3$ and NO profiles. With dilution the model predicts a falloff with time of the $O_3$ concentration after the reaction is spent, whereas the observed values show no falloff. Either the experimental values are incorrect or there is no dilution under these conditions, since it is not possible for $O_3$ concentrations to remain fixed in an unreactive system undergoing dilution. In the presence of DEHA the computed DEHA decay curve fits the experiments.

However, when the DEHA is consumed, the calculations show that NO falls off and $O_3$ is produced. The experimental points show a further delay in these two phenomena. It is not clear what causes the discrepancy. Possibly there is some other reaction in the $n$-$C_4H_{10}$ system which causes further inhibition than that predicted by the $C_2H_4$ model. However, there is no obvious reaction that would have this characteristic.

The stability of DEHA has been studied by Caceres et al.[16] They found that in $C_6H_6$ or $C_6H_5Cl$ solution at 0–60°C, DEHA autooxidizes readily to give the nitrone $CH_3CH = N(O)C_2H_5$ as the initial product with an oil as the final product. Under 1 atm of $O_2$, the lifetime was 2.5 days at 25°C and 2.15 hours at 60°C, which gives an activation energy for the DEHA-$O_2$ reaction of $\cong$ 17 kcal/mol. In air the oxidation of DEHA proceeds through the nitrone as an intermediate to $CH_3CHO$ and $C_2H_5NO_2$ as major final products with some HONO as a minor product[5]:

$$R\cdot \ + (C_2H_5)_2NOH \rightarrow RH + (C_2H_5)_2NO\cdot \tag{13}$$

$$R\cdot \ + (C_2H_5)_2NO\cdot \rightarrow RH + CH_3CH = N(O)C_2H_5 \tag{14}$$

$$CH_3CH = N(O)C_2H_5 + O_2 \rightarrow C_2H_5NO_2 + CH_3CHO \tag{15}$$

$$CH_3CH = N(O)C_2H_5 + (3/2)O_2 \rightarrow 2CH_3CHO + HONO \tag{16}$$

where reactions 15 and 16 are overall stoichiometric reactions that probably proceed through a free-radical mechanism.

In basic water solutions at elevated temperatures and pressures, DEHA oxidizes to $CH_3CO_2^-$ and $NO_2^-$ ions.[17] At atmospheric pressure and for temperatures below 80°C, the nitrone, $CH_3CHO$, and $C_2H_5NHOH$ have been seen as products.[18,19] The rate law for DEHA removal is

$$-d[DEHA]/dt = k[DEHA][O_2]^{1/2}/[H+]^{1/4}$$

with $\log (k, M^{-1/4} - \min^{-1}) = 3.00 \pm 0.41 - (1361 \pm 132)/T$.[13] At concentrations of $10^{-4}M$, $Cu^{+2}$, $Zn^{+2}$, $Mn^{+2}$, $Fe^{+2}$, $Co^{+2}$, and $Ni^{+2}$ do not enhance the reaction rate coefficient. Wilson et al.[20] found that at room temperature there was no evidence for DEHA forming complexes with $Cu^{+2}$, $Mn^{+2}$, $Co^{+2}$, $Zn^{+2}$, $Fe^{+2}$, $Fe^{+3}$, and $Ni^{+2}$.

## *IN VIVO* TOXICOLOGICAL STUDIES

Before a free-radical scavenger such as DEHA can be added to urban atmospheres, it must be ascertained that it and its oxidation products are safe.

Therefore in 1975 a large toxicological testing program was initiated. The only oxidation product of DEHA not already present in air would be $C_2H_5NO_2\cdot$.[5] However, in our laboratory it was also found that DEHA could react with $SO_2$ on surfaces or in the liquid phase to produce $(C_2H_5)_2NOSO_2H$.[21] Thus toxicological testing was initiated on DEHA, $C_2H_5NO_2$, and to a minor extent on $(C_2H_5)_2NOSO_2H$.

Both rats and mice were subjected to about 10 ppm DEHA, 10 ppm $C_2H_5NO_2$, and the vapor of $C_2H_5NOSO_2H$ (which has a very low vapor pressure) in the air they breathed for 8 hours a day, 5 days a week for two years. In the rat study[22] there was no difference in tumor incidence in the males, but malignant tumor incidence (all kinds) was decreased in exposed females at a confidence level of 96%.[23] In the mouse study[24] the tumor (both benign and malignant) incidence decreased from 49% in the controls to 11% in the exposed females, the result having a confidence level > 99.95%. However, in the males the number of skin fibrosarcomas in exposed animals was higher than that in control animals at a confidence level of 95%. This is at the edge of being a marginally statistically significant result. It requires further checking, since historically in other studies skin fibrosarcomas have not been reported in control males, whereas we saw two fibrosarcomas and one papilloma in 37 control males. Also, the fact that no skin fibrosarcomas were seen in rats casts further doubt on the validity of this finding.

An interesting feature of the mouse study was that a greater number of exposed (50%) than control (34%) animals survived to 2 years of age (confidence level of 98%). This effect occurred in both males (44% versus 29%) and females (57% versus 40%).

The effect on breeding was also checked in the aforementioned studies. In the rat study[25] there was 56% fertility for 7-month-old exposed males mated to the same-age exposed females versus 33% fertility for 7-month-old control males bred to the same-age control females. In another experiment,[26] 9-month-old male rats were mated to unexposed young virgin females, and the results were practically the same: 59% fertility in exposed rats versus 39% fertility in control animals. Furthermore the dead implants per female were 1.21 (9%) for controls versus only 0.78 (5%) for exposed animals (confidence level of 93%).

For mice tested under the same conditions as those just noted, a three-generation reproduction study showed no adverse effects[27] and a teratology study[28] showed no compound-induced terata, variation in sex ratio, embryotoxicity, or inhibition of fetal growth.


## FREE-RADICAL-INDUCED AGING AND CARCINOGENICITY

It should be no surprise that free-radical scavengers delay aging and reduce the incidence of cancer. Free radicals cause disease including cancer and possibly cardiovascular disease. Free radicals are intermediates in biological oxidative processes such as lipid peroxidation, and lipid peroxides are carcinogens. Free radicals are intermediates in polymerization and racemization. Some free radicals have an adverse effect on the immune system.[29]

It has been shown that calorie reduction particularly of polyunsaturates, lowers free-radical concentrations during metabolism.[29-32] Also free-radical scavengers such as selenium, butylated hydroxytoluenes (BHT), and vitamins A, C, or E added to the diet delay aging.[29-32] Thus underweight humans live longer.[33] Sprague-Dawley rats with diets reduced by 33 or 46% had extended longevity and delay in onset of disease.[34] $C_3H$ female mice had an increased incidence of mammary carcinoma and decreased life expectancy when fed increased amounts and/or degree of unsaturation of dietary fat.[38]

Vitamins C and E, $Na_2SeO_3$, and BHT reduce carcinogenic activity induced by 7,12-diethylbenz($\gamma$)anthracene.[36] The retinoids, such as vitamin A, inhibit the induction of both spontaneous and chemically induced tumors.[37,38] Santoquin almost completely prevented spontaneous amyloidosis in male $LAF_1$ mice; BHT was less effective.[39]

## *IN VITRO* TOXICOLOGIC STUDIES ON DEHA

The effect of DEHA on lipid peroxidation was studied by Reddy *et al.*[40] DEHA strongly inhibited ascorbate-dependent nonenzymatic microsomal lipid peroxidation. It also completely inhibited nonenzymatic lipid peroxidation of heat-denatured microsomes, as shown in FIGURE 7, thus indicating that the inhibition is protein-independent. DEHA only moderately inhibited NADPH-dependent enzymatic microsomal lipid peroxidation.

Antioxidants in the diet often affect enzyme activity. However, DEHA had no significant effect on hepatic glutathione *s*-transferase, selenium-independent glutathione peroxidase, or selenium-dependent glutathione peroxidase activity in treated CD-1(ICR)Br male mice.[40]

Legator *et al.*[26] performed a number of tests for mutagenic activity of DEHA. *Drosophila* testing showed less than a two-fold increase in mutagenic activity by inhalation of DEHA and about a three-fold increase by feeding. No increase in mutagenic activity was seen in: (1) bacterial testing of DEHA, (2) bacterial testing of the urine of animals fed DEHA, (3) alkylation of hemoglobin and urine of rats treated with DEHA, or (4) micronuclei test of mice fed 200 mg/kg DEHA for five days. However, DEHA did show mild induction of DNA repair.

Additional mutagenicity testing of DEHA was performed by Taylor *et al.*[41] DEHA showed unscheduled DNA synthesis at 37°C, but it did not show thioguanine-resistant mutations in V-79 Chinese hamster cells. At 75°C, DEHA showed mutagenesis of plasmid DNA, but it acted as a very weak mutagen. On a molar basis it was only $3.6 \times 10^{-5}$ as effective as the potent mutagen *N*-acetoxy-2-acetylaminofluorene. Taylor *et al.*[41] also confirmed the earlier findings of Legator *et al.*[26] that DEHA was nonmutagenic to bacteria.

## ADVERSE TOXICOLOGIC EFFECTS

Three adverse toxicologic effects have surfaced with DEHA. The skin fibrosarcomas in male mice have already been mentioned, and the possibility of this

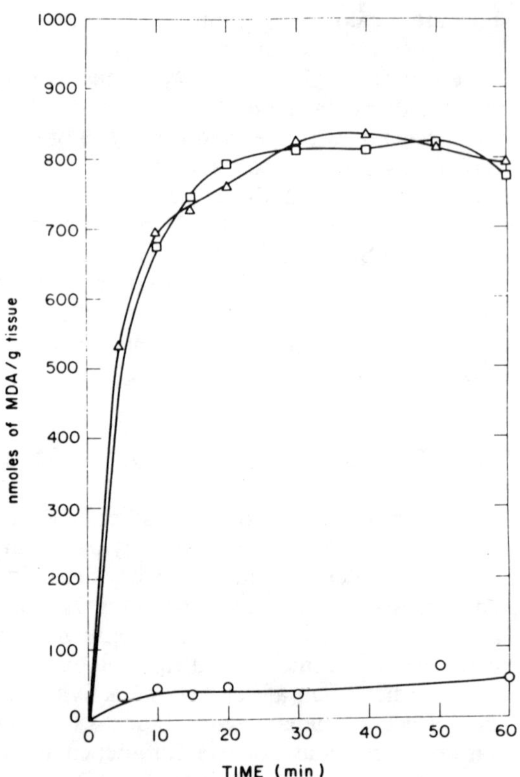


**FIGURE 7.** Time course of ascorbate/ADP-Fe$^{2+}$-dependent lipid peroxidation of heat-denatured rat liver microsomes. △, without DEHA or GSH; O, + 1.0 m$M$ DEHA; □, + 5.0 m$M$ GSH. (From Reddy *et al.*[40] Reproduced by permission of Hemisphere Publishing Corporation.)

being a spurious result has been discussed. The other two effects occurred in a dominant lethal test performed by injection and in a study in which mice were simultaneously fed DEHA and benzo(*a*)pyrene.

The dominant lethal test on male rats injected with up to 180 mg/kg of DEHA, who were then mated to females, gave a significant increase (confidence level > 99.95%) of dead implants in the females if mating was six weeks after injection of DEHA.[26] No adverse effect was seen if mating was four or eight weeks after injection. This result can be compared with that of the dominant lethal test performed by inhalation, discussed previously, in which the dead implants per female were lower in the exposed than in the control animals.

Since phenolic free-radical scavengers in the diet inhibited benzo(*a*)-pyrene-induced neoplasia in the forestomach of ICR/Ha mice,[36] it was decided to see whether DEHA would have the same effect. Groups of 20 male and 20 female CD-1(ICR)Br mice were treated either with 0, 3, 30 or 300 mg DEHA/kg per day (administered via the drinking water) from 57–112 days of age.[42] For the four-week period when the animals were between 70 and 98 days of age, all animals received (via gavage) eight 1-mg doses of benzo($a$)pyrene at 4-day intervals. Contrary to expectation, animals receiving DEHA did not exhibit a reduction in benzo($a$)pyrene-induced tumor incidence. DEHA had no significant effect on lung tumor incidence in either sex or on the incidence of stomach tumor in males. In contrast, DEHA-treated females showed a statistically significant increase in stomach tumors: the $\chi^2$ test gave a confidence level for effect, based on the incidence of females with tumors compared to those without, of 98%. However, on the basis of the dose-response effect (4 × 2 contingency table), the confidence level dropped to 91%, which is only marginally statistically significant. These results suggest that for female mice, DEHA may act as a promoter or cocarcinogen for benzo($a$)pyrene.

## CONCLUSION

DEHA is an effective inhibitor of photochemical smog formation. It can be used efficiently in urban atmospheres without causing second-day or downwind effects. At the concentrations required it should pose no toxicologic threat.

## SUMMARY

Photochemical smog is caused by a free-radical chain mechanism which converts NO to $NO_2$. The $NO_2$ further reacts to produce ozone, nitric acid, and peracylnitrates. This chain mechanism can be inhibited by suitable free-radical scavengers. The chemistry and toxicology of one such free-radical scavenger, diethylhydroxylamine, has been studied in depth. It has been shown to be effective, safe, and practical for use in urban atmospheres to prevent photochemical smog formation.

## REFERENCES


1. HEICKLEN, J. 1976. Atmospheric Chemistry. Academic Press. New York, NY.
2. GITCHELL, A., R. SIMONAITIS & J. HEICKLEN. 1974. The inhibition of photochemical smog I. Inhibition by phenol, benzladehyde, and aniline. J. Air Pollut. Control Assoc. **24:** 357.
3. GITCHELL, A., R. SIMONAITIS & J. HEICKLEN. 1974. The inhibition of photochemical smog II. Inhibition by hexafluorobenzene, nitrobenzene, naphthalene, 2,6-ditert-butyl-4-methylphenol. J. Air Pollut. Control Assoc. **24:** 772.
4. JAYANTY, R. K. M., R. SIMONAITIS & J. HEICKLEN. 1974. The inhibition of photochemical smog III. Inhibition by diethylhydroxylamine, N-methylaniline, triethylamine, diethylamine, ethylamine, and ammonia. Atmos. Environ. **8:** 1283.
5. HEICKLEN, J. 1981. Control of photochemical smog by diethylhydroxylamine. Atmos. Environ. **15:** 229.
6. SCHALL, D., K. PARTYMILLER & J. HEICKLEN. 1978. The inhibition of photochemical smog. VII. Inhibition by diethylhydroxylamine at atmospheric concentrations. Sci. Total Environ. **9:** 209.
7. HEICKLEN, J. 1984. The Inhibition of Photochemical Smog-IX. Model Calcula-

tions of the Effect of $(C_2H_5)_2NOH$ on $C_2H_4$-$NO_x$ Atmospheres. Center for Air Environment Studies Report No. 706–84.

8. HEICKLEN, J. 1985. The inhibition of photochemical smog-IX. Model calculations of the effect of $(C_2H_5)_2NOH$ on $C_2H_4$-$NO_x$ atmospheres. Atmos. Environ. **19**: 597.

9. ATKINSON, R. 1983. Private communication.

10. PITTS, J. N., JR., J. P. SMITH, D. R. FITZ & D. GROSJEAN. Enhancement of photochemical smog by N,N'-diethylhydroxylamine in polluted ambient air. 1977. Science **197**: 255.

11. PITTS, J. N., JR., D. M. GROSJEAN, A. M. WINER, A. C. LLOYD & G. J. DOYLE. 1977. National Science Foundation—RANN Grant ENV 73-02904-A03, Third Annual Progress Report (October 1975–December 1976).

12. OLSZYNA, K. & J. HEICKLEN. 1976. The inhibition of photochemical smog–VI. The reaction of $O_3$ with diethylhydroxylamine. Sci. Total Environ. **5**: 223.

13. SHAFFER, D. & J. HEICKLEN. 1986. The oxidation of diethylhydroxylamine in water solution at 25–80°C. J. Phys. Chem. **90**: 4408.

14. CUPITT, L. T. & E. W. CORSE. 1979. Effect of Diethylhydroxylamine on Smog Chamber Irradiations. Environmental Protection Agency Report EPA-600-379-040.

15. ATKINSON, R., K. M. DARNALL, A. C. LLOYD, A. M. WINER & J. N. PITTS, JR. 1979. Kinetics and mechanisms of the reactions of hydroxyl radical with organic compounds in the gas phase. Adv. Photochem. **11**: 375.

16. CACERES, T., E. A. LISSI & E. SANHUEZA. 1978. Autooxidation of diethyl hydroxyl amine. Int. J. Chem. Kinetics **10**: 1167.

17. HWA, C. M., 1983. Private communication.

18. JOHNSON, D. H., M. A. T. ROGERS & G. TRAPPE. 1953. The autooxidation of aliphatic hydroxylamines. Chem. Ind.: 1032.

19. JOHNSON, D. H., M. A. T. ROGERS & G. TRAPPE. 1956. Aliphatic hydroxylamines. Part II. autooxidation. J. Chem. Soc.: 1093.

20. WILSON, J. R., W. H. WELSH, III, D. M. KNAWBY, J. B. RABENOLD, D. C. ROSER & J. HEICKLEN. 1984. The base-association equilibrium constants for substituted hydroxylamines and the metal-complexing ability of diethylhydroxylamine. Unpublished work.

21. LANDRETH, R., L. STOCKBURGER, III, J. HEICKLEN, J. WEAVER & S. PRESCOTT. 1975. The Reaction of $SO_2$ with N,N'-diethylhydroxylamine. Center for Air Environment Studies Report No. 399-75.

22. HEICKLEN, J., J. F. MEAGHER, J. WEAVER, N. KELLY, K. PARTYMILLER, R. LATT, F. FERGUSON, C. PUTMAN, W. SAPANSKI & L. BILLUPS. 1981. Toxicological testing of rats subjected to inhalation of diethylhydroxylamine, nitroethane, and diethylamine hydrogen sulfite. Environ. Res. **26**: 258.

23. HEICKLEN, J. 1982. Reduction of Malignant Tumor Incidence in Female Rodents Exposed to Diethylhydroxylamine, Nitroethane, and Diethylamine Hydrogen Sulfite. Center for Air Environment Studies Report No. 635-82.

24. HEICKLEN, J., R. LUNDGARD, K. PARTYMILLER, F. FERGUSON, C. PUTMAN, W. SAPANSKI, J. TANKARD & L. BILLUPS. 1982. Chronic inhalation study of mice subjected to diethylhydroxylamine, nitroethane, and diethylamine hydrogen sulfite. Environ. Res. **27**: 277.

25. HEICKLEN, J., J. F. MEAGHER, J. WEAVER, N. KELLY, K. PARTYMILLER, R. LATT, F. FERGUSON, C. PUTMAN, W. SAPANSKI & L. BILLUPS. 1979. Toxocological Testing of Rats Subjected to Inhalation of Diethylhydroxylamine, Nitroethane, and Diethylamine Hydrogen Sulfite. Center for Air Environment Studies Report No. 523-79.

26. LEGATOR, M., R. E. KOURI, A. S. PARMAR, S. ZIMMERING, C. PUTMAN, R. LATT, J. HEICKLEN, J. F. MEAGHER, J. WEAVER & N. KELLY. 1979. Mutagenic testing of diethylhydroxylamine, nitroethane, and diethylamine hydrogen sulfite. Environ. Res. **20:** 99.

27. HEICKLEN, J., K. PARTYMILLER, N. KELLY, W. SAPANSKI, C. PUTMAN & L. H. BILLUPS. 1979. Three-generation reproduction study in mice subjected to inhalation of diethylhydroxylamine, nitroethane, and diethylamine hydrogen sulfite. Environ. Res. **20:** 450.

28. BELILES, R. P., S. L. MAKRIS, F. FERGUSON, C. PUTMAN, W. SAPANSKI, N. KELLY, K. PARTYMILLER & J. HEICKLEN. 1978. Teratology study in mice subjected to inhalation of diethylhydroxylamine, nitroethane, and diethylamine hydrogen sulfite. Environ. Res. **17:** 165.

29. HARMAN, D. 1982. The free radical theory of aging. *In* Free Radicals in Biology, Vol. 5: 255. W. A. Pryor, Ed. Academic Press. New York, NY.

30. HARMAN, D. 1985. Role of free radicals in aging and disease. *In* Relations Between Normal Aging and Disease, H. A. Johnson, Ed.: 45. Raven Press. New York, NY.

31. CUTLER, R. G. 1984. Antioxidants and longevity. *In* Free Radicals in Molecular Biology, Aging, and Disease. D. Armstrong, R. S. Sohal, R. G. Cutler, & T. F. Slater, Eds.: 235. Raven Press. New York, NY.

32. CUTLER, R. G. 1984. Free radicals and aging. *In* Molecular Basis of Aging. A. K. Roy & B. Chatterjee, Eds.: 263. Academic Press. New York, NY.

33. DUBLIN, L. L., A. J. LATKA & M. SPIEGELMAN. 1949. Length of Life. Ronald Press. New York, NY.

34. BERG, B. N. & H. S. SIMMS. 1960. Nutrition and longevity in the rat. II. Longevity and onset of disease with different levels of food intake. J. Nutr. **71:** 255.

35. HARMAN, D. 1971. Free radical theory of aging: Effect of the amount and degree of unsaturation of dietary fat on mortality. J. Gerontol **26:** 451.

36. WATTENBERG, L. W., J. B. COCCIA & L. K. T. LAM. 1980. Inhibiting effects of phenolic compounds of benzo(a)pyrene-induced neoplasia. Cancer Res. **40:** 2820.

37. CALABRESE, E. J. 1980. Nutrition and Environmental Health. The Influence of Nutritional Status on Pollutant Toxicity and Carcinogenicity. Wiley Interscience. New York, NY.

38. BOLLAG, W. & A. MATTER. 1981. Modulation of cellular interactions by vitamin A and derivatives (retinoids). Ann. N.Y. Acad. Sci. **359:**9.

39. HARMAN, D. 1968. Free radical theory of aging: effect of free radical inhibition on the life span of male $LAF_1$ mice — Second experiment. Gerontologist **8:** 13.

40. REDDY, C. C., J. HEICKLEN, R. W. SCHOLZ, C.-Y. HO, J. R. BURGESS & E. J. MASSARO. 1985. Effects of diethylhydroxylamine on hepatic microsomal lipid peroxidation and glutathione *s*-transferases. J. Toxicol. Environ. Health **15:** 467.

41. TAYLOR, W. D., J. S. MUDGETT, N. L. KROEKEL, R. H. BONNEAU, S. G. WILT, P. C. BRONTZ & J. HEICKLEN. 1985. Toxic and Mutagenic Effects on Bacteria and Mammalian Cells of Hydroxylamines: Hydroxylamine, Methylhydroxylamine, Methylamine, Dimethylhydroxylamine, Isoproplyhydroxylamine, and Diethylhydroxylamine. Center for Air Environment Studies Report No. 727-84.

42. HEICKLEN, J., T. WILSON, G. D. MILLER & E. J. MASSARO. 1984. The effects of diethylhydroxylamine on the incidence of tumors induced by benzo(a)pyrene in the mouse. Drug Chem. Toxicol. **7:** 315.

# Air-Cleaning Devices for Home and Office

AMOS TURK

*Department of Chemistry*
*The City College of the*
*City University of New York*
*New York, New York 10031*

The concentration of gaseous contaminants in indoor air depends on various factors that have been changing over the years. In particular, new structural, decorative, and insulating materials have been introduced and ventilation rates have been reduced in recent efforts to make buildings energy-efficient. A consequence of these changes has been the widespread recognition that indoor air quality is a matter for serious concern and that remedial methods are sometimes called for. One response to these felt needs has been the proliferation of home or office-type air cleaners that are available for purchase at the retail level. Two consumer journals have recently tested and rated such devices.[1,2]

First, it should be recognized that air cleaners can be applied in either of two modes. In one mode, the air cleaner is interposed between the source and the indoor environment. An example is the recirculating range hood, which draws cooking vapors directly into its filter media. In the second mode, the air cleaner treats the indoor air after the contamination has been more or less thoroughly dispersed therein. Typical examples of such devices are air cleaners applied to central heating, ventilating, and air-conditioning (HVAC) systems or free-standing air cleaners located somewhere in the building. In the ideally simple case in which both the contaminant sources and the treated air are completely and instantaneously mixed in the entire indoor space, the concentration of contaminants drifts exponentially to a steady-rate value that depends on (1) the volume of space being treated, (2) the rate at which contaminants are being generated, and (3) on the air-handling capacity and contaminant-removal efficiency of the air cleaner.[3]

A simplified form of the general equation is

$$C = C_o e^{-EmQt/V}$$
$$= C_o e^{-EmN}$$

where $C$ = contaminant concentration of any time $t$, $C_o$ = initial concentration of contaminant, $E$ = air-cleaning efficiency of device (fractional), $m$ = mixing factor in room (efficiency of mixing of purified air with ambient room air), $Q$ = air flow rate through device, $V$ = volume of room, $N$ = number of air changes = $Qt/V$.

As an example, consider a 1000-cubic foot room served by a device that recirculates 50 cubic feet per minute (cfm) in a room with a typical mixing factor of about ⅓. The time required for the device to reduce a given contaminant level by 90 percent in a sealed room in which no new contaminant was being generated would be somewhat over 2 hours if the device were 100 percent efficient, about 3 hours at 75 percent efficiency, and 4 hours at 57 percent efficiency. These calculations are given to illustrate the fact that such devices are not expected to act rapidly. The devices could, however, be used to clean the air in a previously contaminated room that is not to be reentered for several hours while the device operates. If the room is not sealed, as most are not, then natural ventilation contributes to the performance and shortens the clean-up time. If generation of contaminant is continuous, then the contaminant level would drift down toward a steady-state concentration, not toward zero. The devices rated by *Canadian Consumer*[1] in 1984 provided air flows ranging from 4 to 23 cfm (low-speed) and 10 to 37 cfm (high-speed) and removed from 4 to 30 percent of smoke haze in 2 hours from a 200-cubic foot chamber.

## AIR CONTAMINANTS

Air contaminants are often arbitrarily categorized as particulate or gaseous (molecular), although a molecule, too, is a particle and the spectrum of particle sizes approaches a continuum. The sizes of contaminant particles are important to the degree that they affect the air-cleaning efficiency of the human respiratory system and of mechanical devices.

The upper respiratory system filters out particles larger than about 5 micrometers in diameter. Particles smaller than about 3 micrometers generally escape all the natural protections and enter the lung. More than 99 percent of all contaminant particles (by particle count, not by mass, and excluding what are considered to be molecules) are smaller than about 1 micrometer. For an airborne particle to be visible to the naked eye, it must be larger than about 10 micrometers. Thus, it is mostly what you cannot see that may harm you. TABLE 1 is a short list of particles of various diameters.

**TABLE 1.** Molecules and Other Particles

| Particles | Diameter | | Settling Rate in Air (cm/sec) |
| --- | --- | --- | --- |
| | Micrometers[a] | ångström Units | |
| Beach sand | 100–1000 | | ⎫ |
| Human hair | 40–1000 | | 1–100 |
| Coal dust | 1–100 | | ⎭ |
| Bacteria | 0.5–15 | | 0.1–0.01 |
| Mold spores | 1–10 | | |
| Tobacco smoke | 0.01–1 | 100–10,000 | ∼0.0001 |
| Viruses | 0.005–0.05 | 50–500 | |
| Gas molecules | | 3–10 | |

[a] One micrometer (micron) = $10^6$ m = 10,000 Ångström units.

# AIR CLEANING

Commercial methods of air cleaning include liquid scrubbing, incineration, filtration, electrostatic precipitation, and adsorption. Liquid scrubbing and incineration are appropriate only for source control of effluents from industrial processes, where contaminant concentrations are much higher than those encountered in ambient air, indoors or out. Odor masking or "air freshening" is not air cleaning, and so will not be considered here. Ozonization produces both sensory and partial oxidation effects, but ozone is toxic and is therefore contraindicated for indoor use.

The remaining methods, filtration, electrostatic precipitation, and adsorption, are embodied, either individually or in pairs, but almost never in a combination of all three, in various types of home or office air cleaners. Each of the methods offers a range of air-cleaning action.

### *Filtration*

Filtration is commonly visualized as a simple straining action, in which particles larger than the filter openings are stopped and small particles pass through. In addition to such straining, there are four other mechanisms by which even a particle smaller than the filter opening can be interrupted. One of these is a direct hit of a small particle onto a portion of the filter. A second means is inertial, in which a small particle that is still much more massive than an air molecule follows a larger radius of curvature in a curved path through the filter medium and thus impinges on the filter surface. A third mechanism involves a diffusional, or Brownian motion effect, where a particle is deflected from the bulk motion of the air stream by its collisions with molecules of nitrogen or oxygen and thus is forced to an impact with the filter medium. Finally, friction effects with filters may cause a particle to become electrically charged and then adhere to the filter electrostatically. Cigarette filters operate to some extent in this way.

The factors to be considered in the choice of filter media are efficiency, resistance, maintenance, and space. These factors are functionally related, so the choices are not independent. High efficiency, which is desirable, is obtained at the expense of imposing high resistance to air flow, which in turn requires more fan power for a given flow rate, as well as demanding more space and energy cost, and which may be noisier. Generally an efficient filter requires a coarse prefilter to screen out larger particles and thus prolong filter life. As a filter becomes loaded, its efficiency increases, but so does its resistance, and it approaches as a limit a condition of 100 percent efficiency with no air flowing through it.

There are various methods by which filter efficiencies are rated. The important criterion is the particle size range of the contaminant against which the filter is judged. One such method is optical; it measures the darkening of a spot on the filter through which a dusty gas is passed. The procedure is described by American Society of Heating, Refrigerating and Air-Conditioning Engineers (ASHRAE) Standard 52-76. Another method is particularly applicable to the rating of high-efficiency particulate-arresting (HEPA) filters of the type used for "absolute" filtration in industrial clean rooms. This method challenges the filter with a fine smoke cloud consisting of essentially uniform 0.3-micrometer droplets of dioctyl phthalate (DOP). TABLE 2 displays the efficiencies of several common filter media according to these two criteria.

The dust spot method tests filters for their ability to remove particles that, when they deposit on surfaces, cause visible soiling. The DOP test assays the ability of a filter to remove particles that are small enough to bypass the protection offered by the human respiratory system.

### *Electrostatic Precipitation*

Electrostatic precipitation is another method of removing particulate pollutants from air; it is not effective for gaseous matter. The contaminated air is blown into a section provided with ionizing wires whose high positive electrical potential, typically about 4000 volts, strongly attracts electrons that are randomly present in air. The rapidly accelerating electrons strike electrons out of air molecules ($M$), creating positive ions:

$$e^- + M \rightarrow M^+ + 2e^-$$

The newly liberated electrons are free to strike other air molecules and produce additional positive ions. This multiplying effect all takes place in the same charging section of the air cleaner. Some of the energy associated with these actions is released in the form of blue light, which accounts for the bluish corona discharge sometimes visible around the ionizing wires.

Within about 0.01 second, these positive ions become attached to the contaminant particles (P) in the air stream.

$$xM^+ + P \rightarrow P^{x+}$$

**TABLE 2.** Efficiencies of Some Common Filter Media

| | Efficiency (%) | |
|---|---|---|
| Filter Medium | ASHRAE Dust Spot | DOP Test |
| Open cell foams | 15–30 | 0 |
| Thin glass or cellulose fibers | 20–35 | 0 |
| Multi-ply glass, cellulose, or wool | 25–40 | 5–10 |
| Thick mats | | |
|     5–10-micrometer fibers | 40–60 | 15–25 |
|     3–5-micrometer fibers | 60–80 | 35–40 |
|     1–4 micrometer fibers | 80–90 | 50–55 |
| HEPA (fibers < 1 micrometer) | NA[a] | 95–99.999 |
| Membranes with holes < 1 micrometer | NA | ~100 |

[a] Not applicable.

The larger the particle, the greater the number of ions that adhere to it and hence the greater the charge. The charged dust particles are then carried by the moving air stream to a collector section consisting of a series of thin metal plates alternately charged positive and negative with a high DC voltage. The charged particles are thus accelerated toward the negative (ground) plates where they adhere to the plates and to each other, and are thus removed from the air stream. In some devices a prepared adhesive is applied to the collector plates to augment the natural adhesive and cohesive forces. Most indoor air contaminants include enough oily or tarry materials from sources such as tobacco smoke and vapors from cooking so that added adhesives are unnecessary. The most important factor in maintenance is periodic cleaning of the dirty collector cell. FIGURE 1 displays collection efficiencies by the dust spot method from a set of commercial electrostatic air cleaners.

## *Adsorption*

Adsorption refers, broadly, to the action of granular media rather than filters for air cleaning. In all cases, the object is the removal of contaminants that are in the form of gas or vapor. In typical indoor air, the gas-phase contaminants of most concern include oxides of nitrogen (from gas flames), radon (from some soils and masonry), formaldehyde (from resins in insulation and particle board), carbon monoxide (from outdoor air intakes), solvent vapors (from lacquers, paints, dry-cleaning solvents, and so on), and a wide range of unpleasant odors, especially from bacterial action on many organic substrates and from tobacco products.

The effectiveness of a granular adsorbent medium for air cleaning depends on two processes that occur in sequence. First, the molecules of a gaseous contaminant make contact with the solid surface of the medium and, second, the contaminant must be retained and thus permanently removed from the air stream or it must be converted to an innocuous product.

The first process, making contact, distinguishes a granular medium from a filter. The kinetic motion of gas molecules is rapid enough (for example, about 400 m/sec for nitrogen at 0°C) compared with the much slower kinetic motion of particulate matter so that it is possible to design granular beds that offer low resistance to air flow while offering effective contact with gas molecules. The adsorbents commonly used for indoor air cleaning are granules of mixed sizes ranging in diameter from about 0.04 to 0.2 inches (1 to 5 mm) packed into beds of uniform thickness. Such arrangements offer relatively low resistance to airflow. Then, if an air stream containing contaminant molecules enters such a granular bed, the contact efficiency, $E$, which is the fraction of molecules that have contacted a solid surface after $n$ seconds of residence time, is:

$$E = 1 - 2^{-100n}$$

From this equation, typical contact efficiencies are 75 percent for a residence time of 0.02 sec, 93.75 percent for 0.04 sec, and 98.44 percent for 0.06 sec.

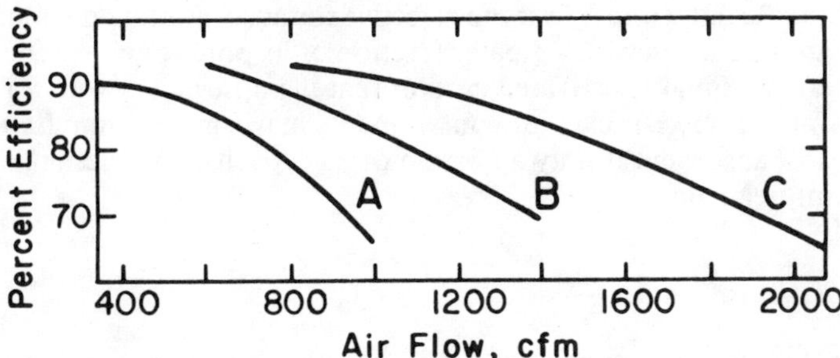


**FIGURE 1.** Efficiency ratings (dust spot method) of a set of commercial electrostatic air cleaners. (A) Face dimensions, 20″ × 12½″; rated for 400-1000 cfm. (B) Face dimensions 16″ × 25″; rated for 500–1400 cfm. (C) Face dimensions 20″ × 25″; rated for 800-2000 cfm. (Courtesy of Honeywell, Inc.)

The second process, which is an action that prevents the return of the contaminant molecules to the air stream, may involve any of the following actions, or a combination of them: (a) physical adsorption, (b) chemical neutralization of an acid or a base, (c) oxidation, (d) addition or complexation, and (e) decomposition.

## Physical Adsorption

Any gas or vapor will adhere to some degree to any solid surface. The factors that are important in practical air cleaning by physical adsorption are the molecular weight and electrical polarity of the gas or vapor and the polarity, surface characteristics, and total capacity of the adsorbent.[4] The molecular weight of dry air (weighted average 29) is sufficiently lower than that of most air contaminants so that physical adsorption can effectively select for contaminants. Exceptions include ammonia (molecular weight 17), carbon monoxide (28), and the lighter hydrocarbons. Water vapor can be a problem because its high polarity gives it a strong affinity for polar adsorbents, which include all those containing oxygen, such as activated alumina, silica gel, and synthetic zeolites. As a result, water vapor renders such adsorbents ineffective for broad-spectrum removal of organic contaminants. Formaldehyde (molecular weight 30), however, can be selectively adsorbed by a synthetic zeolite of appropriate pore diameter, about 13 angstroms. It is because activated carbon lacks electrical polarity that it is the adsorbent of choice for practical use. Activated carbon thus rejects water and preferentially adsorbs organic vapors, while oxygenated adsorbents do the opposite. Other important factors in physical adsorption are the total pore volume and the distribution of pore diameters in the adsorbent. The total pore volume determines the capacity of the carbon to accommodate contaminants by capillary condensation, and the pore diameters influence the preference of the carbon for contaminants of different molecular size or shape. For example, a small-pore carbon, with a large fraction of its total pore volume in the "micropore" range (diameters below about 30 angstroms), is more effective for adsorption and retention of radon than is a carbon with a greater fraction of its pore volume in the "macropore" range.[5] Finally, activated carbon that is impregnated with a reagent for removal of a specific class of contaminants may function simultaneously as a physical adsorbent if only a portion of its capacity is sequestered by the reactive impregnant.

## Chemical Neutralization

Activated carbon impregnated with a caustic substance (NaOH or KOH) is effective for the neutralization of even very weak acids such as hydrogen sulfide or mercaptans. The capacity of the impregnated carbon (about 0.2 g hydrogen sulfide per g of carbon) is greater than can be accounted for by neutralization, and the difference is the result of oxidation of some of the sulfur to form polysulfides. Activated alumina can also be impregnated with a base, but not so strong a base as caustic, which would react with alumina. Alumina impregnated with sodium carbonate will neutralize moderately strong acidic gases such as sulfur dioxide but will not, at the same time, retain a practically useful capacity for physical adsorption of a variety of air contaminants from a moist atmosphere. For removal of basic gases, activated carbon is impregnated with phosphoric acid, yielding a capacity for retention of about 0.1 g of ammonia per gram of impregnated carbon. Such impregnated carbon is also applicable for the removal of light amines.

## Oxidation

Air contaminants may be oxidized to detoxify them, deodorize them, or render them more strongly retained on the granular adsorbent. The oxidizing agent may be a chemical impregnant on the adsorbent, or it may be air, in which case an oxidation catalyst on the surface of the adsorbent is generally necessary. An impregnated oxidant becomes exhausted as it is reduced by the air contaminants with which it reacts. An oxidation catalyst may become inactivated by being chemically poisoned or by being physically blinded by the accumulation of other adsorbed contaminants. When the oxidation pathway is ionic, the presence of some adsorbed moisture is necessary. These various factors dictate the choices for particular applications, such as the following: Activated alumina impregnated with potassium permanganate is used for removal of contaminants that are particularly labile to oxidation, such as formaldehyde. Permanganate oxidation of many other organics, however, is too slow to be completed in a single pass through a thin granular bed.[6]

On the other hand, the reduction product of the permanganate, which is manganese dioxide, can serve as a catalyst for the air oxidation of some of the components of stale tobacco odor. Oxidizing agents do not survive on activated carbon because they are reduced by it. Carbon has been used for many years, however, as a support for oxidation catalysts. Platinized or palladized carbon is a powerful catalyst for air oxidation. Carbons impregnated with copper and/or silver catalysts were developed for the catalytic destruction of World-War-I type poison gases such as Lewisite and arsine. Even unimpregnated carbon is catalytically active, as evidenced by its promotion of the air oxidation of hydrogen sulfide to sulfur and polysulfides.

### Addition and Complexation

Some specific examples are of interest; the substrate in each case is activated carbon: Brominated carbon removes ethylene by converting it to the dibromide.[7] Iodized carbon effectively removes mercury vapor; this action involves both oxidation and addition. Zinc acetate can remove ammonia and some amines by forming a zinc-ammonia or zinc-amine complex; this impregnation is advantageous over phosphoric acid when an acidic medium is to be avoided.

### Decomposition

Activated alumina impregnated with a catalyst such as manganese dioxide, or activated carbon with or without a metallic impregnation, is an effective catalytic medium for the destructive decomposition of contaminants containing oxygen-to-oxygen bonds, such as ozonides, peroxides, hydroperoxides, and ozone itself.

Unless the action is catalytic, adsorbents eventually become saturated and have to be replaced. Unfortunately, there is no universal end-of-service-life indicator for granular media. Since an adsorbent bed is generally preceded by a filter, it does not become physically clogged and therefore its resistance to air flow does not change in a way that can be used to monitor saturation. Adsorbents generally pick up and release moisture in response to changes in ambient humidity, so that gain of weight is not an index of exhaustion. Permanganated alumina does change color, from purple to brown, as it is chemically reduced, but carbon does not. For home and office devices the only practical indicator of saturation is deterioration of performance, which means, in effect, failure to remove odors. A consumer who is inattentive to such change is left with an ineffective air-cleaning device.

## SUMMARY

Methods for purifying contaminated air include liquid scrubbing, incineration, filtration, adsorption, and electrostatic precipitation. The first two of these are appropriate only for concentrations of pollutants much higher than those encountered in occupied spaces. The latter three are embodied, either individually or in pairs, but almost never in combinations of all three, in many types of home or office air-purifiers commercially available as self-contained units.

The effectiveness of these units depends on the following factors: (*a*) the types of contaminants, particulate or gaseous, to be removed; (*b*) the air-cleaning efficiency of the units for the pollutants to which they are addressed; and (*c*) the total air flow through the units relative to the size of the room and its air-mixing characteristics. These factors are discussed with specific references to the various types of commercially available home and office cleaners.

## REFERENCES


1. CANADIAN CONSUMER. **14**(5): 15 (May 1984).
2. CONSUMER REPORTS, **50**(1): 7 (January 1985).
3. TURK A. 1963. Measurements of odorous vapors in test chambers: Theoretical. ASHRAE, **5**: 10.
4. TURK, A. 1977. Adsorption. Air Pollution, Vol. 4, 3rd ed. A. C. Stern, Ed. Academic Press. New York, NY.
5. BRISK, M. A. & A. TURK. 1984. Control of indoor radon by activated carbon filters. Presented at the 77th Annual Meeting of the Air Pollution Control Association, San Francisco, California, June 1984.
6. TURK, A.. S. MEHLMAN & E. LEVINE. 1973. Atmos. Environ. **7**: 1139.
7. TURK, A., J. MORROW, P. F. LEVY & P. WEISSMAN. 1961. Int. J. Air Water Pollut. **5**: 14.

# Air Movement and Vehicular Pollution in Urban Canyons


W. HOYDYSH

*ESSCO, Inc.*
*Long Island City, New York 11101*

M. ORENTLICHER

*Department of Environmental Protection*
*City of New York*
*New York, New York 10013*

W. DABBERDT

*National Center for Atmospheric Research*
*Boulder, Colorado 80302*


This paper will begin with a statement of the practical problem faced by environmental decision makers — estimation of the impact on air quality in the complex urban environment, and a sketch of the physical phenomena that make this estimation so difficult. Atmospheric science issues pertaining to the simulation of urban impacts are detailed, and include considerations such as: thermal stratification, global and local topography, as well as the characteristics of vehicular emission sources. Actual simulation results are shown in conclusion.

## INTRODUCTION

### *The Social and Scientific Context*

Air-quality planning and transportation planning are intimately interrelated, particularly in urban areas. Problems and inaccuracies inherent in transportation planning can have nonlinear impacts on estimates of present or future air-quality conditions. These estimates of the concentrations of ozone or carbon monoxide that are used to measure compliance with local, state, and federal air-quality regulations have major environmental and economic consequences. These consequences follow from the regulatory and policy decisions which are made on the basis of the estimates, and these decisions frequently affect transportation planning.

The concentration of carbon monoxide is highly sensitive to the details of traffic patterns and building configurations. There are several reasons for this: nearly all carbon monoxide (CO) emissions in urban areas can be attributed to vehicular traffic; CO "exceedances" of the air standards occur only in proximity to roadways and intersections; and urban structures act to restrict or modify ventilation and thus to magnify ambient CO levels beyond what they would be in open terrain (out of the city). As a consequence, inaccuracies in our ability to properly describe or project traffic flow can lead to larger uncertainties in air-quality projections: projects run the risk of being denied or approved (partially) on the basis of faulty or inaccurate traffic estimates.

To understand the sensitive nature of the relationship between traffic and air quality, one must consider the three-fold nature of the problem. (1) Motor vehicles generate the pollutant emissions in a complex pattern that varies from hour to hour and block to block. (2) The atmosphere dilutes these emissions according to laws of physics that govern turbulence and atmospheric transport. The presence of buildings near traffic sources grossly modifies the way in which this dilution occurs. (3) Because of spatial variations in traffic flow (and hence emissions) *and* variations in atmospheric properties, it is very important *where* we decide to measure or project pollutant concentrations — such "receptors", if displaced a few tens of feet, can indicate significantly different air quality. For a fixed geometry the dynamic state of the air is determinate; only our ability to accurately represent it is uncertain or inexact. In a planning context, this uncertainty in estimation implies that sensitive areas (parks, schools, hospitals) could be exposed to unacceptable air-quality conditions, or projects could be rejected due to erroneous projections of their air-quality impacts.

Art and science combine when we attempt to incorporate these factors in the tasks of environmental planning and regulation. A schematic description of such a process includes five stages:

(1a)　Develop a model of the traffic network and of appropriate receptor locations.

(1b)　Develop a conceptual and/or physical model of the local topography.

(2)　Estimate and/or measure the local traffic parameters to calculate vehicular emissions.

(3)　Compute emissions with an appropriate method that is recognized by regulatory authorities.

(4)　Use a dispersion model conforming to local conditions of traffic, meteorology, and topography, to estimate pollutant concentrations at receptors.

(5)　Compare estimates to air-quality standards. If there are "exceedances," change traffic flows, land use, and so forth to minimize or eliminate them; and repeat the processes (1–5).

Scientific uncertainties are troublesome in stages (3) and (4), and a paucity of empirical knowledge is prominent in stage (2). Consequently, we ask the transportation planning community to provide us with improved traffic parameters with which we can better estimate vehicular emissions. These parameters include, for example, such link and intersection features as speed, volume, queue length, acceleration rates, idle times and their temporal variability. Second, as atmospheric scientists we set out to improve our capabilities for simulating atmospheric dispersion of emissions from vehicles at intersections and within street canyons. At ESSCO, a unique research and development program develops improved mathematical models of street-canyon dispersion. By utilizing an environmental wind tunnel, ESSCO has been able to make numerous and detailed measurements that would otherwise be prohibitively costly, time-consuming, and difficult to make in the ambient atmosphere. We will proceed in this paper to describe the work of atmospheric scientists, while the reader is asked to keep in mind the traffic inputs that are needed to help yield an improved description of emissions and air quality in the cities.

### *Varieties of Urban Topography and Atmospheric Motions*

The analysis of vehicular air pollution in the urban environment is considerably more complicated than vehicular air pollution analysis in flat rural terrain. As well as considerations of vehicular modal splits and trip generation, some of the special characteristics of the urban environment must be considered.

In flat rural terrain, air pollution diffuses from an "at grade" or "cut section" highway lane as if the lane were a stationary pollution line source instead of a set of dynamic, discrete pollution point sources moving at a constant velocity with constant separation between them. HIWAY-2 is an example of the EPA-approved pollution modeling strategy for such a situation.[1] Mathematical aspects of pollution diffusion from point and line sources have been summarized by Turner.[2]

The presence of a city has major impacts on the air flow both within and around the urban region. This is due to the "heat island" effect that results from both the thermal inertia and heat generation characteristics of a major city as well as the resistance to air flow produced by the physical bulk of the city. These factors are critical for studies of pollution from elevated sources such as utility stacks, but can be assumed to be accounted for in describing microscale impacts by the local atmospheric variables (wind speed and direction, dispersion coefficients) used in the models described in this paper.

The analysis of air flow and pollutant dispersion in deep cuts and urban street canyons is complicated by a number of urban effects that are strictly local. These include vortex formation, channeling of vehicle wakes, convection due to street-level flux of sensible heat and differential heating or cooling of canyon walls. Results of studies of moderate street canyons have been incorporated in the APRAC-1A MODEL.[3] FIGURE 1 presents flow regimes over a shallow sloped cut, a shallow vertical cut, and a moderate vertical cut. The flow in all three of these cases is considered to be perpendicular to the street.

With a shallow sloped cut, the flow parallels the terrain. As the cut walls become vertical, small corner vortices form, although the main flow is (more or less) parallel to the terrain.

As the depth of the notch increases (FIG. 1c) frictional surface drag decreases and the flow accelerates above the notch. This figure shows that the flow decelerates between c and b. An unstable and intermittent helical flow develops

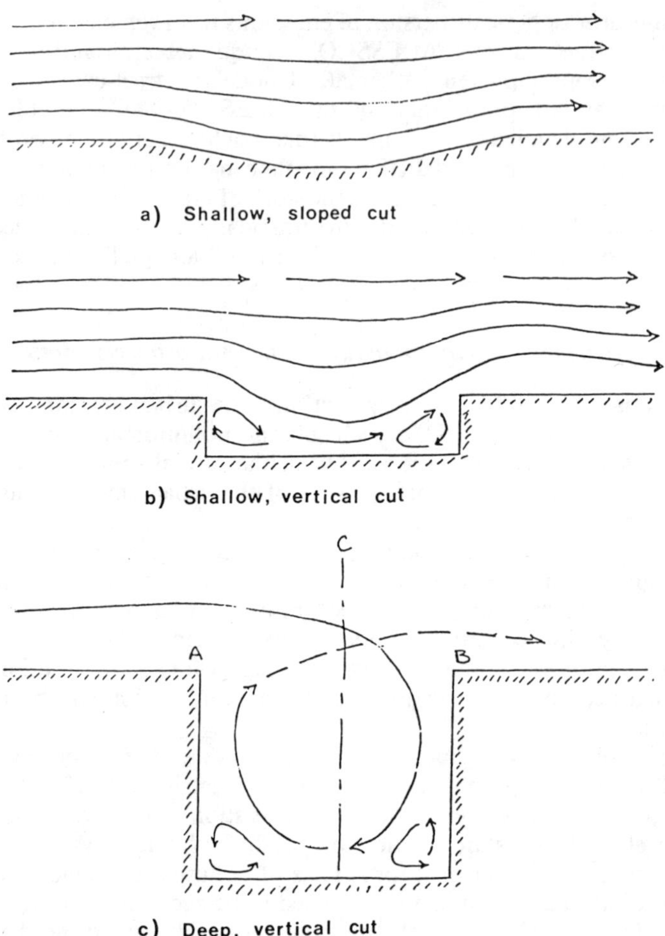



**FIGURE 1.** Schematic illustration of flow regimes as determined by site geometry.

that prevents a monotonic increase in canyon concentration due to street-level emissions.

As the aspect ratio increases, the flow regimes become more complex. At larger aspect ratios the frictional drag of the vertical surfaces prevents vortex penetration to the lower levels of the canyon. Flow regimes within such deep canyons have been treated analytically, but are still under study by ESSCO, and require considerably more research by atmospheric scientists.

An empirical study of deep-canyon dispersion phenomena was conducted for the State of New York by New York City and the Cooper Union. This study demonstrates that "deep canyons" deviate in their dispersion behavior from that predicted by APRAC, and that this behavior is a function of the local building topography.[4] The ratio of CO concentration to intensity of emis-

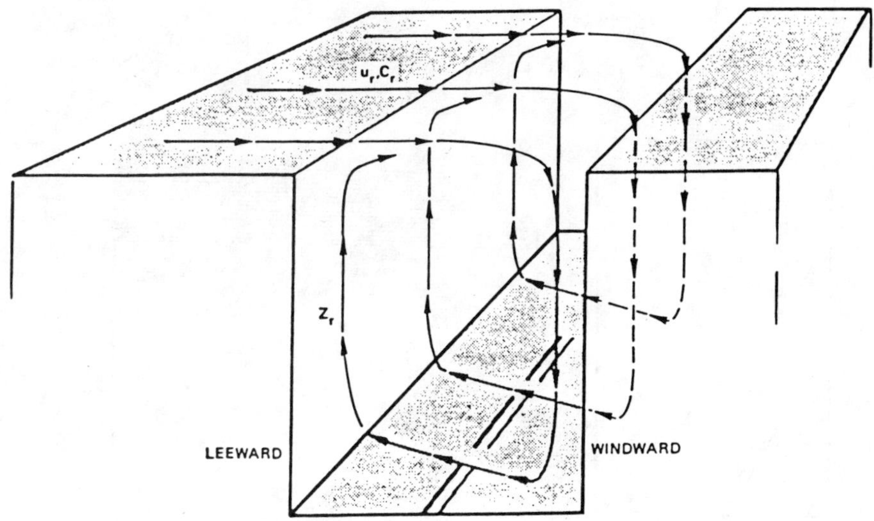


**FIGURE 2.** Indicated typical helical air flow over a street. (Adapted from Georgii *et al.*[5])

sion varied by about one order of magnitude in the deep canyons, presumably because of variations in the local meteorology.

In the next section, we discuss studies of dispersion in the streets of four cities. These streets all exhibit topography appropriate to use of the APRAC model, having moderate ratios of building height to street depths.

## FIELD STUDIES

There have been only four comprehensive atmospheric field studies of air flow and pollutant dispersion in urban street canyons. In these studies, vertical and cross-street pollutant profiles were measured along with local traffic and site-specific meteorology. The four programs are those of Frankfurt am Main,[5] San Jose/St. Louis,[6] Cologne, Federal Republic of Germany,[7] and Lyon, France.[8]

A major finding of the Frankfurt study was that CO concentrations on the leeward sides of streets were considerably higher than those on the windward sides, implying a helical cross-street circulation component near the surface in the opposite direction from the roof-level wind (FIG. 2). The averaged data also showed that the vertical concentration profiles on either side of the street had an exponential form, the mode of air circulation above the street changed when roof-top winds exceeded 2 m/sec, and the concentrations were exponentially related to traffic density.

The San Jose/St. Louis studies by Stanford Research Institute utilized CO concentration monitors mounted on the street-canyon walls, above the street, and in helicopters and vans, as well as instrumentation to measure winds, tem-

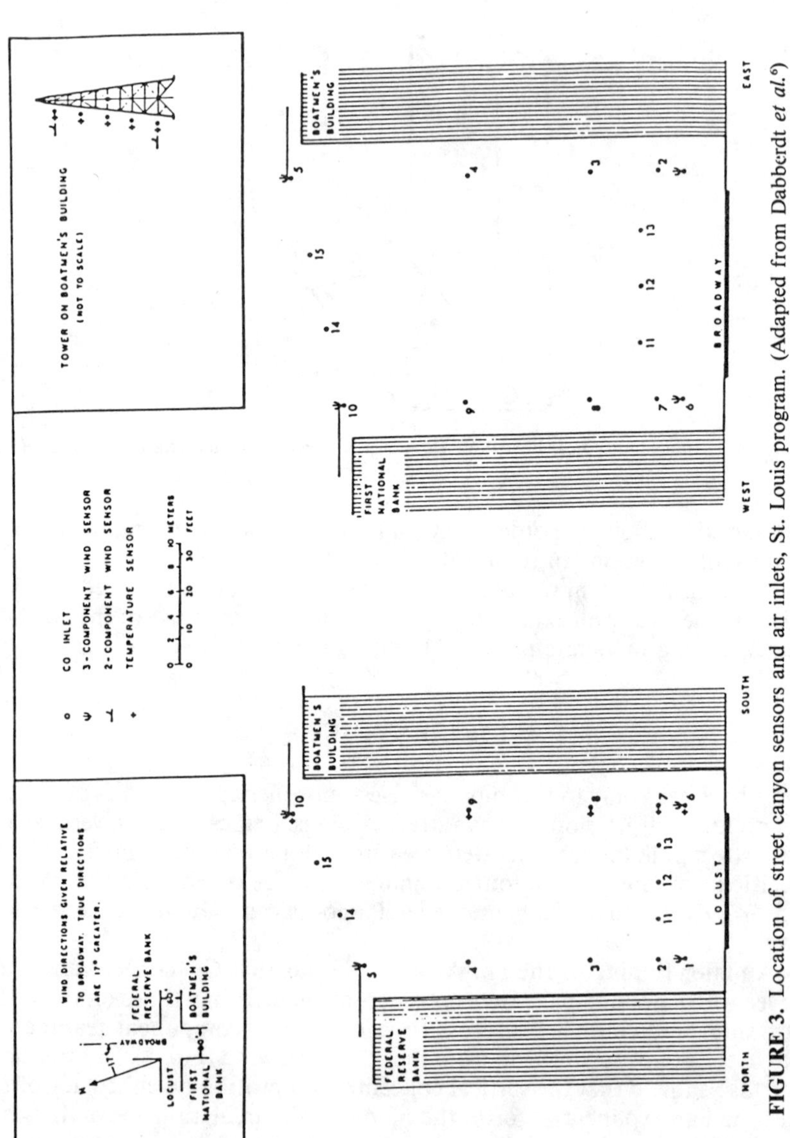


FIGURE 3. Location of street canyon sensors and air inlets, St. Louis program. (Adapted from Dabberdt *et al.*[6])

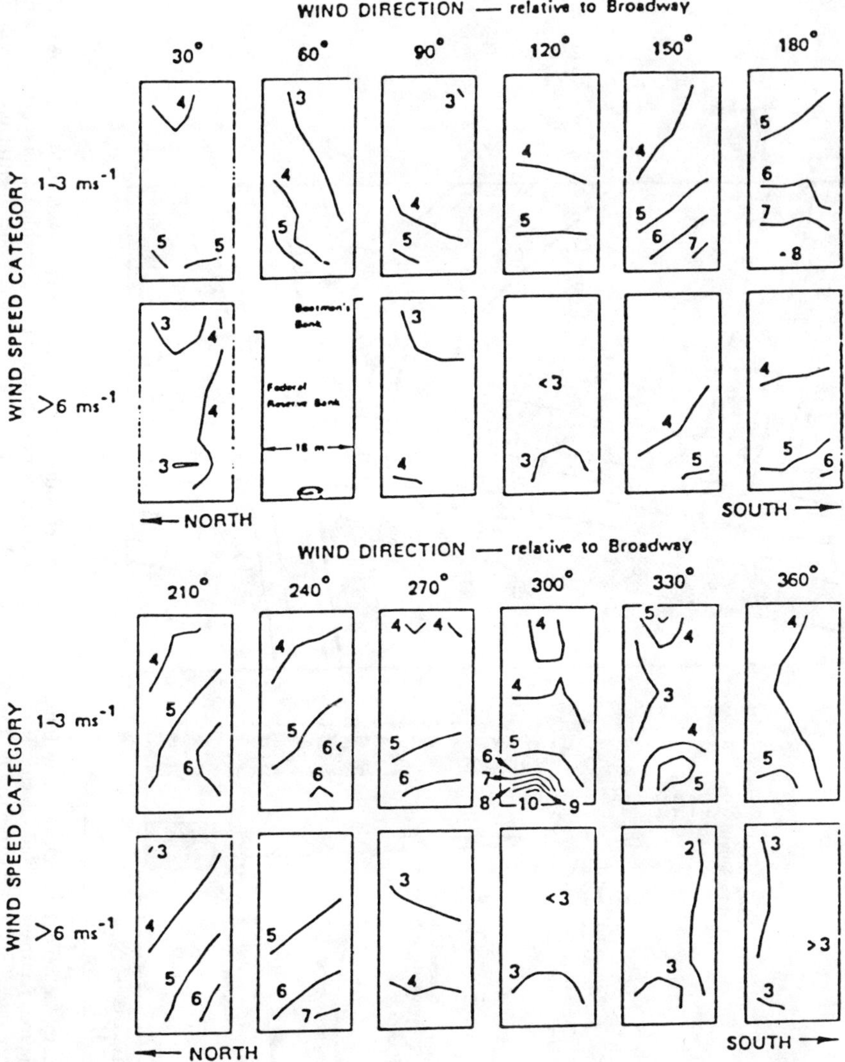



**FIGURE 4.** Distribution of CO concentration (ppm) in Locust Street Canyon, St. Louis Program. (Adapted from Dabberdt *et al.*[6])

perature and traffic. The San Jose/St. Louis studies indicated that a single-helix circulation was present for roof-level winds blowing at an angle of 30 degrees or more to the street direction. The largest cross-street gradient occurs with a wind that is inclined 30 degrees to the longitudinal street axis. The pattern changed abruptly from the low-wind to the high-wind category. FIGURE 3 shows the experimental setup and FIGURE 4 illustrates the cross-street CO

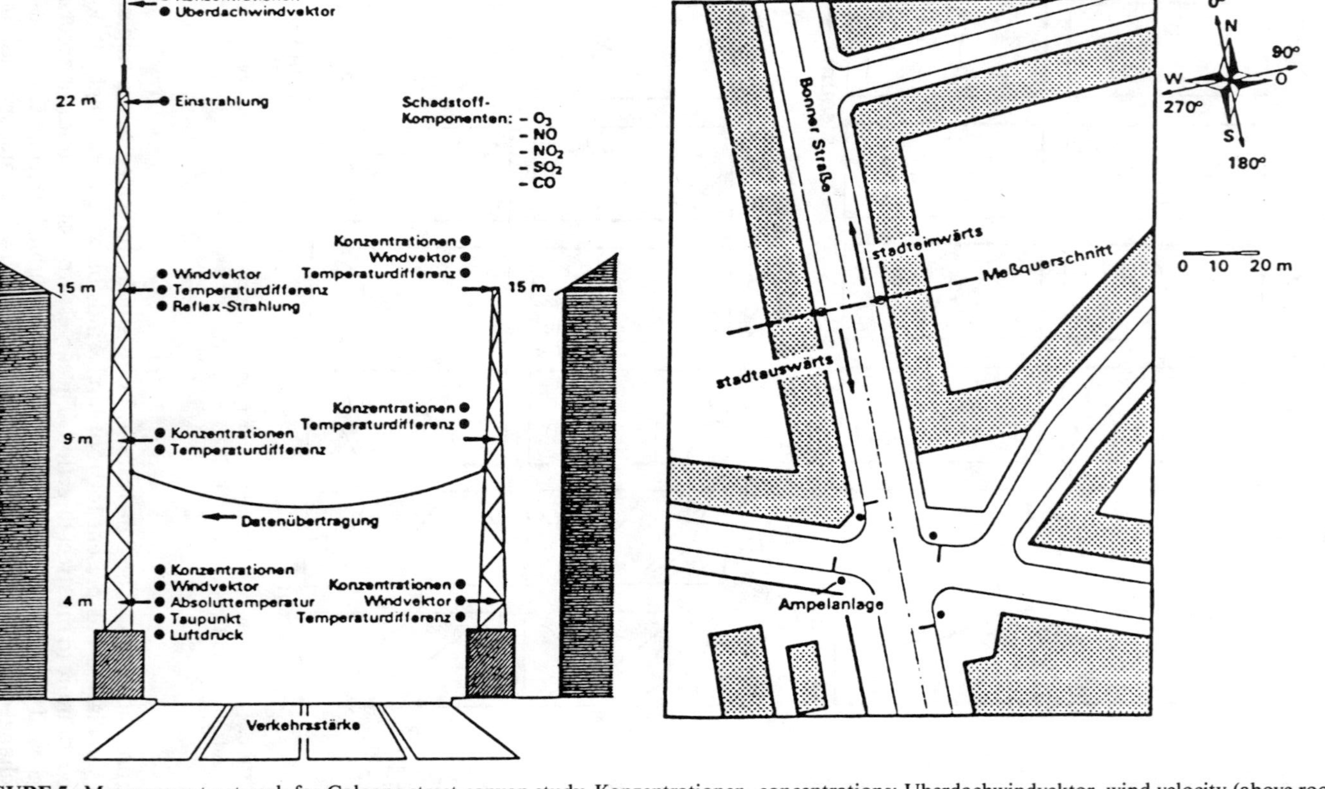


**FIGURE 5.** Measurement network for Cologne street-canyon study. Konzentrationen, concentrations; Uberdachwindvektor, wind velocity (above roof level); Einstrahlung, incoming radiation; Windveletor, wind velocity; Temperaturdifferenz, temperature difference; Reflex-strahlung, outgoing radiation; Absoluttemperatur, absolute temperature; Taupunht, dewpoint; Luftdruck, air pressure; Verkehrsstarke, traffic volume; Datenubertragung, data line; stadteinwarts, inbound traffic; stadauswarts, outbound traffic; Messquerschnitt, measurement cross-section. (Adapted from Waldeyer *et al.*[7])

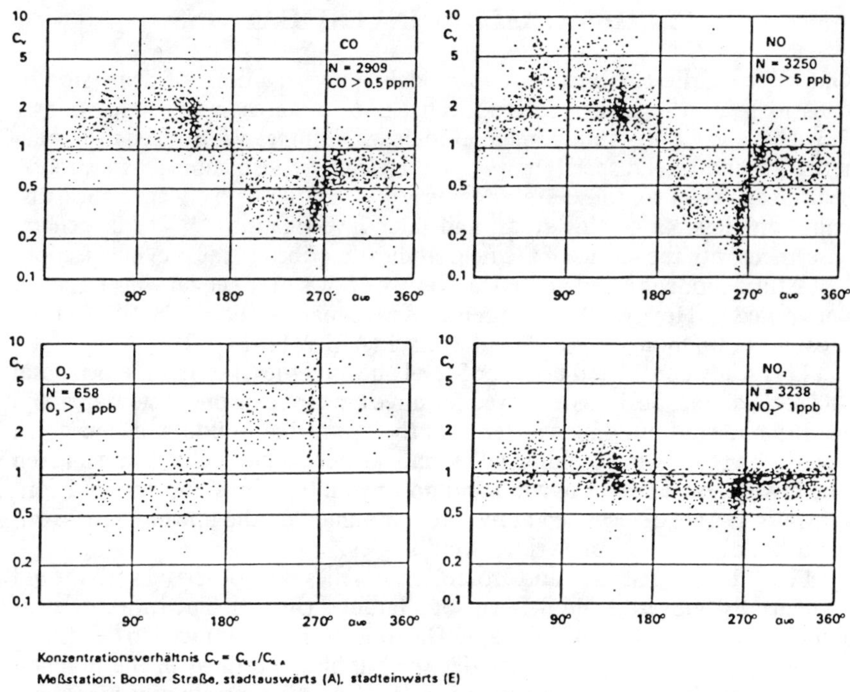



**FIGURE 6.** Ratio ($C_v$) of east-to-west side 4-m concentrations versus rooftop wind direction in Cologne street canyon.

concentration patterns from the St. Louis study, where the patterns are presented according to wind-direction and wind-speed categories.

In Cologne, a network was set up to measure CO, NO, $NO_2$, $O_3$ and $SO_2$ concentrations as well as winds, turbulence, temperature difference, pressure, humidity, and solar radiation. FIGURE 5 shows the experimental set-up. Half-hourly average data were obtained in a 20-m wide, 15-m high (on both sides) urban canyon (street) for the period from January 1 to July 14, 1981. A pronounced cross-street concentration gradient (FIG. 8) was revealed for a wide range of wind directions. A persistent vortex circulation was confirmed.

The Lyon study consisted of one month's CO concentration measurements in a single street canyon. A 20-m meteorologic tower was utilized in the median strip of the roadway. Street-level measurements were taken at both sides of the median and CO monitors were placed at two elevations on the northern building space. A street-canyon vortex air circulation was observed as illustrated in FIGURE 2.

## FLUID-MODELING INVESTIGATIONS

One of the earliest published fluid-modeling investigations of dispersion of emissions from highway sources located in urban areas was carried out by Hoydysh and Chiu.[9] A micrometeorologic wind tunnel was utilized to characterize ground-level dispersion in a symmetrically spaced cubical array, representing an urban complex. Tracer gas was utilized to obtain concentration distribution. Results of this study and later investigations[9-15] are in general agreement with the results of the field studies described in the previous section.

A fluid modeling and analytical study of flows in street canyons has been performed by Hoydysh.[15] Measurements were made utilizing the ESSCO Atmospheric Boundary Layer Wind Tunnel (ABLWT) (FIG. 7).

This study considered both roof-level and street-level sources. Horizontal and vertical transport was observed for three types of canyons: a step-up notch with the upwind building shorter than the downwind building, a step-down notch with the upwind building taller than the downwind building and an even notch with equal-height upwind and downwind buildings. Normalized concentrations were determined at receptors mounted on the upwind and downwind building faces, for various wind angles.

Flow characteristics in and around the notches were observed with the aid of neutrally buoyant helium-filled soap bubbles (FIG. 8). The trajectories of bubbles were recorded on videotape. The trajectories were traced to determine resident times and vertical velocities for bubbles entrained in the notches. Bubble velocities in the notches were compared with free-stream velocities above the notches.

Vertical concentration profiles at mid-block were measured for various wind angles. These empirical profiles were then compared with the predictions of several empirical and analytical models for those cases where the free stream flow is normal to the axis of the street canyon.

The results of this investigation have confirmed some previous research findings and provided several new insights as well:

(i) Street-level emissions contribute about 50 times more to observed concentrations within the notch than do equivalent emissions from the downwind edge of the upwind building.

(ii) Entrainment into the street canyon appears to be primarily the result of horizontal advection that results from vertical-axis corner vortices. FIGURE 8 shows sample trajectories for step-up and even notch configuration.

(iii) The distribution of trace-gas concentration contours is nearly identical on the leeward faces of the "even" and "step-up" notch configurations and is characterized by higher concentrations at mid-block; leeward contours for the "step-down" notch are horizontally stratified throughout the notch, suggesting the absence of convergence (relative to the other configurations).

(iv) Street-level cross-notch gradients indicate that concentrations are generally a factor of two or more greater for the leeward than the wind-

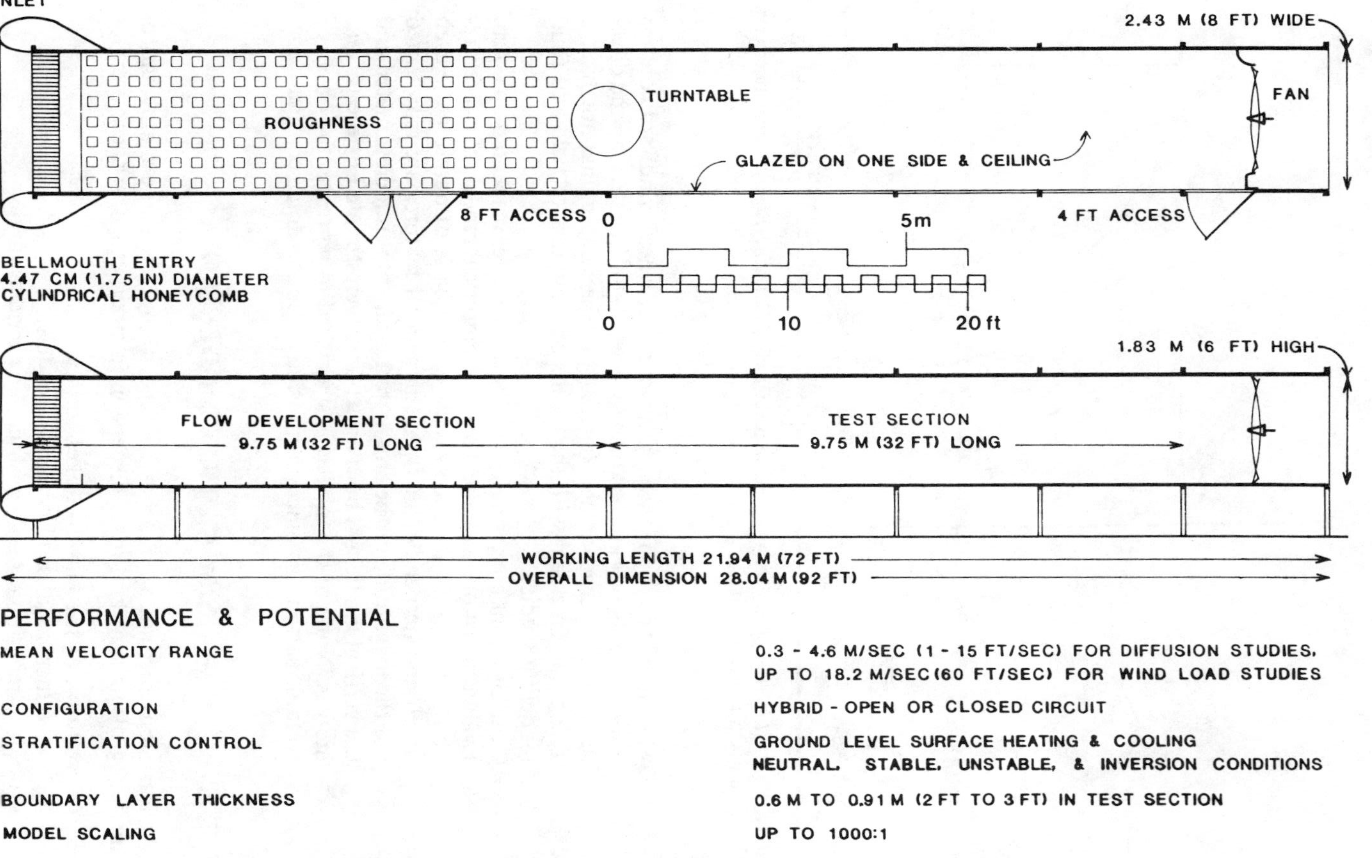


**FIGURE 7.** ESSCO boundary layer wind tunnel: technical specifications.

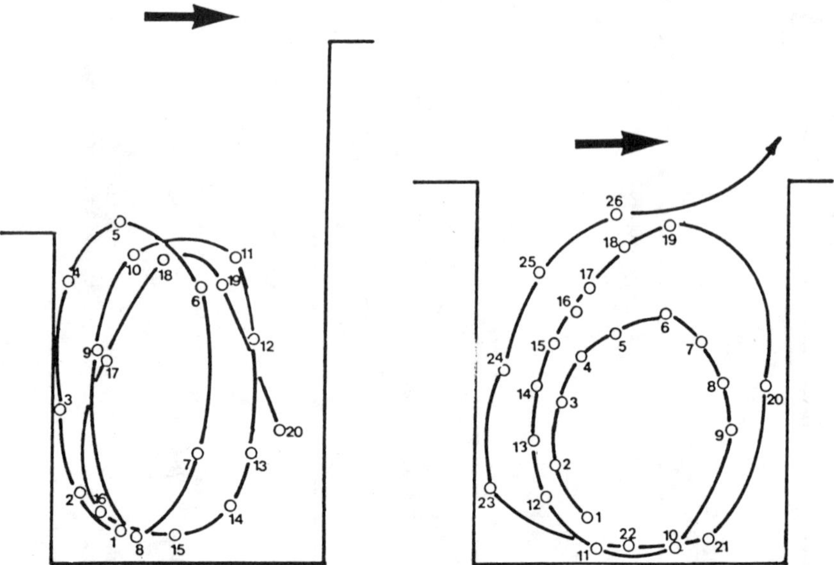

**FIGURE 8.** (*Left*) sample trajectory for step-up notch configuration. (*Right*) sample trajectory for even notch configuration.

ward face, except for the step-down notch where windward concentrations are slightly greater (than to leeward) for free stream wind directions nearly perpendicular ($r \leqslant 30°$) to the longitudinal notch axis. This is in agreement with the results of a number of field and fluid modeling studies.

(v) Concentrations are generally a factor of two lower in the step-up notch.

(vi) Evaluation of one analytical and two empirical models of concentration profile showed mixed results on the basis of data for the even notch and perpendicular flow; the general features are represented but the magnitude of the cross-street gradient is underestimated and the vertical gradient is both under and overestimated by the various models. None of the models is structured to represent the observed contour patterns as functions of wind angle and notch configuration.

## SIMULATION MODELING

At least four distinct approaches have been applied to the mathematical modeling of air flow and mass dispersion in street canyons; these are:

(1) solution of the Navier-Stokes equations;
(2) two-dimensional semi-empirical models;
(3) zero-dimensional semi-empirical box models; and
(4) empirical adaptation of Gaussian line-source dispersion.

The second and third approaches have been used most commonly because they are easily applied and have been evaluated on the basis of field measurements. The 15-year-old APRAC model is a representative example.[3] There are also a number of Navier-Stokes models, including one by Hoydysh *et al.*[16] A recent two-dimensional empirical model (CANY) has been constructed by Santowski and Deland.[4] Dabbert *et al.*[12] have published a recent modification of Gaussian dispersion formulation.

Although rigorous evaluation of various models under a variety of canyon and meteorological conditions is lacking, some of the models produce encouraging results. A comparison of calculations obtained with the CANY model[4] and observation in three Manhattan street-canyon sites revealed that the model performed well for two sites with aspect ratios above two and reasonably even at the third with a moderate aspect ratio of 1:4. Results of this study are currently employed in regulatory review in New York City.

## SUMMARY AND CONCLUSIONS

We have briefly examined some of the approaches to the consideration of pollution dispersion within urban street canyons. None of these approaches is mutually exclusive and all, at best, are approximate. It is surprising that a subject with such a major impact on human welfare has been studied so little from a scientific standpoint. In the absence of a deep and solid scientific foundation, practical problems must be tackled by a combination of approaches that are customized for each problem.

## REFERENCES


1. PETERSON, W. B. 1980. User's Guide for HIWAY-2, A Highway Air Pollution Model, Report EPA-600/8-80-018. Environmental Protection Agency. Research Triangle Park, NC.
2. TURNER, D. B. 1970. Workbook of Atmospheric Dispersion Estimates (Revised). Environmental Protection Agency. Research Triangle Park, NC.
3. MANCUSO R. L., & F. L. LUDWIG. 1972. User's Manual for the APRAC-1A Urban Diffusion Model Computer Program. Report EPA PB-213 091, Environmental Protection Agency. Research Triangle Park, NC.
4. TAN, C. 1984. New York City Carbon Monoxide Hot Spot Study. The Cooper Union Research Foundation and the New York City Department of Environmental Protection.
5. GEORGII, H. W., E. BUSCH, & E. WEBER. 1967. Investigation of the Temporal and Spatial Distribution of the Emission Concentration of Carbon Monoxide in Frankfurt/Main. NAPCA Trans. No. 0477.
6. DABBERDT, W. F., F. L. LUDWIG, & W. B. JOHNSON, JR. 1973. Validation and applications of an urban diffusion model for vehicular pollutants. Atmos. Environ. 7: 603–618.
7. WALDEYER, H., P. LEISEN, & W. R. MUELLER. 1981. Die Aghangigkeit der Immissionsbelastung in Strassenschluchten von meteorlogischen und verkehrsbedingten Einflussgrossen. Proceedings, Abgasbelastungen durch den Kraftfahrzeugverkahr, TUV Rheinland.: 85–114. Koln, FRG.

8.   JOUMARD, R. 1981. Ausbreitungsmodelle fur vevhehrs-Immissionen in Strassensch-
     lucten und Vergleich zu franzosischen Messungen. Proceedings, Abgasbelas-
     tungen durch den kraftfahrzengvezkehr, Tuv Rheinland.: 187–206. Koln, FRG.
9.   HOYDYSH, W. G. & H. CHIU. 1971. An experimental and theoretical investiga-
     tion of the dispersion of carbon monoxide in the urban complex. AIAA Paper
     #71-523.
10.  HOYDYSH, W. G. 1972. Dispersive Properties of Building Wake Flows, a Scale
     Model Study of Automotive Pollution Concentration in the Vicinity of High-
     Rise Structures. New York University, Environmental Engineering Research
     Laboratories. EERL/NYU 112.
10a. HOYDYSH, W. G., Y. OGAWA, & R. GRIFFITHS. 1974. Scale Model Study of Dis-
     persion of Pollution in Street Canyons. Presented at APCA 67th annual meeting,
     Denver, Colorado, June 9–13.
11.  HOYDYSH, W. G. & W. F. DABBERDT. 1983. Dispersion in Deep Urban Canyons —
     A Wind Tunnel Study. ESSCO TM-102. New York, NY.
12.  DABBERDT, W. F., E. SHELAR, D. MARIMONT & G. SKINNER. 1981. Analyses, Ex-
     perimental Studies and Evaluations of Control Measures for Air Flow and
     Air Quality on and near Highways. Federal Highways Administration,
     FHWA/RD-81/051, Final Report. SRI International. Menlo Park, California.
13.  JACKO, R. B. 1972. A Wind Tunnel Parametric Study to Determine the Air Pol-
     lution Dispersion Characteristics in an Urban Center. Ph.D. thesis, Purdue
     University.
14.  LEISEN, P., P. JOST, & K. S. SONNBORN. 1981. Modellierung der Schadstoffaus-
     breitung in Strassenschluchten Vergleich von Aussenmessungen mit rech-
     nerischer und Windkanal-Simulation. Proceedings, Abgasbelastungen durch
     den kraftfahrzeugverkahr, TUV Rheinland.: 207–234. Koln, FRG.
15.  HOYDYSH, W. G. & W. F. DABBERDT. 1986. A wind tunnel study of flow kinematics
     and dispersion characteristics in urban canyons. ESSCO TR-171. New York, NY.
16.  HOYDYSH, W. G., R. PIVA & Y. OGAWA. 1975. A combined experimental and
     numerical study of flow in street canyons. Presented at the 2nd U.S. National
     Conference on Wind Engineering Research, Colorado State University, Fort
     Collins, Colorado, June 23–25.

# Drinking Water Quality Concerns of New York City, Past and Present

GERALD R. IWAN

*Chief, Drinking Water Quality Control Division*
*New York City Department of Environmental Protection*
*Bureau of Water Supply*
*New York, New York 10003*

## THE WATER SUPPLY OF NEW YORK CITY

The City of New York, having a population of more than eight million people inhabiting five boroughs, consumes the incredible amount of 1.5 billion gallons of potable (drinking) water each day, received through approximately 850,000 service connections. This reflects a daily per capita consumption of 200 gallons—one of the highest in the country.

Water to the city's inhabitants is provided through a supply and distribution system which has been considered one of the foremost public works projects of the 19th and 20th centuries. The distribution system consists of a 6000-mile gridwork of water mains ranging in size from 2 to 84 inches in diameter, with roughly 180,000 valves and 100,000 hydrants. Through the system's gravity head and the use of regulating valves, a street pressure of 35–60 psi is achieved. This pressure is sufficient to supply water upward five or six stories without pumping. About 95% of the consumption is provided solely by gravity and head pressure, principles that are similar to those of the classic system of ancient Rome.

With the exception of portions of Queens, which draw less than 100 million gallons per day of groundwater from the last remaining privately owned supply, the Jamaica Water Company, the city's drinking water comes from three great surface water systems: the Croton, the Catskill and the Delaware. Waters from the upstate impounding reservoirs of each system's watersheds are conducted ultimately to the distribution network by an interconnecting series of gravity flow tunnels, aqueducts, and, in certain instances, open streams.

Water enters the North Bronx's Jerome Reservoir from the Croton System's 24-mile-long, 12–14 foot-diameter New Croton Aqueduct. Croton water leaves this 0.8 billion gallon storage reservoir to portions of the Bronx, Harlem, the West Side and portions of the East Side of Manhattan via mains from the one-billion-gallon Central Park storage reservoir, the 135th Street gatehouse, or by pumps at various locations to each elevation of over 40 feet.

The Hillview Reservoir, in Yonkers, New York, just north of the Bronx, acts as the receiving and equalizing reservoir for the Catskill and Delaware waters as they enter the city from the Kensico Reservoir. Although its 0.9-billion-gallon capacity is normally fed via the 92-mile-long, 17-foot-diameter Catskill Aqueduct, there are provisions for blending or isolating Delaware water as it ultimately reaches Hillview Reservoir from its 85-mile-long, 19-foot-diameter aqueduct.

Catskill delivery is made to the Bronx, Manhattan and Brooklyn from Hillview Reservoir via City Tunnel #1, which is a cement-lined pressure tunnel 18 miles long and 15 feet in maximum diameter. Delaware delivery is made through the 15–21-foot-diameter, 20-mile-long City tunnel #2 under pressure to the East Bronx, Queens, portions of Brooklyn, and to Staten Island via the Richmond Tunnel to the buried Silver Lake Tanks, a 100-million-gallon underground storage reservoir. Both city tunnels are 400–500 feet beneath the city and are the predecessors of the city's new Tunnel #3, which will interconnect both tunnels and all boroughs upon completion. It is estimated that construction costs of Tunnel #3 will exceed 25 billion dollars when it is completed in the 21st century.

New York City's water supply sources have a combined storage capacity of 550 billion gallons and consist of 19 reservoirs and 3 controlled lakes within 1950 square miles of watershed situated in nine upstate counties. The city, as landowner of 90,000 watershed acres, pays out more than 15 million tax dollars to these localities annually.

The three upstate supply systems can be briefly described as:

1. **The Croton:** The city's oldest supply, developed as early as 1842, the Croton system is composed of 12 reservoirs and 3 controlled lakes having a storage capacity of 86.6 billion gallons in a drainage area of 375 square miles. This system, also referred to as the "East of Hudson System," provides about 10–14% of the city's daily water needs or a maximum of 200 million gallons. Its contributing reservoirs (or controlled lakes), their completion dates, and maximum capacities are shown in TABLE 1.

In general, the water quality of the Croton supply is categorized as soft (with a hardness of 60 mg/L), neutral (pH 6.8–7.3), and biologically productive. Historic quality concerns have been chiefly aesthetic and related to eutrophic conditions because of high primary productivity. Available treatment consists of chlorination, use of copper sulfate, selective depth withdrawal, and isolation. Relatively high primary productivity associated with occasional taste and order effects and zooplankton population peaks are believed to be related to drainage characteristics and the associated effects of a growing population and real-estate development of the surrounding counties.

2. **The Catskill:** The second oldest supply, started in 1907 and commissioned in 1917, the Catskill system drains 571 square miles of the Esophus and Schoharie watersheds. Its two collecting reservoirs, the Ashokan and Schoharie, have a combined maximum storage of 147.5 billion gallons and satisfy 40% of the city's daily demand (approximately 600 million gallons). These reservoirs are connected via the 18.1-mile Shandaken Tunnel, which diverts the normal northerly Schoharie flow south into the Esophus Creek, Ashokan's major tributary. The Ashokan Reservoir was completed in 1915 and has a total storage between east and west basins of 127.9 billion gallons. The Schoharie,

**TABLE 1.** The Croton System

| Contributing Reservoirs | Date of Completion | Maximum Storage Capacity (billion gallons) |
|---|---|---|
| Lake Gilead (controlled lake) | 1870 | 0.380 |
| Lake Gleneida (controlled lake) | 1870 | 0.165 |
| Lake Kirk (controlled lake) | 1870 | 0.565 |
| Boyd's Corners | 1873 | 1.696 |
| Middle Branch | 1878 | 4.145 |
| East Branch | 1891 | 5.243 |
| Bog Brook | 1892 | 4.400 |
| Titicus | 1893 | 7.599 |
| West Branch | 1895 | 10.427 |
| Anawalk | 1897 | 7.070 |
| Muscoot | 1905 | 5.705 |
| New Croton | 1905 | 28.110 |
| Cross River | 1908 | 10.898 |
| Croton Falls | 1911 | 14.839 |
| Croton Falls Diverting | 1911 | 0.888 |

put on-line in 1927, adds an additional 19.6 billion gallons to the supply. Ashokan effluent feeds the Kensico Reservoir in Westchester County via a 75-mile portion of the Catskill Aqueduct. The Kensico, originally built in 1915 to maintain a then 60-day supply (30.573 billion gallons) of Catskill water for the city, is now recipient of both Catskill and Delaware effluents. Although physically located in the "Croton watershed," essentially all of its waters are "West of Hudson" due to its minor 13-square-mile drainage area.

The water of the Catskill, "West of Hudson" system, is characteristically very soft (20 mg/L), of low alkalinity (10 mg/L) and slightly acidic (pH 6.7). Historical quality concerns focused on the water's potentially corrosive nature, its episodic high turbidities and aesthetic concerns of taste and odor associated with periodic algal blooms, especially of *Synura (sp)*.

Contemporary treatment strategies involve chlorination, selective withdrawal, and flocculation through alum (aluminum sulfate) addition to aqueduct water at Pleasantville, New York and floc settling in Kensico Reservoir.

3. **The Delaware:** This system, our newest and largest source, was completed in 1965 as part of the "West of Hudson" system. Its four drainages—Neversink, East Branch Delaware, West Branch Delaware and Rondout—have a combined area of 1010 square miles. Its supply is collected in the 50-billion-gallon Rondout Reservoir, which was completed in 1944. The Rondout, serving as the central collecting reservoir, impounds waters from three feeder reservoirs: (1) The Neversink, completed in 1953 and possessing a storage capacity of 35.5 billion gallons; (2) the Pepacton, completed in 1955, with a storage capacity of 143.7 billion gallons; and (3) the Cannonsville, completed in 1965, which stores 97 billion gallons. All three impoundments are connected to Rondout by separate tunnels named after their respective drainages. The Delaware, West of Hudson system has a combined storage of 326 billion gallons and through connections with the West Branch and Kensico Reservoirs provides

50% (750 million gallons) of the city's daily requirement. Interconnection with the Hudson River at Chelsea, New York, provides an additional 100 million gallons per day in times of drought, such as in 1966 and 1985.

Treatment available to this system consists of chlorination, use of copper sulfate, diversion, selective withdrawal, and isolation. Characteristic water quality is soft (20 mg/L), acidic (pH 6.7–7.0), poorly alkaline (10 ppm), potentially corrosive, and selectively productive. Historic quality concerns have focused on Cannonsville waters, which are high in nutrients, primary productivity and eutrophication-related turbidities, tastes, odors, and so on. The watershed is currently sparsely developed, with the exception of certain population centers and agricultural districts in the western section.

## HISTORICAL PERSPECTIVE ON QUALITY AND QUANTITY OF WATER IN NEW YORK CITY

New York City's current municipal surface water supply has been in operation since 1842. However, the city's challenge to provide an adequate, pure and wholesome supply began with the founding of New Amsterdam by Peter Minuit in 1626.

Problems about the quantity and quality of water and the solutions to those problems have both shaped the city's character and parallel the story of its development and unequalled growth. Chronologically, there were five major landmark eras and events in the city's quest for an adequate and safe drinking water supply.

### *1626–1671: The Era of Natural Water Sources and Private Wells*

The original Dutch inhabitants depended upon locally available surface water sources such as springs, ponds and streams. The Katch-Hook, a 48-acre pond, was especially favored. By 1658, private shallow wells south of Wall Street supplemented the surface supply for its then 300 residents. In 1667, after the English occupation, the first stone-lined well was dug in Fort New York. The population of the settlement was 1500 and quantity was the only concern.

### *1671–1774: The Era of Public Wells*

The first of six stone-lined public wells was dug at the original site of City Hall (Pearl Street and Coenties Alley) in 1671. These wells were soon inadequate, unpredictable, and foul because of the infiltration of seawater and waste. Of the estimated 1700 pumps and wells in existence during this period, only one, the Tea Water Pump at Park Row, was considered by most to be potable. Foulness of the waters became so increasingly aggravated by waste defilement from privies, cesspools, and street drainage that by 1748 it was claimed that

horses would not drink the water and travelers avoided the city for fear of
the diseases that plagued it regularly. Water quality reflected the general state
of filth and unsanitary conditions that characterized the city at that time.

### 1774–1830: Era of Privately Owned Municipal Supplies

In an attempt to alleviate its primary concern for more water for general
use and fire-fighting, the City supported the efforts of the New York Water
Works in establishing the first distribution system in 1774. This system of
wooden pipes, wells and "Collect Pond" (Katch-Hook) water was never completed because of the Revolutionary War. The city's population of 22,000
swelled to 60,000 in 1800 before the next effort at a formal supply was undertaken. Despite suggestions for an extra-city source by such men as Dr. Joseph
Browne, the city, prompted by memory of the great fire of 1776 and the cholera
and yellow fever epidemics of 1798–1799, commissioned the Manhattan Company to supply a "Pure and wholesome water." An unfortunate era of poor
supply ensued from 1800–1830. Corruption and poor service forced the population of 200,000 into utter dependency on a ground water supply that, according to its scientists, was fouled by 100 tons of excrement daily. Although
sanitary science and the disease theory were not yet formulated, the consensus
was to seek a purer water quality from sources to the north and to cancel the
Manhattan Company's contract with the city for its daily 700,000 gallon service to a mere one-third of the population.

### 1830–1917: Era of the Municipal Supply

The city's water problems were punctuated by the disastrous fire of 1828
and the cholera epidemic of 1832, during which over 100,000 residents im-
migrated to the previously unsettled "north," to the area now known as Greenwich Village, causing further groundwater contamination. In 1835, the
Common Council, encouraged by such city notables as Peter Cooper, approved
plans to bring water to the city from Westchester County by constructing a
dam and the first or "old" Croton Aqueduct to tap the Pristine Croton River.
In this age of innovative engineering, iron pipe, and new technology, the first
great municipal water supply project was conceived. Pure Croton water entered New York City for the first time on July 4, 1842 amidst a massive civic
celebration. The Old Central Park and Murray Hill Reservoirs received the
seemingly inexhaustable supply. Its purity was certified by the methods of
quality determination available at the time. This source of water met the need
only temporarily, for as the 1842 population of 300,000 grew to 1.2 million
by 1883, all available modifications were made to the "Old Croton" system.
The current supply was again unable to meet the demands of growth, leakage
and waste.

In 1883, construction of a new aqueduct and reservoir supply to meet the
high demand was approved. The new Croton Aqueduct was completed in 1890
and the New Croton or Cornell Dam was completed. At 234 feet, it was the

highest dam in the world in 1907. These landmark projects and their appurtenances maximized use of the Croton watershed for the city's population, which, by 1910, was more than 3 million.

Methods used to determine water quality improved between 1835–1910 as the link between disease and contamination by living organisms was being recognized and the agents of typhoid fever, yellow fever and cholera were discovered. It was the era of the microbiological discoveries of C. J. Esberth, Carlos Finley and Robert Koch, and Pasteur's disease theory was being debated, as was Lister's "new" antiseptic technique. Attention was turning to the possible relationship between "impurities" and "animalcules" in water and the occurrence of epidemic diseases. As early as 1831 the Lyceum of Natural History insisted that the Manhattan wells were polluted by sewage. The city of Schenectady was attempting to control its cholera epidemic by sanitation with chlorinated lime. In 1868, Drs. Elisha Harris and C. F. Chandler stated: "The sanitary importance of the excellent water supply which New York City enjoys cannot be overestimated, nor can we sufficiently prize the safeguards by which its continued and constant purity shall be insured . . . [to] prevent every source and possibility of defilement of the stream."[1]

Comparing the organic and inorganic composition of Croton with Manhattan well water, they demonstrated the foulness of the wells and suggested a link between impure water and the epidemics of the time. Charles F. Gissler, in a detailed microscopic study of the Croton supply (1872), advocated filtration of Croton water at Central Park to improve its quality.[2] J. Michales, in 1878, advocated filtration of Croton water because of organisms present from unsanitary sources in the system such as privies, manure piles, slaughterhouse refuse, and so forth. He stated: "Pure water is as necessary for health as food or fresh air yet it is probably more neglected or less understood in modern times than either."[3]

The public and the press, confronted with constant reminders of the effects of disease, advocated that the new technology of filtration be adopted for the Croton supply. Already cities such as London, England (1855), Poughkeepsie (1873) and Hudson, New York (1875), and Providence, Rhode Island showed dramatic reduction of cases of typhoid after the water supply was filtered. In 1905 Thomas Darlington, Commissioner of the New York City Department of Health, expressed concern over the pollution of the watershed, the growth of New York, the suggestion of using Hudson River water, and the contribution of "dead ends" in the distribution system to disease. He advocated filtration, but cautioned: "The plan of filtration herein recommended does not mean that the city should guard any less the source of supply."[4]

The case for filtration was strong, and in 1905, New York, with a population 3.5 million inhabiting all five now legally incorporated boroughs, was about to embark on this enterprise. However, by 1910 filtration was no longer considered a viable alternative, when responsible authorities concluded that it was not necessary to do so at the time. It is likely that certain new developments influenced their ultimate decision. Chlorination was established as a useful disinfectant in Jersey City in 1903, microbiological analyses and chemistry had developed sufficiently since 1882 to be judged reliable indicators of quality, the fermentation tube method was recommended in 1893, and the New York State Health Department was achieving success with water supply legislation enacted from 1880–1892, and again in 1905, which afforded protection to the watersheds. New York established its first calcium hypochlorite disinfection of Croton water in 1910 at Dunwoodie, New York and gas chlorination two years later at the same location. Paralleling similar achievements in the state, it can be seen that it was use of surface waters and disinfection rather than any dramatic abatement of pollution which controlled typhoid, dysentery, cholera, and other fecal-origin diseases in New York City. The typhoid death rate, taken at that time as a quality index, dropped from 15.1 to 6.5 per 100,000 with the first introduction of chlorination. Continuing disinfection without further treatment has resulted in progressive declines, as evidenced by the rates: 5.3 (1916); <4 (1917); 2 (1920); 0.3 (1938), and <0.1 currently. Chlorination remains our major source of protection against bacterial disease today. Consistent with developments in the health area, aesthetic concerns were also being addressed during this period. The use of copper sulfate for algal control, superchlorination for taste and odor mitigation, alum for clarification, etc., were all coming into vogue. Formal monitoring of the New York City water supply began in 1897 with the establishment of the Mt. Prospect Laboratory situated at Mt. Prospect Reservoir in Brooklyn.

### *1917–1984: Modern Era of Municipal Supply*

Even before the Croton reached its maximum extension in 1910, plans for a new New York City water supply were being developed. The momentum for the Catskill system formally began in 1897 with a report to the Brooklyn Merchants Association recommending extension of the existing supply.[5] The final consolidation of Brooklyn's population into the five boroughs in 1898, the ensuing report of the city's consultant, John Freeman,[6] recommending expansion of the supply northward, and passage of the state laws of 1905 opened the way for acquisition of new watershed property and established the State Water Supply Commission. Coincident developments establishing the New York City Board of Water Supply, with a formal laboratory division under George C. Whipple, state repeal of the Ramapo Water Company's claims to various Catskill watershed drainages (1901), increasing concern over the adequacy of the Croton supply and public dissatisfaction over the declining quality of supplementary supplies to Brooklyn, Queens and the Bronx, set the stage for the development of the Catskill (1907–1927) and Delaware systems (1927–1965). The city's activities in the development of the Catskill supply is a classic history of human relations, politics and engineering. Only four years after the final Schoharie segment was completed in 1927, a landmark Supreme Court decision (1931) permitted New York City's tapping the Delaware watershed. With consumption increasing at the rate of 42 million gallons per day (mgd)/year, the existing new Catskill supply could be counted on only until 1932 to meet growth requirements. The Delaware supply, designed to supplement the Catskill system, was begun in 1932. In phases, through World War II, the supply developed. The first phases, the Rondout Reservoir and

Delaware Aqueduct, were begun in 1937, but the last, or third stage—the Cannonsville—did not go online until 1965. The colossal enterprise of constructing the Catskill-Delaware supplies took place during a "new era" of successful public water supply technology. Built on the foundation of the scientific achievements of the turn of the century Croton era, the Catskill/Delaware systems were characterized by further technological advancements in chlorination practices and they incorporated improved New York State sanitary regulations, developments in copper sulfate technology for algal control, use of activated carbon filtration, improved water quality standards and increased Federal- and State-legislated water quality criteria and regulation. During this era, the city maintained, or was a leader in, state-of-the-art water quality practice, having individuals such as Frank Hale, G.C. Whippel,[7,8] and Ben Nesin associated with laboratory operations and contemporary treatment technology. The number of laboratories continued to increase to seven, and functions performed included chemical and biological monitoring, sanitary surveys, treatment and inspections.

## NEW YORK CITY'S DRINKING WATER QUALITY CONTROL SYSTEM

The current system of five laboratories responsible for the monitoring of New York City "raw" and potable waters is under the administration of the New York City Department of Environmental Protection (NYC DEP). Within the DEP's Bureau of Water Supply, the Drinking Water Quality Control (DWQC) laboratories provide the following scientific services:

1. Sampling, monitoring, and analysis of the drinking water supply within watersheds, reservoirs, water plants and the distribution system;
2. Sampling, monitoring and analysis of upstate streams and sewage treatment plants for assurance of New York State Department of Environmental Conservation compliance with SPDES (State Pollution Discharge Elimination System) discharge permits;
3. Investigation of citizen water quality complaints, leaks, new main installation and water quality emergencies;
4. Investigation of new water sources for the city;
5. Recommendations for treatment, purification, sanitation, and protection of the water supply;
6. Investigations and research on the watersheds, supply and distribution systems to insure maintenance of water quality.
7. Water quality documentation as required by the New York State Department of Health and other regulatory agencies to assure the safety and potability of the New York City drinking water.

The current operation employs 60 chemists, microbiologists, and technicians, who collectively participate in conducting physical/chemical and microbiological analyses on the distribution system and on the collective upstate supply. This represents a yearly effort of approximately 400,000 quality analyses conducted on over 30,000 samples.

New York City, as supplier of water, is legally mandated to comply with Part 5 of the New York State Sanitary Code, as revised in 1981, and Part 128, Section 128.1, as revised in 1962. Part 5 contains state rules designated to assure the health and safety of drinking water supplies within the state through implementation of protective measures at water sources and in the treatment and distribution systems. Part 128 of the code specifies rules and regulations for the sanitary protection of the watersheds and reservoirs. The New York City Department of Health, as a regulatory agency, is charged with implementing Parts 5 and 128 in New York City in conjunction with the water supplier, NYC DEP. The responsibility for conducting the potability monitoring and analyses on the distribution system is that of the Central Laboratory staff, located at 51 Astor Place (the Cooper Union School of Engineering Building) in Manhattan.

Under Part 5, Sections 5-1.50 to 5-1.55, the minimum acceptable water quality criteria for potable drinking water is specified for community water systems. The underlying basis for the maximum contaminant level (MCL) criteria is founded in the Federal Safe Drinking Water Act of 1974, which includes the National Interim Primary Drinking Water Standards of Public Law 93-523, effective 1977, and the National Secondary Drinking Water Regulations, effective 1981.

By law, the NYC DEP must conduct analyses and issue reports on the categories shown in TABLE 2.

The parameters and MCLs as established by law and the representative corresponding values of New York City drinking water are shown in TABLE 3.

The city meets the present NYS DOH potability criteria. However, to do so, an annual variance for extension of the turbidity MCL to 5 NTU is necessary. This variance has been granted thus far for the following reasons: experimental demonstration that the nature of the supply's turbidity does not interfere with chlorination, progress mandated by NYS DOH for ultimate filtration of Croton water at Jerome Park Reservoir, and assurance that the city maintain minimum 0.2 ppm free chlorine residual throughout distribution.

Water quality problems within distribution are relatively rare. Local quality problems are more frequent. During one year (1984), approximately 700 water quality complaints were addressed, in the following categories: "Dirty water", "worms", "yellow water", "green water", "blue water", "taste", "lead", "odor",

**TABLE 2.** Analyses Conducted by the New York City Department of Environmental Protection

| Category | Analyzed | Reported |
| --- | --- | --- |
| Inorganic chemicals (Sect. 5-1.51a) | Monthly | Monthly |
| Inorganic chemicals (Sect. 5-1.51c) | Monthly | Monthly |
| Pesticides/herbicides (Sect. 5-1.52a) | Every 3 years | Every 3 years |
| Total trihalomethanes (Sect. 5-1.52b) | Quarterly | Quarterly |
| Entry-point turbidity (Sect. 5-1.53a) | Daily | Monthly |
| Distribution-point turbidity (Sect. 5-1.53b) | 5 times per week | Monthly |
| Microbiology (Sect. 5-1.54) | 500 times per month | Monthly |
| Radioactivity (Sect. 5-1.55) | 4 times per quarter every 4 years | Every 4 years |

**TABLE 3.** Regulated Constituents of Potable Water (NYS DOH Chapter 1, Part 5, State Sanitary Code): Comparison to New York City Drinking Water

| Category | Maximum Contaminant Level (mg/L) | Croton System | Catskill-Delaware System |
|---|---|---|---|
| Arsenic (As) | 0.05 | <0.04 | <0.04 |
| Barium (Ba) | 1.0 | <0.05 | <0.05 |
| Fluoride (F) | 2.2 | 0.96 | 0.94 |
| Lead (Pb) | 0.05 | <0.005 | <0.005 |
| Mercury (Hg) | 0.002 | <0.001 | <0.001 |
| Nitrate (AsN) | 10.0 | 0.28 | 0.36 |
| Selenium (Se) | 0.01 | <0.005 | <0.005 |
| Silver (Ag) | 0.05 | <0.01 | <0.01 |
| Cadmium | 0.01 | <0.005 | <0.005 |
| Chromium | 0.05 | <0.03 | <0.03 |
| Chloride (Cl) | 250.0 | 24.0 | 8.0 |
| Copper (Cu) | 1.0 | <0.02 | <0.02 |
| Iron (Fe) | 0.3 | 0.08 | 0.06 |
| Manganese (Mn) | 0.3 | 0.06 | 0.02 |
| Sodium (Na) | No limit | 11.31 | 4.56 |
| Sulfate (SO4) | 250.0 | 15.7 | 8.6 |
| Zinc (Zn) | 5.0 | <0.05 | <0.05 |
| Color | 15 | 13 | 7 |
| Corrosivity | Noncorrosive | pH 6.8–7.3 | pH 6.7–7.2 |
| Odor | 3 | 0-2 cc (Chlorine) | 0-3 cc (Chlorine) |
| Trihalomethanes (THMs), total | 0.1 | −.039− | |
| Coliform bacteria | (<1/100 ml) | −0.10− | |
| Turbidity | (1 NTU) | 1.2 | 0.9 |
| Endrin | 0.0002 | <0.0002 | <0.0002 |
| Lindane | 0.004 | <0.004 | <0.004 |
| Methoxychlor | 0.1 | <0.1 | <0.1 |
| Toxaphene | 0.005 | <0.005 | <0.005 |
| 2-4-D | 0.1 | <0.1 | <0.1 |
| 2,4,5-TP Silvex | 0.01 | <0.01 | <0.01 |
| Total THMs | 0.1 | 50 | 30 |
| Gross alpha radioactivity | 5 pCi/L | <5 | <5 |
| Gross beta radioactivity | 15 pCi/L | <15 | <15 |

"health problem", "gasoline smell", etc. Of these, only about 10% were substantiated as actual quality problems. Historically, quality complaints have primarily involved episodes of turbidity, color, sediment, taste, and presence of *Diptera* larva. The most recent general complaint was a 2-month episode of "earthy" taste, which plagued the Delaware supply in January-February 1984. The phenomenon was coincident with elevated amorphous matter and the diatom *Synedra acus*. Environmental extremes in 1983 and January 1984 caused perturbations of the Delaware watershed and reservoir basin. Heavy rains increased nutrient availability, accelerating primary production. Subsequent decay and release of decomposition products produced a detectable taste. The incident was strictly aesthetic, but was the most persistent quality "incident" recorded since the Delaware system was placed into service. Aesthetic problems are usually associated with the older Croton system because of its characteristic greater natural productivity associated with a "richer" nutrient profile and eutrophic symptoms. Current projections for increasing development of the Croton watershed will undoubtedly increase the incidence of cultural insults to the system's natural purification processes.

Besides mandated testing, additional analyses are also conducted to secure potability by the Central Laboratory on distribution samples collected from 218 locations in the five boroughs both by its own water quality inspectors and by BWS field crews. TABLE 4 shows the water quality for the New York City distribution system in 1985. It is particularly important to note that the data represent the addition of approximately 80 mgd of pumped Hudson River water into the Catskill-Delaware supply due to drought conditions. Although characteristic parameter concentrations were altered slightly, in no instance was potability compromised nor were regulatory requirements violated.

The Central Laboratory is also participant in contemporary research on water quality topics. Of the many areas being investigated, trihalomethane formation potential, acid precipitation effects, and bacterial screening are currently popular:

*Trihalomethane Formation in New York City Drinking Water:* Concern exists for the presence of the trihalomethanes in chlorinated drinking waters. These potentially carcinogenic compounds originate from the chlorination of humic and fulvic acids characteristic of surface water supplies. The relationship between resulting THM concentration, chlorine dose, natural precursor concentrations, and time in distribution is seen as important for control of these organic compounds. Although New York City is well below the MCL of 0.1 mg/L, operational concerns demand further knowledge of our particular distribution characteristics regarding THMs. Preliminary research conducted at the laboratory has indicated that: (1) THM concentrations are approximately 50% higher in the Croton supply than in the Catskill-Delaware supply; (2) successive chlorination may increase THM levels in certain areas of the city; (3) maximum formation is evidenced in less than one day; and (4) the contribution of total organic carbon (TOC) to THM formation potential of the supply should be investigated. Typical data reflective of this are illustrated in TABLE 5.

*Acid Precipitation/Acidification Potential:* A recent review of historical raw water and distribution water data indicates no classic effects of "acidification" of New York City reservoirs. The stability of ionic concentrations, pH and organism composition as determined by analyses would seem to verify this contention.

Historical analytical data were reviewed and special analyses conducted in cooperation with the NYC DEP Division of Policy and Planning. Results based on characteristic alkalinity values indicated that six Catskill-Delaware system reservoirs, (Ashokan, Schoharie, Rondout, Neversink, Pepacton, and Cannonsville), were moderately sensitive to the potential effects of acidification (average alkalinity = 10 ppm). Criteria generally recognized for acidification classification are as follows:

Acidified             $<0$ μEq/L (1 μEq/L = 1/20 ppm alkalinity)
Extremely sensitive   0–40 μEq/L

**TABLE 4.** Range and Average of Drinking Water Quality Characteristics of the Catskill-Delaware and Croton Supplies from Respective Points in the New York City Distribution System for the year 1985[a]

| Constituent | Standards (MCL) | Catskill-Delaware System[b] Range | Average | Croton System Range | Average |
|---|---|---|---|---|---|
| | | *Chemical and Microbiological Analyses Required by USEPA NDW Regulations, Primary Standards* | | | |
| Arsenic | 0.05 mg/L | < 0.04–< 0.04 mg/L | < 0.04 mg/L | < 0.04–<0.04 mg/L | < 0.04 mg/L |
| Barium | 1. mg/L | < 0.05–0.42 mg/L | 0.15 mg/L | < 0.05–0.83 mg/L | 0.35 mg/L |
| Cadmium | 0.010 mg/L | < 0.005–< 0.005 mg/L | < 0.005 mg/L | < 0.005–< 0.005 mg/L | < 0.005 mg/L |
| Chromium | 0.05 mg/L | < 0.03–< 0.03 mg/L | < 0.03 mg/L | < 0.03–< 0.03 mg/L | < 0.03 mg/L |
| Lead | 0.05 mg/L | < 0.03–< 0.03 mg/L | < 0.03 mg/L | < 0.03–< 0.03 mg/L | < 0.03 mg/L |
| Mercury | 0.002 mg/L | < 0.001–< 0.001 mg/L | < 0.001 mg/L | < 0.001–< 0.001 mg/L | < 0.001 mg/L |
| Nitrate (as N) | 10 mg/L | 0.16–0.48 mg/L | 0.30 mg/L | 0.06–0.46 mg/L | 0.25 mg/L |
| Selenium | 0.01 mg/L | < 0.005–< 0.005 mg/L | < 0.005 mg/L | < 0.005–< 0.005 mg/L | < 0.005 mg/L |
| Silver | 0.05 mg/L | < 0.01–< 0.01 mg/L | < 0.01 mg/L | < 0.01–< 0.01 mg/L | < 0.01 mg/L |
| Fluoride | 1.4–2.4 mg/L | 0.59–1.30 mg/L | 0.95 mg/L | 0.71–1.30 mg/L | 0.96 mg/L |
| Turbidity[c] | 1 NTU | 0.3–3.1 NTU | 1.0 NTU | 0.5–2.9 mg/L | 1.3 mg/L |
| Coliform bacteria[d] | < 1/100 ml (mean) | | < 1/100 ml | | < 1/100 ml |
| Endrin | 0.2 µg/L | < 0.05–< 0.05 µg/L | < 0.05 µg/L | < 0.05–< 0.05 µg/L | < 0.05 µg/L |
| Lindane | 4 µg/L | < 0.05–< 0.05 µg/L | < 0.05 µg/L | < 0.05–< 0.05 µg/L | < 0.05 µg/L |
| Methoxychlor | 100 µg/L | < 0.5–< 0.5 µg/L | < 0.5 µg/L | < 0.5–< 0.5 µg/L | < 0.5 µg/L |
| Toxaphene | 5 µg/L | < 1–< 1 µg/L | < 1 µg/L | < 1–< 1 µg/L | < 1 µg/L |
| 2,4-D | 100 µg/L | < 0.2–< 0.2 µg/L | < 0.2 µg/L | < 0.2–< 0.2 µg/L | < 0.2 µg/L |
| 2,4,5-TP Silvex | 10 µg/L | < 0.1–< 0.1 µg/L | < 0.1 µg/L | < 0.1–< 0.1 µg/L | < 0.1 µg/L |
| Total Trihalo-methanes | 100 µg/L | 20–69 µg/L | 38 µg/L | 39–89 µg/L | 59 µg/L |
| | | *Chemical Analyses Required by USEPA NDW Regulations, Secondary Standards* | | | |
| Chloride | 250 mg/L | 7.0–23.5 mg/L | 12.5 mg/L | 18.0–30.0 mg/L | 26.0 mg/L |
| Color | 15 color units | 1–31 units | 7 units | 2–30 units | 10 units |
| Copper | 1 mg/L | 0.01–0.22 mg/L | 0.05 mg/L | 0.05–0.23 mg/L | 0.11 mg/L |
| Corrosivity[e] | Noncorrosive | Langelier Index (−2.1)–(−3.7) | L.I. −2.6 | Langelier Index (−0.9)–(−2.0) | L.I. −1.5 |
| Dissolved solids | 500 mg/L | 34–115 mg/L | 57 mg/L | 98–156 mg/L | 127 mg/L |
| Foaming agents | 0.5 mg/L | (MB-LAS) < 0.01–< 0.01 mg/L | (MB-LAS) < 0.01 mg/L | (MB-LAS) < 0.01–< 0.01 mg/L | < 0.01 mg/L |

| Iron | 0.3 | mg/L | 0.02–0.18 | mg/L | 0.06 | mg/L | 0.04–0.25 | mg/L | 0.10 | mg/L |
|---|---|---|---|---|---|---|---|---|---|---|
| Manganese | 0.05 | mg/L | < 0.01–0.14 | mg/L | 0.03 | mg/L | < 0.01–0.20 | mg/L | 0.05 | mg/L |
| Odor | 3 threshold | | 0, 1A–2A, 1B, 1Bs, 1CC–3CC 1d, 1M–2M, 1df, 1E | | | | 0, 1A–2A, 1B, 1Bs, 1df–2df 1CC–3CC, 1d–2d, 1M–2M, 2G | | | |
| pH | 6.5–8.5 | | 6.7–7.5 | | — | | 6.8–7.5 | | — | |
| Sulfate | 250 | mg/L | 5.0–12.3 | mg/L | 8.4 | mg/L | 7.0–17.0 | mg/L | 13.9 | mg/L |
| Zinc | 5 | mg/L | < 0.01–< 0.01 | mg/L | < 0.01 | mg/L | < 0.01–< 0.01 | mg/L | < 0.01 | mg/L |
| | | | *Chemical Analyses Not Required by USEPA NDW Regulations* | | | | | | | |
| Alkalinity as CaCO₃ | | | 8–16 | mg/L | 13 | mg/L | 39–50 | mg/L | 44 | mg/L |
| Aluminum | | | < 0.01–0.13 | mg/L | 0.05 | mg/L | 0.01–0.15 | mg/L | 0.04 | mg/L |
| Beryllium | | | < 0.01–< 0.01 | mg/L | < 0.01 | mg/L | < 0.01–< 0.01 | mg/L | < 0.01 | mg/L |
| Boron | | | 0.01–0.11 | mg/L | 0.05 | mg/L | 0.03–0.12 | mg/L | 0.07 | mg/L |
| Carbon dioxide | | | 1.7–2.7 | mg/L | 2.3 | mg/L | 2.2–8.8 | mg/L | 5.2 | mg/L |
| Calcium as CaCO₃ | | | 11.75–20.12 | mg/L | 16.33 | mg/L | 37.50–50.50 | mg/L | 45.21 | mg/L |
| Calcium as Ca | | | 4.70–8.04 | mg/L | 6.53 | mg/L | 15.00–20.20 | mg/L | 18.08 | mg/L |
| Chemical oxygen demand | | | 1.9–9.4 | mg/L | 4.4 | mg/L | 2.3–12.2 | mg/L | 8.0 | mg/L |
| Cyanide | | | < 0.02–< 0.02 | mg/L | < 0.02 | mg/L | < 0.02–< 0.02 | mg/L | < 0.02 | mg/L |
| Dissolved oxygen | | | 8.0–12.8 | mg/L | 10.1 | mg/L | 6.5–14.4 | mg/L | 9.4 | mg/L |
| Hardness as CaCO₃ | | | 16–38 | mg/L | 25 | mg/L | 61–80 | mg/L | 67 | mg/L |
| Iodide | | | < 0.01–< 0.01 | mg/L | < 0.01 | mg/L | < 0.01–< 0.01 | mg/L | < 0.01 | mg/L |
| Lithium | | | < 0.01–< 0.01 | mg/L | < 0.01 | mg/L | < 0.01–< 0.01 | mg/L | < 0.01 | mg/L |
| Magnesium as CaCO₃ | | | 4.92–11.48 | mg/L | 7.09 | mg/L | 20.11–25.42 | mg/L | 23.57 | mg/L |
| Magnesium as Mg | | | 1.20–2.80 | mg/L | 1.73 | mg/L | 4.98–6.20 | mg/L | 5.75 | mg/L |
| Nickel | | | < 0.05–< 0.05 | mg/L | < 0.05 | mg/L | < 0.05–< 0.05 | mg/L | < 0.05 | mg/L |
| Nitrogen-ammonia | | | < 0.03–0.07 | mg/L | 0.04 | mg/L | < 0.03–0.08 | mg/L | 0.05 | mg/L |
| Nitrogen-nitrite | | | < 0.001–0.006 | mg/L | 0.003 | mg/L | < 0.001–0.005 | mg/L | 0.002 | mg/L |
| Phenol | | | < 0.001–< 0.001 | mg/L | < 0.001 | mg/L | < 0.001–< 0.001 | mg/L | < 0.001 | mg/L |

**TABLE 4.** (Continued)

| Constituent | Standards (MCL) | Catskill-Delaware System[b] | | Croton System | |
|---|---|---|---|---|---|
| | | Range | Average | Range | Average |
| Phosphate | | 0.02–0.30 mg/L | 0.07 mg/L | 0.03–0.31 mg/L | 0.08 mg/L |
| Potassium | | 0.51–1.15 mg/L | 0.79 mg/L | 0.93–2.10 mg/L | 1.70 mg/L |
| Silica as silicon dioxide | | 1.3–3.5 mg/L | 2.2 mg/L | 2.0–5.7 mg/L | 3.9 mg/L |
| Sodium[f] | | 3.60–12.89 mg/L | 7.15 mg/L | 9.00–12.90 mg/L | 12.24 mg/L |
| Specific conductance | | 56–155 μmhos/cm | 107 μmhos/cm | 170–225 μmhos/cm | 209 μmhos/cm |
| Strontium | | < 0.05–< 0.05 mg/L | < 0.05 mg/L | < 0.05–< 0.05 mg/L | < 0.05 mg/L |
| Temperature | | 34–71 °F | — | 34–74 °F | — |
| *Microbiological Analyses Not Required by USEPA NDW Regulations* | | | | | |
| Amorphous Matter | | 95–1225 S.U. | 674 S.U. | 325–1250 S.U. | 878 S.U. |
| Heteorotrophic Plate Count (48 hrs. at 35°C)[g] | | 0–300 Bact/ml | 5 Bact/ml | 0–251 Bact/ml | 12 Bact/ml |
| Plankton Organisms[g] | | 25–1200 S.U. | 175 S.U. | 30–4025 S.U. | 577 S.U. |

[a] Prepared by the New York City Department of Environmental Protection, Bureau of Water Supply, Drinking Water Quality Control Laboratory (E. Chen, Laboratory Director).

[b] Because of the 1985 New York City drought, Hudson River water was pumped at Chelsea Upstate New York, and mixed with the New York City Catskill-Delaware System during the period of 7/10–12/11/85. According to the results of chemical analyses, Hudson River water was not shown in the New York City distribution system until the first week of August.

[c] The State has granted the City of New York a turbidity variance with a 5 NTU limit (cf. Part 5 State Code).

[d] TNTC's data are not included in the city-wide coliform (mean) calculation. The city-wide number of the coliforms from the distribution system is 0.10 (mean), number of samples with > 4/100 ml coliforms is 22, and % of samples with > 4/100 ml coliforms is 0.3.

[e] If the Langelier Index (L.I.) is zero, water is in chemical balance; if the L.I. is a plus quantity, scale-forming tendencies are indicated; and if the L.I. is a minus quantity, corrosive tendencies are indicated.

[f] There is no MCL for sodium. Water containing sodium more than 20 mg/L and 270 mg/L should not be used for drinking by those on severely and moderately restricted sodium diets, respectively.

[g] TNTC's data are not included in the calculations.

**TABLE 5.** THM Concentrations (ppb) with Time at Locations/System

| Time | Croton | Croton | Catskill-Delaware | Catskill-Delaware | Catskill-Delaware (Staten Island) |
|---|---|---|---|---|---|
| 0 hr | 39 | 45 | 22 | 23 | 42 |
| 1 day | 55 | 60 | 29 | 38 | 48 |
| <5 days | 55 | 58 | 31 | 35 | 55 |
| Cl residual (mg/L) | 0.7 | 0.3 | 0.5 | 0.5 | 0.6 |

| | |
|---|---|
| Moderately sensitive | 41–200 µEq/L |
| Low sensitivity | 201–500 µEq/L |
| Not sensitive | >1500 µEq/L |

Catskill-Delaware water having alkalinity values of 200 µEq/L is defined by convention as moderately sensitive. To further document water quality-acidification effects, programs in joint cooperation with the USGS are being contemplated, as are further statistical analyses of available data.

*Bacterial Characterization:* The routine microbiological determinations for coliforms and standard plate counts indicate adequate disinfection of the supply as evidenced by the virtual absence of the indicator bacterium *E. coli.* However, the laboratory plans to expand its bacteriological testing. Recent "outbreaks" of seemingly chlorine-resistant coliform bacteria in distribution systems of such cities as New Haven, Connecticut, which filter and chlorinate their supply, have serious implications for a city as large as New York. In light of such episodes, progressive action is indicated.

The Central Laboratory is investigating the merits of establishing a program to elucidate the kinds of bacterial flora, by speciation, that are not revealed in routine coliform operations. This will assist bacteriologists in identifying and cataloging species of atypical flora and relating their incidence to seasonality and system of origin. Potential microbiological concerns will be determined for those bacteria that are now bypassed by routine analyses. This approach will help us to recognize possible latent hazards and recommend necessary corrective action. Such surveillance is thought advisable at this time as supportive of efforts to monitor and preserve supply quality. The future bacterial character of distribution will undoubtedly reflect any increase in degenerative water quality effects from acceleration of watershed/development/related disturbances. Bacterial screening may help to document any such detrimental changes and serve as a basis for effecting corrective or preventive actions in the watershed.

Paralleling the activities of Central Laboratory are the functions of the four upstate laboratories:

- Kensico Laboratory, Valhalla, New York, East of Hudson District; Reservoir monitoring; research; limnology
- Brewster Laboratory, Brewster, New York, East of Hudson District: Waste water treatment plant monitoring; sanitary surveys
- Ben Nesin Laboratory, Shokan, New York, Catskill, West of Hudson District: Reservoir and waste water treatment plant monitoring

• Grahamsville Laboratory, Grahamsville, New York, Delaware, West of Hudson District: Reservoir and waste water treatment plant monitoring

Combined, these laboratories monitor seven New York City-owned sewage treatment plants and over 90 privately owned State Pollution Discharge Elimination System (SPDES) Permittees on the watersheds. Partial protection of the supply under Part 128 of the New York State Sanitary Code and New York City Rules and Regulations is effected in this manner.

To better address concerns for potential deterioration of reservoir quality due to improper waste treatment and to encourage compliance with requirements of SPDES permits, and enforcement if necessary, a formal program of SPDES violation reporting has been instituted. Violations of discharge parameters are substantiated by more than 8000 annual SPDES-related analyses. Results are reported each month by the upstate laboratories for each non-NYC Permittee. It is hoped that such information may be useful in effecting increased surveillance-enforcement of existing SPDES regulations and establishing stricter requirements by the regulatory agency, NYS DEC. It is contemplated that increased NYC activity in this area may produce data substantiating the need to promote enforcement and refinement of New York City watershed rules and regulations and encourage progressive State legislation focused at providing special land use protection for the watersheds of our supply. This is particularly relevant to those watershed areas that are experiencing an acclerated growth and with it increased potential for contamination of the supply.

The routine upstate quality monitoring program covering 180 established points on the three systems is supplemented by a limnology operation which conducts an additional 40,000 yearly analyses. The limnology section, established in 1982, incorporates laboratory analyses with its own field data to effect a more comprehensive understanding of the processes involved in watershed and reservoir dynamics.

There are four major goals for this operation:

1. Develop a comprehensive data base describing each reservoir and its immediate watershed as to chemistry, physics, biology, geology, soils, and so forth.
2. Develop a trophic index for all NYC reservoirs. This index will be designed to provide information as to the longevity of the NYC reservoirs and the extent of their viability as quality water supply sources.
3. Identify major point sources of discharge within the NYC watersheds and assess the contribution of each to total nutrient loading and eutrophication potential of each NYC reservoir.
4. Collect and analyze data on the periodicity, control, treatment, and or removal of problematic organisms within the NYC reservoir system. Develop more efficient monitoring methods, more effective treatment practices, and improve the technical communication between laboratory and engineering operations.

Limnological data collected on thermal profiles, nutrient concentrations, chlorophyll, total organic carbon, and so on, have already begun to be used to qualify the relative quality of the individual sources of supply, the possible cause and effect relationships between quality and the state of watershed development, and the chemical/biological characteristics that can be utilized in indicating treatment for control of water quality problems. Further expansion of efforts in limnology is a logical developmental progression for the DEP laboratories. Combined with a renewed effort toward pollution abatement, it has the potential for refocusing laboratory attention toward more meaningful and contemporary investigative approaches to quality prediction.

The preceeding areas of investigation are incorporated within the laboratories' current program plan. To address the laboratories' responsibility in providing significant and practical water quality information and direction, eight areas have been defined for development.

1. *SPDES Violation Reporting.* Previously described, this system of biweekly reports aids in documenting problematic wastewater discharge situations, helping to focus enforcement efforts.

2. *Limnology Program.* Previously described, this comprehensive monitoring program of the system's supply reservoirs assists in establishing management practice relevant to actual reservoir conditions.

3. *Monitoring Reevaluation.* A reassessment of the applicability of existing monitoring locations both in distribution and supply will increase the practicality of data collected coincident with limnological findings and distribution characteristics.

4. *Training Programs.* Formal programs of training laboratory personnel in contemporary investigative procedures is being formulated.

5. *Analytical Reevaluation.* In-house analytical capabilities will be improved by investigating state-of-the-art instrumentation and methods for analysis of samples.

6. *Report Issue and Data Interpretation.* A combination of reporting requirement review will be incorporated with computerization of data to produce practical periodic water quality reports.

7. *Facilities and Resource Review.* Existing resources and current needs will be assessed to establish programs for upgrading capabilities consistent with changing emphasis in laboratory operations and water quality concerns.

8. *Water Quality Protection Plan.* A total scientific system will incorporate a multidisciplinary approach to assessment of factors affecting ultimate drinking water quality. This includes suggesting areas for policy development on the basis of documentable concerns.

The DEP laboratory's intent by this eight-point approach is to revise traditional approaches to looking at water quality matters, both at the source and in distribution. The intended effect is to gain a better understanding of our water resources, how best to protect them, and in doing so, to protect the public and the future quality of life in New York City.

# THE FUTURE OF NEW YORK CITY'S
# WATER QUALITY

The revisions of the Safe Drinking Water Act (SDWA) and the Amendments of 1986 from the House of Representatives and Senate have recently received considerable scientific and political attention. These long-anticipated enhancements to the original legislation of 1975 reflect a more rigorous and treatment-based approach to water quality health concerns. Although notable issues were covered in the initial proposed legislation of 1984, the process still continues in 1987 with the major areas of focus being the following:

1. Establishment of maximum contaminant levels (MCLs) for Volatile Organic Chemicals (VOCs): trichlorethylene, tetrachloorethylene; carbon tetrachloride; 1,1,1-trichloroethane; vinyl chloride; 1,2-dichlorethane; benzene; 1, 1-dichlorethylene; and *P*-dichlorobenzene.
2. Establishment of the concept of MCLs and Maximum Containment Level Goals (MCLGs) for encouraging maximum mitigation of contaminants by technological development.
3. Establishment of monitoring requirements for unregulated contaminants.
4. Substitution of treatment requirements for MCLs and MCLGs for contaminants where measurement technology is unavailable or poor, such as for *Giardia lamblia.*
5. Establishment of requirements to filter and/or disinfect public water supplies, with emphasis on universal filtration of all surface water sources except where documentation permits exemption.

The proposed legislation has stimulated considerable debate on the issues indicated. Although there is a general consensus of support for improvement of quality standards and for protection of the public, certain aspects have caused polarization between members of the water supply industry and regulators and legislators. An issue of particular interest to New York City is the proposed requirement for universal filtration of surface waters. Essentially, the initial intent of the requirement was to apply filtration technology to all surface supplies to "prevent" public exposure to water-borne contaminants and pathogenic organisms. Superficially, this directive would at first appear a positive contribution to the SDWA, and a laudable endeavor. The concept of universal filtration is, however, contradictory to some very basic surface water quality concepts. Fortunately, in 1986 the fact that certain high-quality supplies did not require filtration was recognized legislatively and proposed criteria in this area are currently being drafted as of Spring 1987.

Current scientific knowledge recognizes the individual characteristics of surface waters. Their degree of quality, as defined by limnology, aquatic ecology, and other "water sciences," are often obvious and quantifiable by both natural and man-made parameters. Violations of regulatory criteria have historically been demonstrated to be on a case-by-case basis. Since there is no universal contaminant scientifically recognized in water, it follows logically that universal application of a particular treatment may be, in many cases, superfluous, wasteful, and damaging to the public's perception of natural water quality.

The future interests of a public served by surface water supplies can best be addressed by legislation that demonstrates an enlightened approach to drinking water quality regulation. At a minimum, such legislation should consider three basic elements:

(1) *Monitoring requirements.* If not previously established, a public surface water supply must have its characteristic limnological parameters accurately monitored and documented to define both individuality and the extent to which health and aesthetic criteria are met.

(2) *Treatment requirements.* Upon review of the supply quality characteristics, a rational evaluation of treatment requirements must be made. Treatment must consider individual supply problems such as chronic violation of quality criteria and be prescribed to meet the particular situations. Specific monitoring requirements must be attached to all treatment mandates to track and otherwise document efficacy.

(3) *Protection requirements.* Suppliers of public drinking water must have the ability to maintain the highest water quality available from their sources. Given the critical concern demonstrated by the proposed SDWA legislation of 1984, it is both inconsistent and detrimental to the utility of the Act to ignore provisions for surface source protection. Any practical approach to protecting the public from water-borne contaminants must have provisions for preventing new introductions and removing established introductions from drinking water surface supplies.

Supplies such as New York City's, which exhibit no chronic water quality problems, would benefit only slightly from filtration. Considerable long-term economic and health rewards would be realized, however, from protection of the quality of the source water.

Rather than cause a disservice to the citizens of New York City, who enjoy an excellent water quality, by mandating expensive, unnecessary treatment, the need for an alternative program for source quality protection is indicated.

A practical program of drinking water quality preservation to provide long-term preservation of supply quality can be briefly presented by the following objectives and associated areas of investigation:

**Objective I:**
—Characterize the primary sources of water in terms of volumes, precipitation temporal patterns, maxima, averages and minima, and short- and long-term trends. Include those factors that effect precipitation composition.

—Identify all available precipitation information relevant to the watersheds and reformat for practical utilization.

—Document meteorology and climatology factors connected with watershed precipitation patterns

—Investigate air pollution chemistry of the atmosphere as it effects the watershed.

**Objective II:**
—Characterize watershed drainages in terms of water quality modification potential.
—Describe and inventory watershed characteristics of topography, soil composition, and geology.
—Describe and inventory vegetation patterns and ecological systems.
—Document active point discharges of contaminants and current controls.
—Document non-point source pollution and controls.
—Develop correlations between specific discharges and impact on receiving waters by cateogory.
—Develop computerized "watershed simulation" models.

**Objective III:**
—Establish a program that provides short- and long-term information on supply quality, including interrelationships between chemical and biological constituents and their effects.
—Develop and implement a standard operating procedures (SOP) manual for laboratory analyses and interpretations.
—Develop and implement computerized information system for water quality and related data.
—Further develop and expand the existing limnology program.

**Objective IV:**
—Document ultimate consumer water quality and distribution system effects on same.
—Develop maps of local distribution water quality.
—Identify causes of potential water deterioration.
—Identify contribution of "local" or residential plumbing to potential deterioration.

**Objective V:**
—Define all operation and treatment capabilities and their contemporary utility in protecting water quality.
—Project costs of increased use of treatment chemicals and methods due to projected raw water qauality decline by eutrophication or pollution.

**Objective VI:**
—Develop recommended strategies for maintaining a high-quality supply for consumers.
—Identify compatibility of existing supply quality and proposed regulations or contemporary concerns.
—Identify main sources of nutrient input aggravating eutrophication effects.
—Assess efficiency of current SPDES treatment requirements.
—Assess technical feasibility and cost of upgrading SPDES treatment facilities.
—Categorize each watershed as to future eutrophic concern potential. Project rates of development and increase of perturbation factors.
—Recommended points for the involvement of New York City in the co-development of land-use policies with other governmental agencies.
—Develop a legislation/regulation package for the preservation of the city's drinking water.

The data produced by this approach, the cause and effect information, and analytical methods generated can all be used in developing drinking water preservation strategies. The combination of strategy implementation, monitoring of critical parameters, and predictive simulation models will serve to afford a long-term preservation of water source and distribution quality.

This multidisciplinary approach is a logical, cost-effective alternative or supplement to implementing additional treatment of NYC drinking water.

The history of New York City's development of an adequate and "potable" drinking water supply, development of disease control, quality criteria, and the dependency of a great population on surface supplies indicate the critical need for maintaining the high quality of the existing drinking water resource, if New York City is to continue to grow or survive.

The City of New York is again entering into a new era in the ongoing history of its water supply development. This era, unlike its predecessors, focuses almost solely on demonstrating, maintaining, and preserving the potable quality of its drinking water. The demonstrable excellence of our drinking water aside, the City is again, because of external pressures at a crossroads.

One direction presents a future of complete dependence on treatment technology and associated economic and technical responsibilities of unimaginable magnitude. With this choice, the reservoir and watershed qualities upon which we are now so dependent will most certainly decline as a result of pressures originating from an attitude of permissive development and multiple use. The resulting continuing spiral of decline of quality followed by increased treatment, followed by yet further degradation and further need for treatment will certainly exert negative effects on the City's overall ability to sustain positive development and quality of life.

The alternative direction, characterized by a policy of resource protection and scientific surveillance, affords advantages in maintaining a cost-effective, high-quality drinking water without the complexities of superfluous treatment technology and source quality degradation. New York, because of its unique high-quality surface supply, again has the opportunity to lead in establishing principles for drinking water preservation that are consistent with the lessons of its historical supply development. The principles of surface water preference, a remote supply, sanitary protection, maximum utilization of natural quality, disinfection, scientific surveillance, and selective application of technology can be synthesized into a policy of reservoir protection that has a utility for filtered and nonfiltered sources alike. The end result afforded by this approach and our contemporary laboratory efforts assure a high-quality source for the future water consumers of New York City and establish the information and confidence enabling the city to meet its next, unknown water supply challenge.

## REFERENCES


1.  HARRIS, E. 1868. Sanitary Qualities of the Water Supply of New York and Brooklyn. Wescott. New York, NY.
2.  GISSLER, C. F. 1872. Contributions to the Fauna of the New York Croton Water: Microscopical Observations. Vogt. New York, NY.

3.   MICHELS, J. 1878. Croton Water, Its Nature, Properties and Impurities, with Original Microscopical Drawings of the Organic Deposit. Metropolitan Water Filtering Company, New York, NY.
4.   DARLINGTON, T. 1905. Report on Filtration of the Water Supply of the City of New York. Martin B. Brown Press. New York, NY.
5.   MERCHANTS' ASSOCIATION OF NEW YORK. 1900. An Inquiry into the Conditions Relating to the Water-Supply of the City of New York. I.H. Blanchard Co., New York, NY.
6.   FREEMAN, J. R. 1900. Report upon New York's Water Supply, with Particular Reference to the Need of Procuring Additional Sources and Their Probable Cost. Martin B. Brown Co., New York, NY.
7.   WHIPPLE, G. C. 1899. The Microscopy of Drinking Water. John Wiley & Sons. New York, NY.
8.   WHIPPLE, G. C. 1907. The Value of Pure Water. John Wiley & Sons. New York, NY.

# The U.S. Geological Survey's Toxic Waste–Ground-Water Contamination Program

STEPHEN E. RAGONE

*Toxic Waste–Ground-Water Contamination Program*
*U.S. Geologic Survey*
*Department of the Interior*
*Reston, Virginia 22092*

## INTRODUCTION

In fiscal year 1982, the United States Geological Survey (USGS) began an interdisciplinary research program entitled the Toxic Waste–Ground-Water Contamination Program (hereafter called the Toxic Waste Program). This program provides both a focus for ongoing studies and an opportunity for new earth science research relevant to toxic waste and ground-water contamination issues. The Toxic Waste Program consists of two primary components: (1) process-oriented research, which consists of the study of the physical, chemical, and microbiological processes affecting contaminant transport in ground water, and field research; and (2) an appraisal of ground-water quality. A secondary component involves the development of general earth science guidelines for situating hazardous waste-disposal facilities. This paper provides a brief description of the objectives, approaches, and status of the components of the Toxic Waste Program. More detailed information concerning the program may be obtained from several reports.[1–4]

## RESEARCH

### Process-Oriented Studies

Knowledge of the physical, chemical, and microbiological processes that affect the movement and alteration of contaminants through the subsurface must be factored into a complex equation to solve the hazardous-waste and ground-water contamination problems of the nation. Laboratory and field studies conducted over the last dozen years or so have helped to characterize the basic processes that affect contaminant transport in the subsurface, but, at the same time, they have also demonstrated how complex natural systems can be. Thus, the objective of the research funded by the Toxic Waste Program is to extend current understanding of earth science processes to better describe the fate of contaminants in the multiphase and multicomponent systems found in nature. Because the unsaturated zone and surface water are hydrologically connected with ground water, studies of processes occurring in these environments must also be conducted in order to evaluate their impact on ground-water transport of contaminants.

For the purpose of description, the research being conducted by the Toxic Waste Program is categorized by the disciplines of ground-water hydrology, organic geochemistry and ecology, and inorganic geochemistry. As should become clear from the project description, considerable overlap exists among these categories. Because the study of basic processes in the earth sciences generally requires information from more than one of the classical sciences, field problems in ground-water contamination require an interdisciplinary approach.

### Ground-Water Hydrology

Ground-water hydrology provides a quantitative understanding of the flow of water and the transport of its constituents through the subsurface. The classical definition of ground water defines it as water that occurs in fully saturated, consolidated, and unconsolidated earth materials. However, it is important that the unsaturated zone also be studied because of the number and complexity of processes occurring there that can affect the transport of hazardous substances to the water table.

As a discipline, ground-water hydrology classically has included the mathematical, chemical, geological, and physical sciences. More recently, biology has become an important discipline in ground-water hydrology. Although the basic mathematical transport models for ground-water systems have been developed, many of the variables in these models have not been adequately investigated. The Toxic Waste Program supports research projects that are intended to improve our understanding of the physics of single and multiphase fluid flow, mass and energy transport, and fluid-phase change. Problems of particular interest include:

* **Hydrodynamic dispersion:** The movement of solutes can be greatly affected by hydrodynamic dispersion, which is a physical process. Dispersion under natural conditions can be 1000 to 10,000 times greater than that measured at laboratory scale. Although it is possible to estimate an asymptotic value for dispersivity (the value approached after a relatively long distance of flow) from the statistical properties of hydraulic conductivity, there is little quantitative understanding of dispersion processes at intermediate field scales, before asymptotic dispersivity is reached.

* **Multiple phase flow:** The transport of fluids and their chemical constituents can occur as a gas or liquid phase separate from the water phase. Many chlorinated hydrocarbons and other organic industrial chemicals are only slightly soluble in water and have a different density than water. Consequently, these fluids can move as separate phases from ground-water flow. Quantita-

tive methods are needed to understand, describe, and predict multiphase transport.

• **Flow through fractured media:** The flow of water through fractured rocks is more difficult to characterize quantitatively than is flow through porous media. However, in many ground-water environments involving wastes or contamination, fractures are the dominant conduit. Although porous-media theory provides a feasible modeling approach, the low permeability and highly heterogeneous nature of fractured rocks require modifications in current theory and field methods. Modeling of flow through fractured rock can be approached by aggregating the flows occurring through individual fractures or by statistical approaches to fracture properties and distributions.

• **Flow and transport in the unsaturated zone:** The quality of ground and surface waters often is significantly influenced by chemical and solute dispersion processes in the unsaturated zone. Physical and chemical processes, and factors controlling them, are considerably more complicated and difficult to measure in the unsaturated zone than in saturated media. Also, our ability to simulate these processes with numerical models is generally inadequate for all but the most simplistic field conditions.

• **Chemical and microbiological reactions:** Simple chemical reactions currently can be included in transport equations and numerical models used to solve the equations, but only after making simplifying assumptions. Considerable work remains to incorporate more complete and realistic chemical and microbiological reactions in transport equations and models. Studies of the chemical and microbiological processes affecting ground-water transport are described in later sections.

Traditionally, geohydrologic information used to determine the physical and chemical properties of ground-water systems was obtained by analyzing the aquifer material collected when wells were emplaced and from water-level measurements made through the fininshed well. Although analyses of samples from wells and other well data remain an essential part of ground-water studies, additional indirect methods are being sought that will help determine aquifer flow properties, chemical characteristics of ground water, and the pathways and locations of contaminant plumes. Surface geophysical methods provide a relatively quick, inexpensive, and indirect way to characterize some aspects of the hydrogeologic setting and may provide a means for determining the three-dimensional distribution of leachate plumes under some conditions. However, the potential utility of this methodology will have to be evaluated through additional field demonstration and verification studies conducted in a variety of hydrogeologic settings. There is also considerable potential for further research and development of these techniques.


## Organic Geochemistry and Ecology

Disposal of a large number of hazardous organic compounds derived from industrial, agricultural, and domestic sources has contaminated ground water.

The problem is compounded by the complex geochemical and microbiological processes affecting the behavior of organic contaminants in the subsurface and a lack of understanding of these processes. In some cases, organic geochemical processes also directly or indirectly affect the transport and fate of inorganic contaminants such as heavy metals.

Research is currently under way in the following areas:

- **Characterization of natural organic solutes in ground and surface water:** Naturally occurring organic polyelectrolytes are highly reactive materials that are present in practically all water systems. They interact with both organic and inorganic pollutants and nutrients and may influence or control the toxicity, rate of movement, and rate of degradation of the pollutants and nutrients in aquatic environments.

- **Study of adsorption phenomenon:** The adsorption characteristics of organic compounds on soil, sediment, and aquifer materials are only qualitatively understood. The role of soil/sediment organic matter, mineralogy, and soil moisture content on the sorption and the transport of organic substances also is not clear. Quantification of these processes, and their rates and partition constants is an important step in defining the level of contaminants in the environment and the pathways of chemical exposure to human beings.

- **Biological partitioning of trace metals:** The same biota in different environments may differ widely in their susceptibility or their response to trace elements. These differences may, at least partly, be related to the differences in the availability of metals in sediments. However, little is known about the geochemical and physiological factors that influence the transport of metals from sediments to organisms.

- **Development of methods to evaluate sublethal concentration of toxic substances in aquatic ecosystems:** Standard methods for evaluating the effect of sublethal concentrations of toxic substances in aquatic ecosystems are not available. Field and laboratory data on the effects of low concentrations of toxicants on function and structure of aquatic communities are largely lacking.

*Inorganic Geochemistry*

The movement and concentration of inorganic solutes in the subsurface may be affected by a variety of reversible or irreversible chemical reactions, such as ion exchange, mineral solution or precipitation, ion filtration, adsorption/desorption, oxidation/reduction, and complexation. Inorganic reactions with aquifer materials can change the aquifer's hydraulic characteristic as well as the surface chemistry of aquifer materials. Research in this subject area is aimed at advancing our understanding of these reactions and include projects addressing the following topics:

- **Clay-solute reactions, particularly with regard to the interaction of clay minerals with contaminants:** Clay minerals are constituents in soils, sediments, and sedimentary rocks. Clays also are among the most chemically complex of sedimentary minerals and are often sensitive to changes in the chemical (fluid) environment. Interaction between migrating solutes and clay minerals results in both modification of the host pore-fluid chemistry and in modification and concomitant change of physical and chemical properties of the clay-mineral assemblage. Characterization of the interaction of clays with waste fluids and rates of their reactivity are required to evaluate and predict behavior of waste disposal and containment and to assess the effects of migration through clay-bearing strata.

• **Adsorption behavior of trace organic and inorganic solutes on particulate materials such as aluminosilicate minerals and colloids, and solids of biogenic origin in natural systems:** The chemical composition of natural water can be influenced by sediment surface reactions. This is attributed to sorption on the sediment surface of trace metals, pesticides, and salts. These sorbed species are interactive across the solid-liquid interface and may move in either direction in response to changes in chemical potential set up on either side of the interface. Knowledge of the processes taking place between the solid and liquid phases by which sediment particles acquire coatings, the chemical composition of the coating, and the chemical reactivity of the coatings when exposed to various bulk water compositions is necessary to predict water quality.

• **The relationship of chemical quality of natural water and human health:** In recent years there has been increasing interest and study concerned with the relationships between the chemical quality of natural water and human health and disease. After nonenvironmental factors are excluded, it appears that local and regional differences in water quality may have some effect on health and disease.

### *Site Studies*

To evaluate how well the current state of knowledge of earth science processes can be used to quantitatively describe real-world problems and to advance that state of knowledge, teams of scientists from the USGS and universities representing all major earth science disciplines began studies in fiscal year 1983 at selected ground-water contamination sites. Sites were selected to study ground-water contamination resulting from a crude oil pipeline break near Bemidji, Minnesota; from the infiltration of creosote and pentachlorophenol from waste-disposal pits at Pensacola, Florida; and from the infiltration of sewage treatment effluent on Cape Cod, Massachusetts. The types of contaminants at the three sites are commonly associated with many ground-water contamination problems throughout the United States. Thus, although the field studies are intended to test and develop the understanding of the processes affecting contaminant transport rather than to solve the specific problems at the sites, the knowledge gained through these studies should have transfer value to other sites where similar contamination has occurred and hydrogeologic conditions are similar.

Although predominantly ground-water studies, the studies at all three sites include examination of the unsaturated zone and surface water. Because ground-water contamination generally occurs in shallow systems, the unsaturated zone and surface-water system often become impacted by contaminants in ground water. Inclusion of surface water and unsaturated studies is necessary as a "sink term," or, in some cases, a "source term." Loss of organic constituents to the unsaturated zone through such processes as volatilization or by entrapment in interstitial pore space, or to the surface water through ground-water discharge, must be understood in order to understand the fate of these constituents in ground water.

The site investigations began with the development of an integrated work plan to aid in the exchange of information and ideas among the individual research studies and to ensure that the major processes affecting contaminant transport were being considered. The following sections provide a brief description of the sites and the major research goals being addressed.

### Infiltration of Sewage-Plant Effluent: Cape Cod, Massachusetts

Otis Air National Guard Base is located in the western side of Cape Cod and is situated on a glacial outwash plain that is underlain by an unconfined aquifer. Since 1938, the Base's sewage-treatment plant has treated more than 8 billion gallons of sewage and discharged the effluent to the aquifer through infiltration beds. As a result of this long-term discharge, a plume of sewage effluent formed in the sand and gravel aquifer that is currently 2500 to 3500 feet wide, 75 feet thick, and more than 11,000 feet long.

The sand and gravel aquifer is 90 to 140 feet thick and is underlain by silty sand and till. Recharge from precipitation, estimated to be 21 inches per year, has depressed the plume 20 to 50 feet below the water table. Ground water flows laterally through the sand and gravel at a rate of 0.7–1.5 feet per day.

An objective of the study is to develop a better understanding of the spatial variation in hydraulic conductivity and dispersivity. Hydraulic conductivity can be measured by variety of well-known techniques, but the results generally are average values for a volume of the aquifer. Small-scale variation in hydraulic conductivity, not accounted for in the spatially averaged values, may significantly affect the rate of contaminant movement.

Another objective of the study is to better understand the geochemical and microbiological processes affecting the transport and fate of contaminants in the sewage plume. Information on trace metals, nutrients, and organic solutes, and mineralogical composition of the aquifer is a necessary part of the research on solute transport, but little detailed information is available. This information will be used to delineate the extent of the plume and also will be used in chemical modeling of the plume. Microbiological processes in the sewage plume can affect the plume's organic and inorganic geochemistry. Detailed knowledge of these processes is necessary in order to explain the fate of many inorganic and organic constituents in the plume.

### Petroleum Pipeline Break near Bemidji, Minnesota

The study site is at a major pipeline break (10,000 barrels) of crude petroleum that occurred near Bemidji, Minnesota on August 20, 1979. Clean-up efforts recovered 6434 barrels, and some of the oil was burned off. Regulatory and remedial action has been completed. Oil percolated through the unsaturated zone, resulting in as much as 4 feet of oil floating on the water table. Ground water contaminated by the oil moves through the glacial outwash aquifer towards a shallow lake.

The objectives of the project are to obtain a more complete understanding of the mobilization, transport, and fate of petroleum derivatives in ground water and to use this understanding to develop predictive models of the behavior of this type of contaminant in the subsurface. A large part of ground-water contamination problems in the United States results from the discharge or leakage of liquid petroleum products such as oil, fuel oil, and gasoline. Organic substances from the liquids can move through the subsurface as vapors, dissolved constituents of water, a separate fluid phase, or as aggregates of individual compounds that are neither truly dissolved nor a distinct fluid phase. A substantially improved understanding of the physical, chemical, and biological processes controlling the dissolution and movement of organic liquids is needed in order to better predict contaminant behavior.

## Infiltration of Creosote and Pentachlorophenol, Pensacola, Florida

An abandoned creosote plant is located in northwest Florida within the city limits of Pensacola. The 18-acre plant site is situated approximately 600 yards north of Pensacola Bay near the entrance to Bayou Chico. The plant was operated from 1902 until December 1981. During the 80 years of continuous operation, wastewaters, generated from the use of creosote and pentachlorophenol (PCP) in the wood-treatment process, were discharged into two unlined surface impoundments which are in direct contact with the sand-and-gravel aquifer. Data collected from the disposal site indicate that the areal extent of ground-water contamination is confined to approximately 26 acres downgradient from the source. The approximate vertical extent of contamination of the aquifer extends to 90 feet at its greatest depth, just downgradient from the ponds.

Despite the 80 years that the plant had been in continuous operation (1902–1981), ground-water contamination has remained confined to a much smaller part of the aquifer than would be expected for conservative (nonreactive) contaminant at the calculated rate of ground-water flow. Laboratory and field studies at the Pensacola site indicate that methanogenic fermentation of selected phenolic compounds may significantly contribute to attenuation of concentrations of phenolic compounds in ground water, but other processes related to hydrogeologic factors and geochemical properties of the aquifer material may also contribute to the attenuation of the contaminants.

The objective of the research at the Pensacola site is to quantitatively define the physical, chemical, and biological processes that occur in ground and surface water that have been contaminated by creosote and PCP and to determine the extent to which they contribute to attenuation of contaminants in the various hydrologic systems.

# REGIONAL GROUND-WATER QUALITY STUDIES

About half of the population of the United States relies on ground water as a source of drinking water. In some places, like Florida, New Mexico, and Long Island, New York, ground water supplies more than 90 percent of the population with water.

Hundreds of billions of gallons of wastes leak into the ground annually from a variety of industrial and domestic point sources.[5] Some of these waste waters contain synthetic organic compounds that were unknown at the turn of the century. Many of these compounds are toxic or can cause degradation of the environment through indirect means. Pesticides have been found in ground waters in 18 states because of widespread agricultural usage.[6] Many new agricultural chemicals are more effective by virtue of their persistence and solubility in water. Unfortunately, these very properties account for their presence in drinking water supplies and foodstuffs throughout the country. Although there is abundant evidence that ground-water contamination has occurred and is occurring in virtually every state of the nation, our knowledge of these problems has come after the fact and in a piecemeal fashion. There are several reasons for this: (1) the inaccessibility and complexity of the resource; (2) the vastness of the resource and our inability to collect samples in sufficient quantity to be considered as representative; and (3) the relative newness of sampling and analytical methodologies.

The objective of the regional ground-water quality studies is to assess the present quality of the nation's ground-water reserves and the nature and extent of the ground-water contamination problem. Because it is impossible to study every aquifer system in every part of the nation, the assessment will be done by intensive study of individual areas that are representative of wider regions in terms of climate, hydrogeologic conditions, and human activities.

The objectives of the individual studies are to develop and test methods (1) to provide information on ambient ground-water chemistry with emphasis on organic substances and trace metals in order to explain the chemistry in terms of the local hydrology and human activities, and (2) to identify the causes, effects, and processes which are applicable to similar areas elsewhere.

Fourteen ground-water quality studies began this fiscal year (TABLE 1). Study areas vary in size from a few tens of square miles to a few thousand square miles. Regardless of the scale, each area is characterized by relatively uniform climatic and geohydrologic conditions. The general hydrologic and ground-water flow system in the study area are reasonably well understood, and there is a reasonable body of available data on inorganic chemical quality of ground water and land use.

Each study area includes more than one type of land use. The major recognized land classifications include: irrigated cropping, nonirrigated cropping, livestock, sewered residential areas, unsewered residential areas, light and heavy industry, active and abandoned mining areas, petroleum production, and undisturbed natural areas. The transitions from one land-use subarea to another are well defined so that correlations between ground-water quality and land use can be established.

**TABLE 1.** Fourteen Areas Selected for Ground-Water Quality Studies

|  |
|---|
| Atlantic and Gulf Coastal Plain[a]: |
|   Louisiana and Mississippi |
|   Long Island, New York |
|   Philadelphia, Pennsylvania |
|   Potomac-Raritan-Magothy outcrop area, New Jersey |
|   Houston, Texas |
| Southeast Carbonate[a]: |
|   Floridian Aquifer, Florida |
| Non-Glaciated Central[a]: |
|   Edwards Aquifer, Texas |
| Piedmont and Blue Ridge[a]: |
|   Combined Regolith, North Carolina |
| Alluvial Systems[a]: |
|   San Joaquin Valley, California |
|   Arkansas River Valley, Colorado |
|   Albuquerque-Belen Basin, New Mexico |
| High Plains[a]: |
|   Kansas |
|   Nebraska |
| Glaciated Central[a]: |
|   Connecticut River Valley, Connecticut |

[a] Major ground-water system as defined by Heath[8].

Each study begins with a reconnaissance phase to assess and refine understanding of the ground-water system by assembling existing hydrologic, geologic, and water-quality information, and pertinent land use and related cultural information. The scale of the effort will vary according to what has been done in the past. In some cases, ground-water models may have to be constructed or updated, and simulations carried out to define ground-water flow patterns in sufficient detail to support water-quality studies. Existing data on ground-water quality and contamination will be assembled, evaluated, and entered in automated data systems as appropriate to the needs of the study. This information will generally include chemical analyses of water from wells and springs, analyses of water from surface-water bodies which recharge or drain the ground-water system, and information on nonpoint and point sources of contamination.

Using the assembled water-quality information in conjunction with information on ground-water flow patterns, the studies will assess the factors affecting ground-water chemistry. Water samples will be collected for analysis of the major classes on organic compounds, trace metals, microbiota, common constituents and other relevant properties. Spatial and temporal variability of the data will be analyzed using graphical and statistical methods. Changes in the concentrations of organic substances and other constituents will be defined whenever possible.

Emphasis will be placed on interpretations concerning the current state of the shallow aquifer systems according to types of land use. Hypotheses concerning the correlation between land use and ground-water quality will be evaluated through a variety of statistical methods. In conjunction with ground-water models, interpretations will be made of the avenues for contamination and processes affecting changes of contaminant concentration. Potential future impacts on both ground and surface waters in the system will be analyzed. Deeper aquifers also will be sampled to determine their vulnerability to contamination from surface-water or shallow ground-water sources. To the extent feasible, results from the areal studies will be related to other similar areas to make regional and national interpretations.

On the basis of results of the first appraisals, the sampling networks and data collection activities will be modified to improve results and cost effectiveness. Additional sampling and analyses will be scheduled for future years to determine long-term changes and trends, if possible. The frequency of future samplings will depend on such factors as hydrogeologic setting, land-use types, and changes in land use.

## DEVELOPMENT OF GUIDELINES FOR SITUATING HAZARDOUS WASTE FACILITIES

Currently, about 90 percent of the solid waste produced in this country is disposed of in landfills.[7] Secured landfills, those designed to isolate hazardous wastes from the environment, represent the state of the art in land-based disposal. However, the long-term integrity of these sites is not yet established. A potential problem with regard to maintaining the integrity of landfills is that, if engineering measures fail in the future, many of the sites may not have the necessary earth science characteristics necessary to maintain waste isolation. Earth science criteria, along with economic, engineering, and social factors, should be considered when selecting sites for disposal of hazardous chemical wastes. The geologic, hydrologic, and geochemical characteristics of the site should be identified and used to characterize the impact of the particular type and form of waste on the environment. Nondegradable wastes that present a hazard to the environment require permanent isolation from the biosphere. Burial or injection in a geologic environment, if it is to provide adequate safety with a minimum of continuing operational or maintenance efforts, must take advantage of geologic, hydrologic, and geochemical barriers which prevent migration of the wastes to the biosphere. Criteria used to site systems of this type should consider the following attributes: site stability, low ground-water velocities and long flow paths to discharge areas, the potential for chemical retardation of waste movement, conditions favorable for chemical or biochemical degradation of waste, and, for some wastes, conditions suitable for deep burial.

Information is currently being assembled concerning general earth-science considerations for long-term containment of hazardous wastes.

## REFERENCES

1. RAGONE, S. E. 1984. U.S. Geological Survey Toxic Waste–Ground-Water Contamination Program—Fiscal Year 1983: U.S. Geological Survey Open-File Report 84-474.

2. MATTRAW, H. & B. J. FRANKS, Eds. 1986. Movement and fate of creosote waste in ground water, Pensacola, Florida: U.S. Geological Survey Toxic Waste–Ground-Water Contamination Program. U.S. Geological Survey Water-Supply Paper 2285.
3. LEBLANC, D. R., Ed. 1984. Movement and fate of solutes in a plume of sewage-contaminated ground water, Cape Cod, Massachusetts, U.S. Geological Survey Toxic Waste–Ground-Water Contamination Program. *In* U.S. Geological Survey Open-File Report 84-475.
4. HULT, M., Ed. 1984. Ground-water contamination by crude oil at the Bemidji, Minnesota research site: U.S. Geological Survey Toxic Waste–Ground-Water Contamination Program. *In* U.S. Geological Survey Water Resources Investigation Report 84-4188.
5. MILLER, D. W. 1980. Waste Disposal Effects on Ground Water. Premier Press. Berkeley, CA.
6. COHEN, S. Z., S. M. CREEGER, R. F. CARSEL & C. G. ENFIELD 1984. Potential for pesticide contamination of ground water resulting from agricultural uses. *In* American Chemical Society Symposium Series: Treatment and Disposal of Pesticide Wastes. Preprint, February, 1984.
7. NATIONAL RESEARCH COUNCIL, NATIONAL MATERIAL ADVISORY BOARD, COMMITTEE ON DISPOSAL OF HAZARDOUS INDUSTRIAL WASTES. 1983. Management of Hazardous Industrial Wastes: Research and Development Needs. Publication NMAB-398. National Academy Press. Washington, DC.
8. HEATH, R. C. 1982. Classification of ground-water systems of the United States. Ground Water **20(4):** 393–401.

# The NIMBY Syndrome

## Its Cause and Cure

C. P. WOLF

*Director, Social Impact Assessment Center*
*Box 2087*
*Canal Street Station*
*New York, New York 10013-9998*

## INTRODUCTION

In 1985 I was asked to present testimony to the Subcommittee on Investigations and Oversight of the House Committee on Science and Technology for hearings on hazardous waste cleanup. Although the issue before Congress at the time was Superfund reauthorization, since approved, my charge was to explore a longer-term proposition: "The roles of community participation in the determination of acceptable risk." At that time I was also beginning work on the socioeconomic impact assessment of siting a high-level nuclear waste repository at Richton and Cypress Creek Domes, Mississippi. Since then I have continued my nuclear excursions, next in low-level waste (LLW) siting and now again in high-level waste (HLW), this time at Hanford Site, Washington.

While HLW may represent a "worst case" of toxic and hazardous materials, routes, sites, and facilities,[1] it is by no means an isolated one. The problem has literally surfaced to public view in the "Valley of the Drums" (Bullitt County, Kentucky), and elsewhere in the country; hardly a week passes but that some new toxic sites are unearthed. (Last week it was a string of 51 sites of polychorinated biphenyls (PCBs) in 12 states along the Texas Eastern Gas Pipeline corridor.) Nor is risk of public exposure confined to land pollution; facilities for handling and treating "people pollution"—halfway houses for deinstitutionalized mental patients, drug addicts, parolees and shelters for AIDS victims and the homeless—have likewise encountered widespread local opposition. As experience with these conditions and proposals for their correction become more familiar, so has an acronym describing the public's typical response: NIMBY (Not-In-My-Backyard). This paper will examine the "NIMBY syndrome," diagnose its symptoms, and prescribe a possible cure.

## THE NIMBY SYNDROME

Though the NIMBY syndrome was first diagnosed as recently as 1981, by Farkas,[2-4] it is clearly a concept whose time has come. (Lately it has been joined by another engaging acronym, LULU—"locally unwanted land use"[5,6]—which

216

identifies the objects of aversion.) There have been close approximations before and since, e.g., "not on my block"[7] and "any place but here."[8] What is telling about NIMBY, and what that acronym tells, is that perceived risk of environmental and technological hazard has been brought close to home, where people live, in families and communities. Whereas pockets of local resistance have punctuated land use controversies previously, the condition NIMBY describes is now thought to have become general, spanning diverse population segments and project types. The superficial attraction of the term masks a deep structure of conflict in personal and public choice. A representative example is contained in this account of efforts to site LLW in Texas[9]:

> We have learned quite a bit in our 30 months of siting efforts. We have held over ten town meetings, and all of them have been almost identical. The basic concerns of people throughout our state are the same. The same questions and concerns have been raised at all of the meetings. The overriding statement of the people is clear: "We need a site, but not here." NIMBY (Not In My Backyard) has followed us wherever we have gone.

The pervasiveness of this response has led to the frustration of siting efforts nationwide as well. According to Kasperson, between 1979 and September 1984 thirty-two hazardous waste siting attempts were made, of which five treatment and two storage facilities were approved (though not yet built) and only one disposal facility was actually constructed.[10] Reviewing this dismal record, Susskind[11,12] (pp. 157–158) concludes:

> Traditional approaches to the siting of regionally necessary but locally noxious facilities are not working. The record is unambiguous. It has been practically impossible to site much needed hazardous waste treatment facilities anywhere in the United States. Almost every attempt to site new oil refineries, power plants, resource recovery facilities, radioactive containment areas, prisons, and power lines has been stymied. The siting puzzle appears to be more difficult than ever to solve.

Perversely enough, locational disputes are intensifying because environmental protection measures are *improving,* not because they are deteriorating. According to Regens[13] (pp. 125–126), unless means of overcoming local opposition are found, implementation of the Resource Conservation and Recovery Act and the Comprehensive Environmental Response Compensation and Liability Act

> . . . will result in the closing of inadequate facilities now receiving hazardous wastes and increase the demand for proper disposal sites . . . remedial action for abandoned sites will identify even more hazardous waste requiring proper disposal. If public opposition continues to frustrate attempts to site properly managed disposal facilities, the nation's effort to regulate hazardous waste may collapse.

The failure to site and regulate approved facilities will result in continued mismanagement of toxic and hazardous materials and illegal practices in their disposal, e.g., "midnight dumping." What accounts for this uniformly negative response?


## THE CAUSE


We can identify three linked causes of the NIMBY syndrome whose understanding may indicate a possible cure: (1) loss of control, (2) loss of community, and (3) equity concerns. The first suggests that the locus of effective decision and action has shifted beyond the control of those most likely to be affected. The second implies the disruption of personal networks and the depletion of cultural resources. The third depicts a resulting situation of distributive injustice in which there is a dissociation of costs and benefits among the parties at interest. A short comment on each must suffice.

### Loss of Control

Siting proposals evoke a perceived loss of control over circumstances deeply affecting personal and social well-being, and a sense of powerlessness and alienation from the forces and processes that shape these basic conditions of existence.[14] This perception is heightened by the involuntary nature of the risk imposed; as Slovic and Fischhoff[15] (p. 135) observe, the public accepts much greater risk from voluntary than from involuntary activities. Shift in the locus of control to nonlocal authorities has been especially marked in nuclear development, i.e., "federal preemption." On the local level, implementation of national policy comes down to a question of alternative sites, not whether a facility is wanted at all. It does no good to remind "locals" that their elected representatives participated in formulating the policy. Community people will contend *they* were not consulted, did not approve, and will not accept the proposed action.

### Loss of Community

Mary Douglas[16] has called attention to the collective (institutional, cultural) dimension of perceived risk. In community context, loss of control implies a weakening of local ties and a loosening of the social fabric. Residents perceive a threat to a settled way of life and a set of stable relations that constitute the sense of community and the satisfactions that provides. These social supports and informal services also sustain community viability by providing access to and control over local resources. In extreme cases of sudden loss, such as the Buffalo Creek disaster, the risk is a physical "destruction of community."[17] Chronic exposure to toxic substances and chronic stress resulting from perceived risk may produce the same result over extended periods, as at Love Canal, New York and Times Beach, Missouri.

### Equity Concerns

The class of problem we are dealing with may thus be characterized as one of perceived low benefits and high disadvantages to potential host communities and regions. For example, Carnes and others[18] (p. 17) note "the general lack of conventional benefits associated with such activity" as HLW repositories, and Voth and Herrington[19] (p. 261) state, "In the case of repositories, the gainers (for example, users of electricity generated by nuclear power) are highly diffused, whereas the losers (residents of the affected communities) are concentrated." According to Regens[13] (p. 125), one source of contention has been that

> Facility sponsors rarely, if ever, addressed the question of equity. Instead, developers tended to define the question in terms of regulation and technology. Opponents of the facility siting, on the other hand, primarily focused their arguments on the potential for risks.

In risk-benefit terms, it is also noted[20] that experts and different segments of the public diverge on the respective weights assigned to probability and consequence; the former emphasize (low) probabilities of a risk event actually occurring, while the public more heavily weighs the likely (adverse) consequences of such an occurrence.

# THE CURE

If this diagnosis is accurate, the conditions of community risk acceptance are those that meet and mitigate the concerns and losses identified above. Loss of control must be offset by empowerment which allows for "meaningful" public and community participation through increased capacity for self-governance. Loss of community must be countered by social reconstruction; community risk assessment and management are required to exercise community control over technological and environmental risks. Equity concerns can be allayed by a fairer apportionment of benefits and risks, e.g., by weighted decision input from potentially affected segments of the public, and by upholding social acceptability as the decisive siting criterion. Only by such means can natural, and rightful, community opposition be surmounted.

That is on the "receiving" side. Equally, on the "sending" side, conspicuous and determined efforts must be undertaken to remove the sources of personal and community risk, whether by policy choice, technology development, or management practice.[21] For example, toxic program design and site development cannot be defended unless a serious effort is mounted at the same time towards source reduction and onsite retention, recycling, and reuse of waste materials. Along with this, an orderly but steady transition to environmentally benign policies and technologies should proceed so that source generators are eliminated altogether, such as commitment to develop alternative energy sources and systems that drastically curtail emissions. By this standard, conventional fossil and nuclear energy technologies can be justified only as transitional, e.g., to solar photovoltaics.[22,23]

### *Taking the Public Seriously*

Most of all, public and local concerns must be taken seriously by agency officials and project proponents; until this "mindset" is altered no real prospects exist for ending the current impasse. Evidence of such disregard and disrespect is found in allegations of "irrational fear" and similar derangements, such as "nuclear phobia."[24] A specimen of this attitude is the notion that gaining community acceptance requires overcoming ignorance and error by means of programs of public information and education: "Sadly, public response to nuclear-waste disposal is often emotional and poorly-informed. This state of affairs cries out for better public information, so individuals can deal knowledgeably and reasonably with the problem."[25]

This statement reflects a conception of risk that Timmerman[26] labels "instrumental rationality," a form of bureaucratic utilitarianism in which persons are treated as means to some higher social end such as hazardous waste siting or, more abstractly, the "general welfare." The contrasting risk conception, widely held by members of the public, is "coherent rationality," an appeal to "reasonableness" rather than to strict rationality. In this view, personal experience and social context are highly relevant to judgment; reason and emotion are united rather than divorced. The calculus for determining public preference and choice employs a "weighted input" scheme,[27] not unlike the "intensity" measure of multiattribute utility theory, whereby decision units — families and communities, not disparate individuals — are weighted as a function of probability of exposure to risk and its consequences. Above all, persons are treated as ends in themselves rather than as instrumental means, and as such possessing of inherent rights and responsibilities they cannot relinquish even willingly. If this position appears "irrational" from the proponent's viewpoint, it is hardly surprising that they in turn are regarded as "immoral" by members of the public.

Another polarity familiar to the risk assessment field is that between "actual" and "perceived risk." The former is associated with experts' risk estimates, the latter with public perceptions (or misperceptions, as risk analysts might view them). The larger question is how risk should be regarded — whether from the standpoint of the analyst or from that of the public. In general, risk professionals expect the public to conform to their technical standards and to deny the reality of perceived risk insofar as it deviates from them. One may very well question the "rationality" of analysts who deny this social reality, however, and who fail to conceive a form of rationality that includes values rather than excluding them. Just as there is need for greater public understanding of science, so there is a counterpart need for greater scientific understanding of the public. The persistent and pernicious disjunction of "technical" and "political" factors in many decision contexts suggests that a broader risk conception is required for their higher resolution.


### Information and Decision

This is not to say that technical issues are irrelevant or that the public would not benefit from improved access to information resources. Nor does it imply that siting decisions should not be evaluated on their geotechnical merits. For example, three-quarters (77%) of Tennessee respondents, compared with 41% of the state legislators surveyed, favored leaving siting decisions to "scientific experts."[10] The authors interpret this as a desire on the public's part to remove siting issues from the arena of partisan politics and also to confirm public confidence in finding a rational basis for decision. If public information has the effect of increasing dependency on external authority, however, then information itself becomes a factor in loss of control and will be mistrusted and resisted accordingly. This is entirely rational from the community's standpoint, and underscores the non-neutrality of so-called "objective" information. Payne[28] (p. 205) puts it this way:

> . . . in many cases, local residents and officials have little technical knowledge about radioactivity and radioactive waste management. In general, they cannot acquire the knowledge quickly and cannot afford to hire advisors they trust. Thus, when the community is initially informed that its region is being evaluated as a possible respository site, local citizens and officials immediately feel at a technical disadvantage. No amount of effort and information from siting officials can erase these initial feelings.

Recognition that such technical expertise is outside the control of the community and may operate to the detriment of its interests can result in a general negativism. Forcade[29] (pp. 112–113) observes that public opposition is not confined only to poor hazardous waste sites, but extends to good sites as well. He goes on, "While this opposition to all sites may, at first, appear irrational or unfounded, it is not. . . . A citizen who is technically incapable of distinguishing good sites from bad ones, and good regulatory control programs from bad ones, may logically and rationally decide to avoid the risk by opposing all sites." Such uninformed and unfocused opposition is often ascribed to deficiencies in scientific and technological literacy, but until recently public information on which to base reasoned technical judgments has not been available and usable. Late in 1985, EPA Administrator Lee M. Thomas declared, "For the first time, people will be able to identify where chemical risk exists in their communities."[30,31] What, and on what basis, were they supposed to believe up till then?

In earlier days, nuclear development was accepted as a continuation of the general technological progress of the age; the curiosity then was resistance to innovation[32] and the drama was how the prestige of science would overcome that benighted opposition. Far from "informed consent," such acceptance was taken on faith and trust in public institutions and officials. As late as 1973 an EPA survey found that ". . . most people have a positive attitude toward a hazardous waste facility, would accept one in their county, and believe such a facility would be beneficial to their area"[29] (p. 111). By 1980, hazardous waste facility siting had become an issue of major public concern, however. A Council on Environmental Quality survey found that more than half of the public would want "to move to another place or actively protest" the siting of such a facility within 100 miles of their home, even if "disposal could be done safely and . . . the site would be inspected regularly for possible problems."

Part of the change in public attitudes may be attributed to a dreary succession of events, such as the 106-T leak at Hanford Reservation,[33] which has eroded confidence and betrayed trust, and in light of which public apprehension seems entirely rational. Farkas[2] (p. 47) points to managerial and technical defects as triggering this response:

> The factual situation is that the major industrial liquid waste proposals made in Ontario to date have been inadequate and incomplete on technical grounds. Thus it is no wonder that they have been rejected on a combination of technical, social, political, and environmental grounds. Those who blame the NIMBY syndrome for the failure of these proposals are . . . using NIMBY as a scapegoat to cover the inadequacies of these proposals.

This and many similar instances reaffirm the need for community-based approaches to risk assessment and management.

Effective means for asserting local control are at hand in the application of impact assessment methods and procedures at the community scale.[34,35] It is necessary that they be employed, comprehensively and consistently, in order to assure the community that its interest is being recognized and protected. Community impact assessment has been practiced mainly on the "back end" of the assessment cycle—mitigation, monitoring, and management[36]— but can serve as well for purposes of project and technology selection.[37,38] By developing and exercising the capability for impact assessment, communities can reliably inform themselves of the potential risks and consequences of site acceptance. The basis on which a prospective host community or region will evaluate these assessed risks and benefits leads further to the question of compensation.

### Compensation

Now and again, someone will ask, "Why don't they just give them a couple of billion dollars?" and forego the lengthy site screening and selection process that many critics dismiss as a charade. The answer is, you can't pay them enough legally, and morally it is a corrupt form of payment. By the terms of the Nuclear Waste Policy Act of 1982, individuals are compensated at fair market value for tangible property losses incurred in the siting process. The attempt to add further monetary incentives to coerce unwilling residents is perceived as bribery, sacrificing the welfare of future generations for selfish personal gain. Moreover, were it legal to overcompensate that would create another situation of perceived inequity and generate pressures for redistributive justice. There are real property concerns that can and must be arbitrated and adjusted through the market, but this addresses only a narrow and comparatively minor range of issues in the overall siting process. These limits of economic incentives are noted by Solomon and Cameron[39]:

> . . . an "economic" approach to the incentives and compensation problem does not address those nuclear waste disposal burdens which cannot be quantified, and may not be subject to alleviation through monetary payments. These burdens include psychological stress, continued uncertainty over the risks of facility accidents and negative social changes accompanying any large-scale, long-term changes in the local economic structure.

Short-run economic benefit cannot circumvent the issue of intergenerational equity that arises over long-term risks to public health and safety. Indeed, one pressure for waste isolation now is precisely to remove that burden from future generations. Moreover, economic "incentives" cannot avoid the appearance of bribery or coercion; only if residents were willing to accept a facility *regardless* of economic benefits would we consider their decision a principled one.

A "pork barrel" refinement[40] on the bribery question has lately been offered by Luther Carter,[41] that of linkage between a noxious facility such as a HLW repository and a desirable one, e.g., the Superconducting Super Collider. (Parenthetically, breakthroughs in superconducting materials may have the effect of drastically scaling down this project, and hence its attractiveness.) "World class" nuclear research centers are another option,[42] especially if directed to health physics and nuclear medicine. Co-located nuclear industries such as reprocessing are a further possibility, carrying, however, their own consequences for increased risk of chemical and radiologic accidents and subsequent population exposure. In the case of Hanford Site, the basic resource is already in place: the storage, seepage, and spillage of liquid waste throughout the area. The appropriate nuclear industry here would seem to be one for decontamination and cleanup, whether or not repository-connected, thus converting waste to resource as Buckminster Fuller urged. A novel linkage has been advanced by the Oak Ridge-Roane County citizen task force (TF) in its negotiations with the U.S. Department of Energy (DOE) over siting a proposed Monitored Retrievable Storage (MRS) facility at Clinch River. As described by Peelle,[43]

> One of the . . . major concerns was how to deal with distrust of DOE, viewed as a principal major impediment to acceptance of the proposed facility. Since media revelations about extensive past pollution of the DOE reservation and DOE's past record of indifference to local concerns were part of the local context, the TF sought to build local trust through performance requirements upon DOE. This included tying the MRS schedule to cleanup of the DOE reservation.

These examples point to the efficacy of nonmonetary incentives in securing community acceptance.

Community acceptance can also be gained through a variety of nonmonetary (environmental and sociopolitical) incentives whose aim is to remedy any aversion to risk that the community might have. These include:

- Access to information on site development and management;
- Funds for hiring independent experts;
- Community monitoring of site operations;
- Representation on the facility governing board; and
- Local power to shut down the facility until treatment or management deficiencies are corrected

Together these items form a scale of community control measures believed by members of the public to enhance site acceptability far beyond monetary payments.[44] Survey research findings from Wisconsin, Pennsylvania, and Tennessee are consistent in endorsing these measures for site selection and regulation. It seems obvious that community control is the condition of community risk acceptance and that informational resources and procedural guidelines for exerting that control must be developed and deployed. At the same time, this calls into question the very nature of "community" itself and its capacity for self-management.

### *The Limits of Community Control*

As well as benefits, there are perils to empowerment, not the least being the matter of community accountability. If nonlocal institutions and officials should be made accountable to the community, the question remains to whom the community itself is accountable. Answers heard so far are unsatisfactory. If it is said that the community is accountable only to itself, that appears self-serving and violates the principle of balancing rights with responsibilities. If it is said the community is accountable to future generations, enforceability becomes problematic. If the community is held accountable to higher governmental authority, then we revert to the original position empowerment is seeking to alter.

A related concern is the matching of community responsibilities and resources. It is all very well to argue that communities should assume responsibility for managing their own affairs and the forces that impinge on them, but what resources are required for discharging these tasks and what is their availability? The resources in question may be social and cultural as well as physical and fiscal, e.g., interpersonal support systems for community treatment whose creation entails stable relations over extended periods. It may well be that resource scarcity is a contributing factor to the very existence of the need. Moreover, if external (nontraditional) resources are imported into the situation, legitimate and effective means for their mobilization and allocation must be devised as in the case of "community control" over neighborhood schools in New York City.[45] In any case, community response capability[46] should weigh heavily in siting decisions.

Other cautions attach to exclusive reliance on community control. "Neighborhood power"[47] is not always exercised judiciously, as repeated incidents of housing discrimination attest.[48] Community control measures may intensify such patterns of inequality and inequity. While community cohesion and consensus may be required for effective control, factionalism and conflict may instead prevail. As the surveys in Pennsylvania and Tennessee disclosed, local officials and agencies may be no more respected and trusted than nonlocal ones. For these and similar reasons, the community approach to siting decisions cannot stand alone.

## THE LARGER COMMUNITY

Local community acceptance, while necessary to successful efforts at siting, is not and cannot of itself be sufficient. For that, a larger conception of community is required, one that accommodates and coordinates special interests among diverse portions of the public. For example, in asserting its own right of interest, the local community must at the same time adopt a "national perspective" which takes seriously the generic problem of nuclear waste disposal and puts forward constructive suggestions for its resolution. If a site is determined to be the best for HLW repository development, then the adjoining community would have to seriously question whether local preference should dominate the siting decision at the expense of the national interest.[49] The assumption here is, of course, that the procedure for site screening and selection is acknowledged to be fair and it is exactly on this point that critics denounce the entire DOE program as politically motivated and "fatally flawed."

The prior "constitutional" question of the community's asserting that its right of interest takes precedence over the national interest must be addressed, however. Lyons, Freeman, and Fitzgerald[10] (p. 14) observe, ". . . time and again, siting attempts typically result in anguished local protests that frequently lead to project vetoes." Indeed, their own Tennessee sample strongly affirms (89%) the community's right to prevent a hazardous waste treatment plant from being sited nearby. The authors continue, "It is nonetheless clear that the general societal interest must eventually prevail over protesting local interests if essential facilities are to be constructed." Carter,[50] (p. 28) to the contrary, argues that ". . . given the right strategy for picking (and not picking) sites, no repository ever need be imposed on an unwilling locality."

Yet another issue centers not on what the community *will* accept but what, as a matter of principle, it *should* accept. In several instances, local willingness to host a facility has been overridden by opposition at the state level. For example, a proposed LLW site near Edgemont, South Dakota supported by the local community was defeated in a state referendum. Voters in Sioux Falls, at the other end of the state, feared exposure to transportation risks and to public liability in the event of default by the site operator, as happened at West Valley, New York.[51] Similarly, despite the achievement of the Clinch River MRS Task Force in its negotiations with DOE, the site was rejected by the Tennessee governor and legislature.

What constitutes "the community" in these situations is open to question. In the case of Hanford Site, Tri-Cities residents have been generally favorable to repository development in the past. Some persons residing outside that area have objected, however, that the "community" of interest extends far beyond those boundaries, to encompass statewide and interstate concerns, e.g., downwind Idaho and downstream Oregon. One empirical finding bearing on community scale is the response of Bord's[52] (p. 26) Pennsylvania sample: "Most respondents, 66 percent, place the acceptable boundary within a 10-mile radius of the site. However, 29 percent would go beyond a 10-mile radius." If a risk measure of community is applied, that distance might be further expanded. Evidently, the concept of community is not a terribly exact or operational one.[53] Yet the social reality it expresses may have more to do with inclusion and integration than with exclusion.

The real issue of risk acceptance may reside less with a prospective host community and region than with the proponent agency, however. For instance, in the MRS case the chief difficulty may well turn on obtaining DOE's acceptance of the conditions for site monitoring and management demanded by the local community. Agency officials may respond that such conditions not only infringe on their authority, but also abridge or abrogate their responsibility under the laws of the land. The community impact agreement negotiated between the Township of Atikokan and Ontario Hydro and implemented over a nine-year period[54] suggests, however, that institutional arrangements can be achieved which are conducive to effective consultation and concurrence between community and agency. As these authors remark, ". . . the main contributing factor to the success of the program was the sense of trust and good faith that was developed and maintained. . . ." Indeed, it is just because of the decline in public trust that the demand for community intervention arises in the first place.

The NIMBY syndrome has come to prominence just because communities have become more potent in pursuit and defense of their interests. The macropolitical reality is that in the nuclear field, and perhaps in general, the principle of "federal preemption" is yielding to a more equal balance of powers throughout the federal system. Under provisions of the Nuclear Waste Policy Act, states may veto DOE site selection, subject to ratification by Congress. In reactor siting cases such as Seabrook, Shoreham, and Perry, state and community noncooperation with federal emergency plans has paralyzed the licensing process. Morrell and Magorian[55] (p. 55) conclude that "It seems safe to assume that if a locality is vehemently opposed to a proposed hazardous waste facility, the project will not be successful, regardless of the existence of state preemptive . . . powers."

In fact, the NIMBY syndrome is symptomatic of a general trend toward political and institutional decentralization.[56] If the slogan, "There goes the neighborhood!" was once uttered in bemusement or resignation, those are now fighting words. Whether the NIMBY syndrome constitutes a "neighborhood revolution,"[57] it does seem to point toward a restitution of the federal system in which the local public is accepted as a more equal partner. Until now, community interest and influence have largely been left out of the siting equation. Such greater parity of community power may actually prove beneficial to fostering genuinely cooperative intergovernmental relations. For all the negativism that NIMBY connotes, there is this constructive side as well. Indeed, the fundamental proposition before our country and people is community-building on and across all levels of self-governance. The prognosis is uncertain, but NIMBY may yet make a positive contribution to this end.

## ACKNOWLEDGMENTS

A number of people have encouraged and assisted me in the preparation of this paper: Audrey Armour, Justine Cullinan, Thecla Fabian, Joyce Henderson, Joel Hirschhorn, William Lyons, Glenn Paulson, Elizabeth Peelle, Arthur Purcell, Michael Simpson, Barry Solomon, and Frances Sterrett. I wish to acknowledge my gratitude to them and to assume full responsibility for any misuse of their contributions.

## REFERENCES

1. JAKIMO, A. & I. C. BUPP. 1978. Nuclear waste disposal: Not in my backyard. Technol. Rev. **180**(5): 64–72.
2. FARKAS, E. J. 1981. The NIMBY syndrome. Alternatives **10**(2/3): 47–50.

3. ARMOUR, A., Ed. 1984. The Not-In-My-Backyard Syndrome. York University. Downsview, Ontario.

4. FREUDENBERG, N. 1984. Not in Our Backyards! Community Action for Health and the Environment. Monthly Review Press. New York, NY.

5. POPPER, F. 1985. Nobody loves a LULU. Environment **27**(2): 5–11, 37–40.

6. INSTITUTE FOR ENVIRONMENTAL NEGOTIATION, Ed. 1985. Not-In-My-Backyard! Community Reaction to Locally Unwanted Land Use. University of Virginia. Charlottesville, VA.

7. O'HARE, M. 1977. Not on my block you don't: Facility siting and the strategic importance of compensation. Public Policy **25**(4): 407–458.

8. SMITH, C. J. & R. Q. HANHAM. 1981. Any place but here! Mental health facilities as noxious neighbors. Professional Geographer **33**: 326–34.

9. BLACKBURN, III, T. W. 1985. Low-level radioactive waste activities in Texas. *In* The 1985 Washington Conference on Low Level Nuclear Waste Disposal & Cleanup.: C-7. Center for Energy and Environmental Management, Ed.: C4–8. CEEM. Alexandria, VA.

10. LYONS, W., P. FREEMAN & M. R. FITZGERALD. 1986. Public opinion and the legislative response to the hazardous waste challenge. Paper presented at the annual meeting of the American Political Science Association. Washington, DC.

11. SUSSKIND, L. W. 1985. The siting puzzle: Balancing economic and environmental gains and losses. Environmental Impact Assessment Review **5**(2): 157–63.

12. ANDERSON, R. F. 1986. Public participation in hazardous waste facility location decisions. J. Planning Lit. **1**(2): 145–61.

13. REGENS, J. L. 1983. Siting hazardous waste management facilities. *In* Public Involvement and Social Impact Assessment. G. A. Daneke, M. W. Garcia & J. D. Priscoli, Eds.: 121–128. Westview. Boulder, CO.

14. LEFCOURT, H. M. 1982. Locus of Control: Current Trends in Theory and Research. (2nd ed.) Lawrence Erlbaum Associates. Hillsdale, NJ.

15. SLOVIC, P. & B. FISCHHOFF. 1983. How safe is safe enough? Determinants of perceived and acceptable risk. *In* Too Hot to Handle? Social and Policy Issues in the Management of Radioactive Wastes. C. A. Walker, L. C. Gould & E. J. Woodhouse, Eds.: 112–50. Yale University Press. New Haven, CT.

16. DOUGLAS, M. 1985. Risk Acceptability According to the Social Sciences. Russell Sage Foundation. New York, NY.

17. ERIKSON, K. T. 1976. Everything in Its Path: Destruction of Community in the Buffalo Creek Flood. Simon & Schuster. New York, NY.

18. CARNES, S. A., E. D. COPENHAVER, J. H. REED, E. J. SODERSTROM, J. H. SORENSON, E. PEELLE & D. J. BJORNSTAD. 1982. Incentives and the Siting of Radioactive Waste Facilities. Oak Ridge National Laboratory. Oak Ridge, TN.

19. VOTH, D. E. & B. E. HERRINGTON. 1983. Community development in nuclear waste isolation. *In* Nuclear Waste: Socioeconomic Dimensions of Long-Term Storage. S. H. Murdock, F. L. Leistritz & R. R. Hamm, Eds.: 251–265. Westview. Boulder, CO.

20. COLGLAZIER, JR., W. & M. RICE. 1983. Media coverage of complex technological issues. *In* Uncertain Power: The Struggle for a National Energy Policy. D. S. Zinberg, Ed.: 112–133. Pergamon Press. New York, NY.

21. PEELLE, E. & R. ELLIS. 1987. Hazardous waste management outlook: Are there ways out of the "not-in-my-backyard" impasse? Forum of Applied Research and Public Policy. In press.

22. HAMAKAWA, Y. 1987. Photovoltaic power. Scientific American. **256**(4): 86–92.

23. MOORE, T. 1987. Opening the door for utility photovoltaics. EPRI Journal **12**(1): 5–15.
24. THE MEDIA INSTITUTE. 1980. Nuclear phobia—phobic thinking about nuclear power: A discussion with Robert L. DuPont, M. D. The Media Institute. Washington, DC.
25. LA SALA, JR. A. M. *et al.* 1985. Radioactive Waste: Issues and Answers (revised edition).: 25. American Institute of Professional Geologists. Arvada, CO.
26. TIMMERMAN, P. 1984. Ethics and Hazardous Waste Facility Siting.: 35. Institute for Environmental Studies. University of Toronto. Toronto, Ontario.
27. KRIMSKY, S. 1984. Beyond technocracy: New routes for citizen involvement in social risk assessment. *In* Citizen Participation in Science Policy. J. C. Petersen, Ed.: 55–56. University of Massachusetts Press. Amherst, MA.
28. PAYNE, B. A. 1984. Estimating and coping with public response to radioactive waste repository siting. *In* Waste Management '84: I. Waste Policies and Programs, High-Level Waste. R. G. Post, Ed.: 205–209. University of Arizona. Tucson, AZ.
29. FORCADE, B. S. 1984. Public participation in siting. *In* Hazardous Waste Management: In Whose Backyard? M. Harthill, Ed.: 111–122. Westview. Boulder, CO.
30. DIAMOND, S. 1985. U.S. names 403 toxic chemicals that pose risk in plant accidents. The New York Times. 18 November.
31. SHERRY, S. *et al.* 1985. High Tech and Toxics: A Guide for Local Communities. Golden Empire Health Planning Center. Sacramento, CA.
32. WHITNEY, V. H. 1950. Resistance to innovation: The case of atomic power. Am. J. Sociol. **50**(3): 247–54.
33. GILLETTE, R. 1973. Radiation spill at Hanford: The anatomy of an accident. Science **181**(4101): 728–730.
34. RUNYAN, D. 1977. Tools for community-managed impact assessment. J. Am. Inst. Planners **43**(2): 125–34.
35. SANFORD, F. L. & S. B. FAWCETT. 1977. Community impact analysis: A manual for assessing the possible effects of planned environmental change. Institute of Public Affairs and Community Development. University of Kansas. Lawrence, KS.
36. KING, E. 1981. Citizens' monitoring network. Environmental Impact Assessment Review. **2**(2): 198–201.
37. OLSEN, M. E., P. CANAN & M. HENNESSY. 1985. A value-based community assessment process: Integrating quality of life and social impact studies. Sociol. Methods Res. **13**(3): 325–361.
38. BRONFMAN, B. H., S. A. CARNES & R. S. AHMAD. 1983. Community-based technology assessment: Four communities plan their energy future. *In* Integrated Impact Assessment. F. A. Rossini & A. L. Porter, Eds.: 202–218. Westview. Boulder, CO.
39. SOLOMON, B. D. & D. M. CAMERON. 1985. Nuclear waste repository siting: An alternative approach. Energy Policy **13**(6): 563–580.
40. KOSHLAND, JR., D. E. 1986. Dealing in hot property. Science **232**(4758): 1585.
41. CARTER, L. H. 1986. Nuclear imperatives and public trust: Dealing with radioactive waste. Issues in Science and Technology **3**(2): 46–61.
42. CREIGHTON, J. L. 1985. The U.S. Department of Energy's Implementation of the Consultation Provisions of the Nuclear Waste Policy Act.: 35. Creighton & Creighton. Saratoga, CA.
43. PEELLE, E. 1987. The MRS Task Force: Economic and non-economic incentives for local public acceptance of a proposed nuclear waste packaging and storage facility. Paper presented at Waste Management '87 Conference. Tucson, AZ.

44. Pushchak, R. & I. Burton. 1983. Risk and prior compensation in siting low-level nuclear waste facilities: Dealing with the NIMBY syndrome. Plan Canada. **23**(3):68–77.

45. Altshuler, A. A. 1970. Community Control: The Black Demand for Participation in Large American Cities. Pegasus. New York, NY.

46. Waterstone, M. 1983. Toxics and groundwater: The development and application of net risk analysis. The Environmental Professional **5**(1): 46–56.

47. Morris, D. & K. Hess. 1975. Neighborhood Power: The New Localism. Beacon. Boston, MA.

48. Cass, J. 1986. The Elmwood incident. The Philadelphia Inquirer Magazine. 4 May: 12–20, 28–32.

49. Power, M. A. 1986. Radioactive waste in my backyard? The Nuclear Waste Policy Act and Washington's use of scientific analysis. Paper presented at the annual meeting of the American Association for the Advancement of Science. Philadelphia, PA, May 26, 1986.

50. Carter, L. J. 1985. Review of Panel on Social and Economic Aspects of Radioactive Waste Management, Ed., Social and Economic Aspects of Radioactive Waste Disposal. Environment **27**(2): 25–29.

51. The New York Times. 1984. Plan for nuclear dump divides Dakota town. 4 November.

52. Bord, R. J. 1985. Opinions of Pennsylvanians on Policy Issues Related to Low-Level Radioactive Waste Disposal.: 26. Institute for Research on Land and Water Resources. The Pennsylvania State University. University Park, PA.

53. Minar, D. W. & S. Greer, Eds. 1960. The Concept of Community: Readings with Interpretations. Aldine. Chicago, IL.

54. Hancock, S., M. Smith & J. Lockhart-Grace. 1986. Community Impact Monitoring Program: Final Report. The Township of Atikokan and Ontario Hydro. Atikokan and Toronto, Ontario.

55. Morrell, D. & C. Magorian. 1982. Siting Hazardous Waste Facilities: Local Opposition and the Myth of Preemption.: 95. Ballinger. Cambridge, MA.

56. Naisbitt, J. 1982. Megagtrends: Ten New Directions Transforming Our Lives, chapt. 5. Warner Books. New York, NY.

57. Boyte, H. C. 1980. The Backyard Revolution: Understanding the New Citizen Movement. Temple University Press. Philadelphia, PA.

# Exposure of Sheet-Metal Workers to Asbestos during the Construction and Renovation of Commercial Buildings in New York City

## A Case Study in Social Medicine

ERNEST DRUCKER,[a] DEBORAH NAGIN,[a] DAVID MICHAELS,[a] MARGOT LACHER,[a] AND STEPHEN ZOLOTH,[b]

[a] *Department of Epidemiology and Social Medicine*
*Montefiore Medical Center*
*Albert Einstein College of Medicine*
*Bronx, New York 10467*

[b] *Hunter College School of Health Sciences*
*City University of New York*
*New York, New York 10010*

## BACKGROUND

In 1982 the Occupational Health Program at Montefiore Medical Center was asked to provide assistance to a New York City sheet-metal contractor who had received a citation from the Occupational Safety and Health Administration (OSHA) for exposure of his workers to asbestos. At the time of the citation these workers had been engaged in renovation of the interior office space of a 45-story office building in lower Manhattan. A clerk-typist working in the building had noticed a fine white powder collecting on the surfaces of her work area. Upon testing, the material was found to contain asbestos, and was traced to a renovation site at the other end of the floor on which she worked. When OSHA visited the site they found a sheet-metal work crew engaged in replacing ductwork as part of an overall renovation of the area. Air monitoring in the breathing zone of several of the sheet-metal workers revealed asbestos levels in excess of the OSHA action level (0.1 fiber/cc). In the citation and following its usual practice OSHA called for medical monitoring of the sheet-metal workers and education of these workers about the hazards of asbestos. The citation made no mention of the source of the asbestos nor did it address remedial measures that would protect other workers (both construction and clerical) or visitors to the building.

## Asbestos Spraying of Office Buildings during Construction

The building in which this incident took place was constructed in 1968 and, like many other high-rise office towers built in New York between 1958 and 1972, contained structural members—beams, girders, floors, and decking—coated with asbestos insulation compound. Heat insulation is required by law to prevent the steel structure from losing strength in the event of fire and, until 1972, asbestos was the material of choice. The asbestos, usually 1–3 inches in thickness, was applied either as a slurry, spread onto the girders to create a fibrous mat, or, more frequently, as a mixture of fibers and binders blown under pressure from large hoses onto the superstructure. As this phase of the construction was under way, members of other building trades worked on the partially completed lower floors. Workers interviewed about this period describe the presence of clouds of asbestos fibers which often made it difficult to breathe or indeed to see. Wet handkerchiefs and goggles were often the only protection used. In addition, smoking, eating and drinking took place routinely on the job site.

The practice of using sprayed asbestos to insulate high-rise structures was known and utilized as early as the 1930s, but its use became widespread with the large building boom taking place in many urban centers in the 1960s. It is estimated that, nationally, 50 percent of the high-rise buildings constructed before 1970 were insulated with asbestos.[1] We now know that this practice was a source both of substantial environmental contamination and of massive occupational exposure during construction. Indeed, a 1970 study by the New York City Department of Air Resources, in cooperation with the Mt. Sinai School of Medicine, Environmental Sciences Laboratory found elevated levels of asbestos in the ambient air of New York City. Samples taken throughout the five boroughs and Manhattan, where extensive asbestos spraying was taking place, revealed significantly increased contamination levels. In some instances, asbestos levels were ten times "background" levels.[2]

## Renovation of Sprayed Buildings

In addition to the occupational and environmental asbestos exposures associated with the construction of these buildings we learned that their renovation, which takes place regularly, also results in exposure of construction workers and building occupants. Several studies have found that the soft asbestos insulation is easily disrupted and entrained in the air during the course of routine renovation activities.[3,4] During the renovation, insulation material is scraped away to gain access to those fixtures that are being removed or replaced, such as ductwork, wiring, and piping. This work is generally performed by workers untrained in asbestos handling and leads to substantial exposure of all those on the renovation site.[3,4] Further, because the majority of high-rise commercial buildings are constructed using the ceiling plenum (that space between the dropped ceiling and the structural decking) as the return duct of the air conditioning and ventilation system, asbestos-contaminated air may be carried to the adjacent areas on the same floor. Finally, many of these buildings utilize a single, large-scale ventilation and air conditioning system that may encompass the entire building on a common duct. These systems are generally left on-line during renovation work, and asbestos fibers generated on one floor can easily be carried throughout the remainder of the building. Thus, during the renovation of those buildings, significant quantities of asbestos may be disturbed, made airborne, and distributed through the ventilation system, potentially exposing office workers and building visitors.

### *Sheet-Metal Workers' Exposure to Asbestos*

Because of these methods of construction and renovation, sheet-metal workers were exposed to extremely high levels of asbestos during the period of asbestos spraying and, subsequently during the renovation of these same buildings. Asbestos fiber (f) concentrations as high as 100 f/cc were measured in the spray zone during construction. On-site samples taken 30 minutes after spraying had been completed, showed counts as high as 4 f/cc of asbestos. Workers not directly associated with spraying operations, but working in the same building were significantly exposed.[1] Sheet-metal workers, installing ductwork within the superstructure, often worked adjacent to spraying operations and were among those heavily exposed. While such spraying was banned in New York City in 1972, frequent renovation of these buildings guarantees a continued asbestos exposure for these workers.

In Manhattan alone, approximately 150 office buildings were built in the period between 1958 and 1972.[5] In the majority of these commercial buildings, renovation and repair is a full-time activity due to frequent tenant turnover and reconfiguration of office space. The level of this continued activity can be gauged by the fact that in 1982, 50 percent of the activities of the members of the Sheet-Metal Workers Association in New York involved renovation, much of it in asbestos-sprayed buildings. Measurements of air levels during such renovation demonstrate that sheet-metal workers may have the heaviest exposure of any members of the construction crews with 8-hour time-weighted averages of 0.19 f/cc,[3] with past exposure probably significantly higher.

Work practices that would limit the generation of asbestos fibers in the work area or restrict its enclosure to the renovation area itself are not commonly employed. It has been our experience that workers generally do not wear protective equipment and, until quite recently, showed almost no cognizance of the dangerous nature of asbestos. During the years of construction one hears of "snowball fights" with the insulation material, and workers regularly brought significant amounts of asbestos home on their clothes, shoes and in their hair with no awareness of the risk to their families, which has now been well documented.[6]

A total of more than 27 million U.S. workers were exposed to asbestos during the period 1940–1979, with the largest single group of these being construction workers. It has been estimated that 50 percent of all construction workers employed between 1958 and 1972 had significant exposure to asbestos.[7] Given the process of insulation described above and the continuing exposure to asbestos of those involved in renovation, it is not surprising that with the passage of time high rates of asbestos-related disease should emerge among these workers.

### *Asbestos-Related Disease among Sheet-Metal Workers*

Our initial examination of the 19 sheet-metal workers whose employer had been cited by OSHA revealed the problem. Of the 19 men examined, most were involved in construction during the period when asbestos was being sprayed and all had had extensive experience in renovation of asbestos-sprayed commercial buildings. While work histories yielded two cases where other sources of asbestos exposure had occurred (i.e., shipyard work during World War II), the remainder of the subjects had only been occupationally exposed to asbestos during their years as sheet-metal workers in the construction trades. In this small sample we found lung abnormalities in five workers (27 percent), predominantly those with 25 or more years in the trade.[8]

These findings were consistent with the earlier work of Selikoff on senior New York area sheet-metal workers (I. J. Selikoff, personal communication) and the recent Norfolk, Massachusetts study of sheet-metal workers which found similarly high rates (60–75 percent) of lung abnormalities in older sheet-metal workers.[9] Recognizing the extent of sheet-metal workers' historical exposure to asbestos as well as the potential for continuing exposure during renovation, the Sheet Metal Industry Promotion Fund and Local 28 of the Sheet Metal Workers Association asked the Montefiore Program in Occupational Health to conduct a program to assess and reduce the risk of asbestos-related disease among members of that trade. The remainder of this report outlines that program and our recommendations for future action.

## EPIDEMIOLOGIC STUDIES

To understand the dimensions of the health impact of this history on sheet-metal workers in New York City, we investigated the patterns of asbestos-related mortality and morbidity among the membership of Local 28. We attempted to determine whether this population's exposure to asbestos during construction activity and its continuing exposure during building renovation had resulted in the elevated disease levels typical of asbestos exposure.

### *Mortality*

To investigate mortality, a proportional mortality ratio analysis was conducted on 385 deaths reported to the union pension plan between 1976 and 1983. This analysis compared proportional cause-specific mortality in the population with that of an age-, race-, and sex-matched segment of the United States population. The results of this analysis are summarized in TABLE 1.[10]

The data indicate that sheet-metal workers have suffered an increased rate of mesothelioma and of cancer of the lung, and of the digestive tract. Elevated risks in these cancer sites, particularly the mesotheliomas, are consistent with the pattern of mortality expected in a population with significant exposure to asbestos. Further, these increases cannot in all likelihood be attributed to smoking patterns among sheet-metal workers, because the rate of other diseases normally associated with smoking were not elevated. Clearly, past exposure to asbestos during construction has led to increased mortality in this population.

### Morbidity

In order to measure the prevalence of nonmalignant lung disease among currently working sheet-metal workers, we examined the medical records of more than 800 workers who utilized a free medical examination program sponsored by the Sheet-Metal Industry Association. We reviewed the medical records of all participating sheet-metal workers and abstracted relevant information. The primary sources of information on asbestos-related lung disease were chest X-ray interpretations, and pulmonary function tests. At the time of our anal-

**TABLE 1.** Proportional Mortality among Sheet Metal Workers[a]

| ICD[b] | Cause of Death | Observed | Expected | PMR[c] | 95% CI[d] |
|---|---|---|---|---|---|
| 140–209 | All cancer | 118 | 77.8 | 152[e] | 123–186 |
| 150–159 | All digestive organs | 29 | 20.9 | 139 | 95–202 |
| 150 | Esophagus | 1 | 1.7 | 59 | 8–407 |
| 153–154 | Colorectal | 18 | 10.0 | 181[e] | 113–288 |
| 160–163 | All respiratory system | 44 | 26.2 | 168[f] | 123–250 |
| 161 | Larynx | 1 | 1.0 | 97 | 14–690 |
| 162 | Lung | 40 | 25.0 | 160[f] | 116–222 |
| 163 | Mesothelioma | 3 | —[g] | —[g] | —[g] |
| 185–189 | All genitourinary | 9 | 8.0 | 113 | 58–218 |
| 186–187 | Testes | 2 | 0.3 | 746[f] | 230–2,432 |
| 188 | Bladder | 2 | 2.9 | 68 | 17–270 |
| 193 | Thyroid | 2 | 0.1 | 1,704[f] | 615–4,721 |
| 200–209 | All lymphatic and hematopoietic | 12 | 7.1 | 170 | 96–299 |
| 202–203 | Non-Hodgkin's lymphoma | 5 | 2.1 | 236[e] | 100–556 |
| 204–207 | Leukemia and aleukemia | 7 | 3.0 | 232[e] | 112–479 |
| 390–458 | All diseases of circulatory system | 205 | 215.6 | 95 | 81–113 |
| 406–519 | Nonmalignant respiratory disease | 25 | 29.3 | 85 | 57–128 |
|  | Emphysema | 2 | 6.8 | 29 | 8–109 |

[a] This table is reprinted from the *American Journal of Industrial Medicine* 7:315–321
[b] International Classification of Disease, 8th revision.
[c] Proportionate mortality ratio = observed divided by expected multiplied by 100.
[d] 95% confidence interval.
[g] The expected number of mesothelioma deaths was too small to permit the calculation of an accurate proportionate mortality ratio.
[e] $p < 0.05$.
[f] $p < 0.01$.

ysis, 757 sheet metal workers had received a chest X-ray with the results read by a radiologist and reviewed by a specialist in occupational lung disease.

In general, exposure to asbestos can result in two types of nonmalignant lung disease. The most well-known of the two is asbestosis, characterized by the appearance of small opacities on the X-ray film, indicating the presence of fibrosis. Pleural disease, characterized by bilateral fibrotic plaques or diffuse thickening of the chest walls or diaphragm, is also frequently observed and may be an indicator of early impairment of pulmonary function.[11] Using the medical data, in combination with the number of years an individual was employed in the union, we investigated the relationship between years of asbestos exposure and asbestos-related X-ray changes.

Preliminary data analysis based on the specialist's X-ray interpretation suggests that the prevalence of asbestos-related chest X-ray findings increases with the number of years employed in the sheet-metal industry (TABLE 2).

Among asbestos workers with 30 or more years of employment, almost 40 percent have some sign of asbestos-related nonmalignant lung disease. Specifically, 10.4 percent of the sheet-metal workers examined show some degree of asbestosis and the prevalence rises to almost 22 percent in those employed 30 years or more. Overall 8.6 percent of the entire group show signs of pleural disease, and this figure rises to 26.7 percent for those employed 30 years or more.

Recently, a similar pattern of results was demonstrated in a second population of sheet-metal workers in the Boston area.[9] In the Boston workers, the prevalence of pleural disease was even higher — 70 percent in individuals with more than 30 years of exposure. A similar study in Sweden[12] found pleural changes in 62 percent of the construction workers examined. Taken together the results of the morbidity and mortality studies provide convincing evidence of the presence of asbestos-related disease in this group of construction workers.

## HEALTH EDUCATION AND INTERVENTION

In designing an asbestos intervention program for the sheet-metal industry, we chose a multidimensional approach. Program components included the epidemiologic analyses already discussed; a survey of knowledge and attitudes about asbestos among sheet-metal workers; a comprehensive health education and training program aimed at all levels of the industry as well as the general public; industrial hygiene control measures; and, finally, support of regulatory and legislative action.

**TABLE 2.** Prevalence of Asbestos-Associated Lung Abnormalities Relative to Years Worked

| Years Worked | Number With Abnormalities | Number Without Abnormalities | Percentage With Abnormalities |
|---|---|---|---|
| 0–9 | 5 | 183 | 2.7 |
| 10–19 | 20 | 190 | 9.5 |
| 20–29 | 51 | 149 | 24.3 |
| 30 and over | 40 | 61 | 39.6 |
| Total | 116 | 591 | 16.4 |

Our long-term goal is to eliminate risk of further exposure through the safe systematic removal of asbestos from New York City buildings. Recognizing that such a removal strategy could only be achieved over the course of years, most likely through legislation, we developed more immediate goals directed at lowering current asbestos levels for sheet-metal workers engaged in renovation, and enhancing the medical screening program to more effectively detect and manage the developing asbestos disease in this large population.

Looking at the age and exposure history of the members of Local 28, these goals seemed appropriate. Almost 30 percent of the membership is under 34 years of age, with 10 percent between the ages of 20–24. This group also comprises the heaviest smokers in the union. Developing a control strategy to immediately lower exposure levels in this younger population as well as in the overall union, thereby reducing risk, was a priority. In addition, we encouraged effective medical screening and medical management, particularly for more senior workers whose exposure placed them at a high risk of developing disease. Finally, we educated members about the importance of stopping smoking and of limiting inadvertent exposure of family members to asbestos through contaminated work clothes.

### Educational Program

A pivotal part of the program is education. Our efforts were directed at a wide constituency, extending beyond the sheet-metal industry to include other construction trades, health professionals, and the general public. This strategy seemed appropriate since our long-term goal — removal of asbestos from New York City buildings — was complex and would require broad-based support within the labor, real estate, and health communities in New York City.

In an effort to reach these populations and stimulate discussion on the subject, we participated in a variety of public information events including radio and television programs; held a press conference reaching the construction, office worker and broader labor community; and did several professional presentations. In addition, details of our program appeared in several major newspapers throughout the New York City area as well as in regular articles in both the union newspaper and a special contractor's newsletter. Our participation in these events enabled us to communicate this information about the problem, its potential health consequences, and prudent public health measures to eradicate the problem. However our short-term goal was to reduce exposure levels on renovation sites. This required a comprehensive education effort reaching contractors, union leaders, and the broader union membership. Seminars were scheduled for each of these groups. The emphasis of our overall training program was to provide a solid understanding of the health consequences of continued exposure and to develop and present the potential control measures suitable to sheet-metal work.

### Asbestos Awareness Survey

As an initial step we surveyed 100 sheet-metal apprentices to determine their degree of familiarity with the asbestos problem. Workers were generally aware that asbestos presented serious health consequences to their trade. However, they lacked basic knowledge that would enable them to determine whether they were being adequately protected at the worksite.

Further, few understood the purpose of protective clothing, assuming it was required to prevent absorption of asbestos through the skin. Not surprisingly, none took preventive steps to separate their work clothes from normal family laundry or to change into street clothes before coming home. Most understood that regular clean-up was necessary to reduce dust levels on the site. However, a large percentage assumed that a conventional industrial vacuum or frequent sweep-up were effective in controlling dust levels. In terms of dust-reduction measures few were aware of the effectiveness of wetdown techniques. Finally, most had a vague notion that there were a variety of respirators, some providing more protection than others. However, few knew of the need for proper fitting and maintenance of respirators, distinctions between filters, (e.g., toxic versus nontoxic dust, organic cartridge, and so forth), or the need for periodic replacement of filters.

The survey results underscored the need for regular educational efforts regarding asbestos. Presumably, these workers were representative of the membership of Local 28. Our informal discussions with management, union officials, and journeyman members reinforced this assumption. On the basis of these data, we designed our educational program to inform workers about safe work practices and respiratory protection. The results of our morbidity and mortality studies in New York sheet-metal workers were used in our educational program to directly illustrate the magnitude of the problem in sheet-metal work and proved to be a particularly persuasive source of information. Additional findings from this survey will be reported elsewhere.

### Safe Work Practices

Our educational efforts outlined and recommended a safe work practice protocol aimed at reducing asbestos exposure levels on renovation sites. The protocol includes wet handling techniques along with the use of a HEPA vacuum (high efficiency particulate absolute) for surface and floor clean-up. Wetdown techniques have proven to be extremely effective, reducing dust levels as much as 75–90 percent.[13] A HEPA vacuum provides filter efficiency down to 0.3 microns and is necessary for effective asbestos clean-up. The conventional industrial vacuum is an inappropriate choice for asbestos clean-up.[13] The protocol also specifies clean-up and disposal procedures and guidelines to prevent ingestion of asbestos as well as to prevent family exposure. In addition, general dust-reduction precautions are specified.

Many of these work-practice measures are identical to those utilized on asbestos-removal projects, but there is little experience in application of these methods to general building renovation or to the sheet-metal industry. As a consequence, we will introduce and monitor these work-practice measures on a sheet-metal site in order to evaluate their effectiveness in reducing dust levels and their feasibility in sheet-metal work. In the next phase of this program we plan to work with a group of senior sheet-metal workers, utilizing their expertise in evaluating the practicality of these methods for use in their trade.

### Protective Equipment

Respirators and protective clothing represent inferior solutions to the asbestos problem and have serious limitations which are well established in the literature of industrial hygiene.[14,15] Problems associated with respirators include quality and predictability of fit, discomfort in wearing, and breathing resistance, which presents potential health problems for breathing-impaired workers. Likewise, the disposable protective clothing recommended for asbestos work is hot and uncomfortable.

Despite these limitations, protective equipment along with safe work practices currently represent the only alternative immediately available to reduce asbestos exposure levels on renovation sites short of properly removing the asbestos before renovation begins. Therefore as part of our education efforts we provided the industry and union with specific information on these protective modes. We examined a variety of respirators, evaluating their protective capability, price (respirators and replacement filters), and manufacturers' servicing reputation. In addition, we looked at the suitability of these respirators for sheet-metal work. Sheet-metal workers must climb scaffolding, enter dropped ceilings, move and install bulky ducts and carry heavy tools on their belt. These work characteristics are considerations in choosing respiratory protection. We evaluated the single-use disposable respirator; the half-face, reusable air-purifying respirator with a high-efficiency cartridge; the powered air-purifying respirator; and the Type "C" supplied-air respirator. We immediately judged the single-use, disposable paper mask to be inadequate. While the filter medium from which the mask is constructed is effective for asbestos, the mask itself provides poor facial fit. As a consequence, unfiltered air can readily leak through the sides of the mask and be inhaled or ingested.[16]

Several reusable, half-face air-purifying respirators were evaluated. These molded rubber masks are generally manufactured in three sizes and, if properly fitted, guarantee a better facial seal than the disposable mask. As negative-pressure respirators, facial seal is essential since any breach in the seal permits the contaminant to enter the mask, thereby exposing the worker. A respirator-fit testing program, where individual workers are fitted with their own personal respirators, is necessary to guarantee optimal fit.

We also assessed the powered air-purifying respirator and the Type "C" supplied-air respirator. The Type "C" provides the highest degree of protection, but was judged to be unfeasible and potentially dangerous in sheet-metal work since these workers move around on scaffolding, climb ladders, and handle awkward ducts. Entanglement with the respirator's air-hose presented a potentially hazardous condition. The powered air-purifying respirator (PAPR) provides a higher protection factor than the normal air-purifying respirator and is cooler and more comfortable to wear since the positive pressure reduces breathing resistance. These factors make the PAPR an appealing option for construction worker protection. There are, however, several limitations to its use. These include bulkiness, a requirement for regular maintenance, and costliness (about $450 per unit). In addition, recent research has questioned their protective capability, suggesting that the protection factor be lowered. Nonetheless they offer significant advantages, particularly for breathing-impaired workers.

All of this information was presented to the trade in various educational meetings in order to further understanding of control options and to facilitate decision-making in protecting sheet-metal workers. As an immediate first step, a pilot program was initiated to fit workers with reusable, half-face air-purifying respirators.

### Respirator-Fit Testing Program

Using two brands of respirators equipped with high-efficiency filters, we initiated a pilot fit-testing program for 75 apprentice and journeyman sheet-metal workers. Workers received training in proper use, care, and donning of respirators as well as information on respirator limitations. Respirators were fitted by qualitative tests including negative pressure, positive pressure, and irritant smoke challenge. Future information will be collected from these workers regarding frequency of filter replacement (a function of contaminant concentration) and suggestions for future fit programs. A mechanism for coordinated replacement filter distribution is also being developed.

## AUDIT OF MEDICAL SCREENING PROGRAM

Under the OSHA standard in effect at the time, contractors were required to provide annual medical examinations for workers exposed to asbestos levels of 0.1 fibers/cc or greater. As indicated earlier, these levels are frequently found on asbestos-sprayed renovation sites. OSHA requires pre-placement, annual, and termination physical examinations which must include, at a minimum, a chest X-ray film (posterior-anterior), medical history (with an emphasis on respiratory disease), and a pulmonary function test (FVC and $FEV_1$). Complete and accurate records must be maintained for at least 20 years.[19]

As part of the 1982 negotiated agreement, Local 28 members are provided with access to an annual medical examination. The examination is comprehensive, including specific tests recommended for diagnosis of asbestos-related disease. The union health and welfare fund pays for this examination as well as compensating employees for one-half day of work. We examined patient medical records, evaluating the practices and procedures utilized in conducting medical examination, managing and organizing records, and communicating results to affected patients. In addition, we studied patterns of this program's utilization by the union members. Our objective was to evaluate the program's effectiveness in detecting and responding to the already demonstrated risk of Local 28 members for developing asbestos-related disease. The overall evaluation was based on a review of 833 medical records generated over a 1-year period, as well as the utilization patterns since the inception of the program. Specific details of the evaluation have been published elsewhere.[20]

Overall we found a series of deficiencies requiring immediate action. Only one-third of the Local 28 membership made use of the service in any one year. We recommended more visible publicity efforts on the part of the union and the health-care provider to encourage greater utilization of this program.

We evaluated the effectiveness of the "in-house" radiologist in detecting lung abnormalities as compared to the "second-opinion" physician, a specialist in occupational lung disease. We found a marked difference in detection patterns, underscoring the necessity of utilizing a specialist in this area.

We also examined recordkeeping and patient follow-up procedures. We found a pattern of incomplete charts, with a large percentage of second-opinion reports missing. Timely notification of patients regarding health problems was also deficient.

We found that there were no procedures in place to notify retiring workers that they are entitled to and, more importantly, should have this physical examination. In addition, neither the union, management, nor health-provider seemed to be aware of the 20-year recordkeeping requirement. Finally, no occupational history was elicited and the smoking history was inadequate. In each of these areas we provided programmatic recommendations to remove these deficiencies.

## LEGISLATIVE AND REGULATORY ACTION

As we noted above, systematic removal of asbestos from New York City commercial buildings is the only long-term solution to eliminate the asbestos problem. Currently, where removal is occurring, it is often done haphazardly by untrained and unprotected workers. As a consequence a portion of our efforts were targeted to legislative and regulatory agencies to which we presented testimony and contributed expert information, particularly as it related to the sheet-metal industry.

In view of the growing concern about the asbestos problem in public and commercial buildings, as well as concern about the inadequacies of current Federal regulations, several states have developed legislation regulating asbestos abatement activities. Maryland, New Jersey, and Rhode Island are among the states that have passed laws requiring licensing of all asbestos-removal operators, mandating comprehensive training, and establishing guidelines on safe asbestos abatement. In New York, efforts to regulate asbestos removal are being made on both the state and city level. In New York City legislation passed by the City Council in 1985 provides for systematic evaluation of all commercial buildings constructed between 1950 and 1972. When renovation is being planned building owners must hire a trained "asbestos evaluator" to survey the location and condition of asbestos in a building. The legislation mandates comprehensive training and certification of asbestos abatement workers. In addition, it outlines detailed requirements for asbestos renovation and removal work. In New York State, licensing legislation for asbestos abatement firms was passed in 1986.

## CONCLUSIONS AND RECOMMENDATIONS

The Montefiore Asbestos Control Program is now entering its fifth year of work with the New York City sheet-metal industry. In that time we have been able to identify and intervene in the significant public health problem that asbestos-sprayed buildings pose for those who first constructed and now renovate them. Further, we have helped to call attention to the fact that these buildings continue to be a locus of environmental and occupational asbestos contamination which, when disturbed, exposes large numbers of construction workers, office workers, and building visitors to this hazard. Asbestos exposure has already taken its toll on the health of the workers we studied and, given the characteristic latency of asbestos disease, we can anticipate a mounting toll over the next 10–15 years.[17] Finally, the persistence of this hazard is virtually assured by the current unregulated practices employed in the routine renovation of these buildings.

Both construction and office workers are, in our experience, unaware of the nature or extent of this hazard and are ill-equipped to deal with it in a safe and effective manner. Hence the need for a program of public disclosure about the hazardous character of asbestos-sprayed commercial buildings — a need that would be best met by the requirement for a public registry listing all such buildings and the condition of the massive amounts of asbestos that they contain. Such a registry should be part of the comprehensive legislative approach needed to assure the safe management and eventual containment or removal of this asbestos.

While the renovation of asbestos-sprayed office buildings represents a significant problem because it generates and distributes airborne asbestos into the work space, it also represents a significant opportunity for cost-effective and permanent abatement procedures. Because the rip-out and reconstruction work taking place during renovation is similar to that which takes place during asbestos abatement, the costs of such work need not be duplicated. Therefore, if asbestos abatement is done as an integral part of renovation, the cost of asbestos removal is sharply reduced. During such renovation, the additive costs for the asbestos abatement are only a fraction of the cost if abatement were done *de novo*. Thus renovation is the ideal time, from both the health protection and cost-saving perspective, to deal with the asbestos hazard in commercial office buildings.

Finally, it is our view that legislation to regulate and require asbestos abatement in such buildings represents a crucial step. The formal assumption of governmental responsibility for addressing this public health problem is in all likelihood the only way to achieve the goal of reduced exposure over the next decades. There may be disagreement about the best strategies for asbestos abatement, but these should not impede progress. Clearly, the determination of these methods and practices should utilize the best technology available and be upgraded over time as we learn more. But what cannot and must not be delayed is the formal recognition of the problem and the resolve to deal with it immediately. The history of human exposure to asbestos is a long one, full of negligence and deliberate deceit.[21] The consequence has been death and disease on a scale akin to that of modern warfare. Knowing what we now know about the potential risks that many of New York's most prominent buildings pose, we can no longer turn a blind eye to the problem.

## SUMMARY

New York City sheet-metal workers have a history of significant exposure to asbestos. Prior to 1972 when the use of sprayed asbestos insulation was banned in New York City, sheet-metal workers involved in building construction were exposed as they worked adjacent to spraying operations. Subsequent to that date, exposure continued as they renovated these same buildings. In 1982 the Occupational Health Program of Montefiore Medical Center and the Albert Einstein College of Medicine initiated a multidimensional asbestos evaluation and intervention program for the sheet-metal industry and union in New York. The long-term goal of the program was to eliminate asbestos exposure through the safe, systematic removal of asbestos in New York City buildings, most likely a legislated solution. In the short term, we attempted to assess and reduce asbestos exposure in the sheet-metal trade by a series of steps consisting of: mortality and morbidity studies; a medical audit of clinical screening services provided to sheet-metal workers; a comprehensive health education program; development of safe work practices; evaluation of personal protective equipment; and investigation into and support of legislative and regulatory solutions to the problem of asbestos contamination of commercial buildings. This intervention can be seen as a case study in the practice of social medicine.

## ACKNOWLEDGMENTS

This work could not have been possible without the support and active cooperation of the Sheet-Metal Promotion Fund and Local 28 of the Sheet-Metal Workers' International Association of New York and New Jersey. Our particular gratitude goes to Arthur Moore, Dan Wilton, and the officers of Local 28; William Rothberg, Paul Tone, and the members of the Promotion Fund; and to the sheet-metal workers themselves, who have taught us so much.

## REFERENCES


1. REITZE, W. B., W. J. NICHOLSON, D. A. HOLADAY & I. J. SELIKOFF. 1972. Application of sprayed inorganic fiber containing asbestos: Occupational health hazards. Am Ind. Hyg. Assoc. J. **33**: 178–191.
2. NICHOLSON, W. J., A. N. ROHL & E. F. FERRAND. 1970. Asbestos pollution in New York City. Second International Clean Air Congress. :136–139.
3. PAIK, N. W., R J. WALCOTT & P. A. BROGAN. 1973. Worker exposure to asbestos during removal of sprayed material and renovation activity in buildings containing sprayed material. Am. Ind. Hyg. Assoc. J. **44**: 428–432.
4. ONTARIO MINISTRY OF THE ATTORNEY GENERAL. 1984. Report of the Royal Commission on Matters of Health and Safety Arising From the Use of Asbestos in Ontario. Vol. 1–3. Ontario, Canada.
5. ROBBINS, Y. & H. ROBBINS. Manhattan Office Buildings—Midtown, 1984, Downtown, 1985. Yale Robbins Inc., New York, NY.
6. ANDERSON, H. A., R. LILIS, S. H. DAUM, *et al.* 1976. Household-contact asbestos neoplastic risk. Ann. N.Y. Acad. Sci **271**: 311.
7. NICHOLSON, W. J., G. PERKEL & I. J. SELIKOFF. 1982. Occupational exposures to asbestos: Populations at risk and projected mortality, 1980–2030. Am. J. Ind. Med. **3**: 259–311.
8. DRUCKER, E., K. MILLER & R. RULIN. 1981. Asbestos Hazards in the Sheet Metal Trade: An Evaluation of the Health of 22 Sheet Metal Workers in New York City. Unpublished manuscript.
9. BAKER, E. L., R. DAGG & R. E. GREENE. 1985. Respiratory illness in the construction trades—The significance of asbestos-associated pleural disease among sheet metal workers. J. Occup. Med. **27**(7): 483–489.
10. ZOLOTH, S. & D. MICHAELS. 1985. Asbestos disease in sheet metal workers: The results of a proportional mortality analysis. Am. J. Ind. Med. **7**: 315–321.
11. NIOSH-OSHA ASBESTOS WORK GROUP. 1980. Workplace Exposure to Asbestos, Review and Recommendations. DHHS (NIOSH) Publication No. 81–103.
12. HEDENSTIERANA, G., R. ALEXANDERRSON, B. KOLMODIN-HEDMAN, *et al.* 1981. Pleural plaques and lung function in construction workers exposed to asbestos. Eur. J. Resp. Dis. **62**: 11–122.
13. SAWYER, R. N. & C. M. SPOONER. 1978. Sprayed asbestos containing materials in School Buildings: A Guidance Document, Vols. 1 & 2. EPA-450/2-78-014.
14. SCHULTE, H. F. 1973. The Industrial Environment, Its Evaluation and Control. Personal Protective Equipment. U.S. Department of Health, Education and Welfare-NIOSH.
15. A Guide to Industrial Respiratory Protection, DHEW (NIOSH) Publication No. 76–198. Reprinted April 1979.
16. LOWRY, P. L., P. R. HESCH & W. H. REVOIR. 1977. Performance of single-use respirators. Am. Indus. Hyg. Assoc. J. **38**: 462–467.
17. MYERS, W. R. & M. J. PEACH. 1983. Performance measurements on a powered air purifying respirator made during field use in silica ragging operation. Ann. Occup. Hyg. J. **27**(3): 251–259.
18. MYERS, W. R., M. J. PEACH, K. CUTRIGHT & W. ISLANDER. 1983. Workplace Protection Factor Measurements on Powered Air Purifying Respirators at A Secondary Lead Smelter. (NIOSH field research project, unpublished study).
19. U.S. Department of Labor (OSHA), General Industry Standard (29 CFR 1910), Asbestos Standard (1910.1001) 545–500.
20. ZOLOTH, S., D. MICHAELS, M. LACHER, D. NAGIN & E. DRUCKER. 1986. Asbestos

disease screening by non-specialists: Results of an evaluation. Am. J. Publ. Health **76:** 1392–1395.
21.   BRODEUR, P. Annals of law: The asbestos industry on trial. The New Yorker, June 8, 15, 22 and July 1, 1985.

# Index of Contributors